*Hawks,
Owls &
Wildlife*

RED-TAILED HAWK. FROM A PAINTING BY WALTER A. WEBER. SEE ALSO FRONT COVER. *(Courtesy Frederic C. Walcott Memorial Fund, North American Wildlife Foundation)*

HAWKS
OWLS
AND
WILDLIFE

BY

John J. Craighead

Biologist, U. S. Fish and Wildlife Service;
Leader, Cooperative Wildlife Research Unit,
University of Montana

AND

Frank C. Craighead, Jr.

U.S. Fish and Wildlife Service

Drawings by Jean Craighead George
Photographs by the Authors

Dover Publications, Inc.

New York

Published in Canada by General Publishing Company, Ltd., 30 Lesmill Road, Don Mills, Toronto, Ontario.
Published in the United Kingdom by Constable and Company, Ltd., 10 Orange Street, London WC 2.

This Dover edition, first published in 1969, is an unabridged republication of the work originally published by the Stackpole Company and Wildlife Management Institute in 1956.

This editions contains a new Foreword by the authors.

Standard Book Number: 486-22123-7
Library of Congress Catalog Card Number: 74-81670

Manufactured in the United States of America
Dover Publications, Inc.
180 Varick Street
New York, N.Y. 10014

DEDICATED
to
MARGARET and ESTHER

Foreword to the Dover Edition

Arrangements for this reprint edition were completed a number of months ago, but republication was delayed in the hope that we would be able to incorporate into the new edition some of the results of our recent field work. However, we have now concluded that our research duties will not allow us time in which to prepare such new material in the foreseeable future, and that further indefinite delay in republication is unwarranted. Current researches confirm rather than alter or invalidate the conclusions of this volume, which are already well supported by the data provided in Appendices A and C. Hence this edition conforms exactly to the original.

J. J. C.
F. C. C., Jr.

March, 1969

Preface

The relationships between predators and prey have produced about as much heat and as little light as any topic in the wildlife management field. Formerly, predators were widely if not universally condemned, and every man's hand was against them. It is only within the past few decades that efforts have been made to understand the actual role of predation in the ecology of living things.

Hawks and owls of all species are among those that were and still are persecuted, although it has been known for some time that depredations against game birds and poultry usually were the work of a limited number of species.

Laboratory studies and field observations continually provide new information, but many facets of the dynamics and ecology of raptor predation remained unknown, largely because of the difficult and tremendous task of studying those highly mobile, wary, and partially nocturnal bird populations.

Frank and John Craighead, in the studies on which this book is based, undertook to find the answers to some of these questions. It required endless hours of patient watching, hundreds of climbs up and down huge trees and treacherous cliffs, a wide variety of skills, a sympathetic understanding of the habits of the subjects, and a boundless enthusiasm for the work.

No pair of biologists is better qualified for the task than the Craighead brothers. Both have an unusual background for the study of raptorial birds. In conducting this study, they found it necessary to devise some new techniques and to modify others to meet difficulties encountered in securing and consolidating the enormous amount of data on which they now are reporting. The results of their work should prove interesting and informative to anyone interested in predation or in the species involved in their study.

IRA N. GABRIELSON

Acknowledgments

In conducting this research and in preparing the results for publication, the help of numerous colleagues, associates, and close personal friends has been valuable.

For inspiration, so essential to any prolonged effort, we are grateful to our father, Frank C. Craighead, Sr., formerly Chief of the Division of Forest Insect Investigation, U. S. Department of Agriculture, and to the late H. M. Wight, formerly professor of Wildlife Management at the University of Michigan.

Inspiration is kept alive by encouragement and this was supplied in generous measure, through sixteen years of interrupted labor, by our wives Margaret and Esther.

This task was started as partial fulfillment of the requirements for the degree of Doctor of Philosophy in the University of Michigan. For helpful advice and constructive suggestions during early stages of the work, we are especially indebted to S. A. Graham. Josselyn Van Tyne, William Burt, Earl C. O'Roke, and S. T. Dana reviewed the manuscript.

Financial aid was received from the Horace H. Rackham School of Graduate Studies, University of Michigan, and the Wildlife Management Institute.

For help in preparing maps and drawings we are grateful to Alvina M. Woodford, George Brasher, William H. Lawrence, and Dwight Stockstad. Alvina K. Barclay contributed by typing portions of the manuscript.

The wash drawings contributed by Jean Craighead George are most gratefully acknowledged.

The time consuming task of critically reviewing the final manuscript fell to Ira N. Gabrielson and C. R. Gutermuth of the Wildlife Management Institute, and to Daniel L. Leedy of the U. S. Fish and Wildlife Service. Their valuable advice and suggestions were much needed and appreciated.

We wish to thank the U. S. Fish and Wildlife Service for allowing John Craighead to complete the manuscript while serving as leader of the Montana Cooperative Wildlife Research Unit.

To Harrison F. Lewis, formerly Chief of the Canadian Wildlife Service, we extend special thanks for skillful editing of the manuscript and for many helpful suggestions.

Table of Contents

PART I. ECOLOGY OF RAPTOR PREDATION IN FALL AND WINTER

PART II. ECOLOGY OF RAPTOR PREDATION IN SPRING AND SUMMER

LIST OF PLATES

Frontispiece—The Red-tailed Hawk

LIST OF TABLES

LIST OF MAPS

LIST OF FIGURES

Introduction

Nearly a thousand feet above the valley, the wind hurled glazed snow particles over the lip of a great snow cornice. Four adult mule deer and two fawns stood on the precipitous edge, hidden intermittently by the gusts of flying particles. All heads were lifted skyward silhouetted against the purple cast of a wintry sky. Directly above, wings half closed, hung a Golden Eagle. The wings pumped two, three, four strokes, closed tightly and the eagle plummeted toward the closely grouped deer. The lead one almost in slow motion rose on hind legs, lifted her head and forefeet to the sky and pawed the air. Sharp hoofs and needle talons nearly met as the eagle with spread tail and bent pinions broke her dive and rose gracefully to her former pitch. Five times the eagle dived and five times the deer rose in unison to ward off the attack with flailing hoofs. A magpie, the only other spectator, trailed the eagle in every dive and catching the wind, each time climbed safely above the great bird. Three deer moved off the ridge when a final plunge carried the eagle behind the cornice. Had she made a kill?

It was an hour later and a thousand feet higher before we learned that the deer were unharmed; another time the outcome could be different. The action had been dramatic, thrilling and aesthetic in its rhythmic cadence; yet purposeful and harshly realistic. Not always so spectacular, such incidents are, nevertheless, familiar to the biologist attempting to understand nature. The killing and eating of one form of life by another form is known as predation. It is familiar to the farmer who glances up from his plowing to see a Marsh Hawk strike a mouse, to the hunter who glimpses a Cooper's Hawk carrying off a wounded Bob-white, and to the sportsman whose tramping leads him to the fresh kill of a coyote, fox, Horned Owl, or Golden Eagle. It is an everyday occurrence experienced in this form by all. Predation, however, is by no means thoroughly understood. It is a normal biological phenomenon of very great antiquity and has long been a subject of study, particular attention being paid to the conspicuous, mobile raptors —the hawks and owls.

If it is considered that, in the time it has taken to read this paragraph, predation has eliminated thousands of individuals, its influence is better realized.

The data to be presented show that it is a precise and powerful force operating continually to keep within limits the fluctuating numerical relations among animal populations. Though only one of numerous regulating forces, it is characterized by a steady and proportionate pressure that tends to lower the persistent increase of prey species, before more drastic but less steadily functioning forces, such as starvation and disease, run rampant.

The killing of a Blue Jay by a Cooper's Hawk may, for example, appear as mere chance—the jay happened to move into the open when the hawk was overhead. The thousands upon thousands of daily incidents of hawks and owls catching their prey are not, however, a hit-or-miss affair, but are governed by definite natural laws, whose operations are part of the web of life. This continual killing is not cruel nor wasteful, but a necessary biological function that has evolved simultaneously with the development of life. To understand predation—and more specifically the role of hawks and owls in nature—we must avoid the mistake of the blind men who, upon feeling different parts of the elephant, got divergent and erroneous impressions of the animal. It is planned to present in the text sufficient data about various aspects of predation so that the reader can visualize it as a whole.

One cannot understand predation by judging the Goshawk that strikes a grouse nor the coyote that kills a sheep. It is necessary to think, not in terms of individuals and species, but in terms of populations. To do this one may visualize a farm and surrounding country, a favorite hunting area, or a stretch of familiar woods and fields and imagine these multiplied many times. Over this great area there are populations of many prey species, such as cottontail rabbits, meadow mice, small birds, Bob-whites, pheasants, and squirrels. These, taken as a whole, are the prey population. On the same area are populations of Cooper's Hawks, Red-tailed Hawks, Sparrow Hawks, Horned Owls, Barn Owls and many others, making up the population of raptors. Predation can be understood and evaluated by determining the effect of this aggregate population of raptors on the aggregate population of prey over an extended period of time.

This means that factual data pertaining to interrelated factors necessary for an analysis of this subject must be dealt with quantitatively. It is essential to know the year-round numbers and composition of the raptor population, as well as its movements, the numbers or densities of prey species, the food habits of the raptors as related to the available prey and a complex of variables affecting the vulnerability of prey species. These data have been gathered and are presented in this work. As these data are analyzed and as conclusions are drawn, it will be seen that on the study area, as elsewhere, each member of the population of raptors is adapted to take a limited number of prey forms—not all of them. The Marsh Hawk feeds largely on meadow mice, the Cooper's Hawk on small birds, the Horned Owl kills rabbits, the Goshawk takes squirrels. Moreover it will be seen that the raptors are adapted to take almost all prey species present and that the predation coverage is even more inclusive when the mammalian predators, also present, are taken into account. The population of raptors changes from season to season and from year to year, for each species is present in numbers more or less proportionate to the numbers of its major prey.

The result is that the collective prey population, which includes all the familiar wildlife forms—rats around the barn, mice in the fields, pheasants in the corn, rabbits in the brier patch—is taken by raptors in approximate proportion to the relative densities of the prey species. In other words, if mice, as is usual, are much more abundant than pheasants —"mouse hawks," such as Marsh Hawks and American Rough-legged Hawks, will be more common than "bird hawks," like the Cooper's Hawk or Goshawk, and the raptors as a population will take proportionately more mice than pheasants. If there are fewer fox squirrels than there are pheasants, a proportionately smaller number of them will be taken, for the vulnerability of fox squirrels is affected not only by their own numbers but also by the numbers of pheasants and other prey. Despite some impressions to the contrary, hawks and owls do not gang up to wipe out a few squirrels, rabbits, or Bob-whites. Their average kill over any large area and for any substantial period is determined chiefly by the relative densities of the prey species around. They will, of course, take some game species, but these actually are a fraction of the animals they kill and eat. Such victims may be animals that a hunter otherwise might have harvested. On the other hand, while taking this toll, they are reducing greatly the very numerous undesirable species, such as mice, rats, pocket gophers, and ground squirrels. Furthermore, their combined activities result in a force that is essential to the continuous and proportionate control of wildlife populations, a force that under most conditions and in the long view of things outweighs in value the game killed. Such a universal and basic natural phenomenon as predation must be understood thoroughly by both sportsman and naturalist before wildlife management practices will be sound and efficient.

For those who enjoy a day spent among wild creatures how much fuller will be the appreciation of nature if the beautiful and the pleasant and the seemingly cruel and unpleasant—the song of the Meadowlark and the deadly pursuit of the hawk—can be interpreted more correctly!

PART I

Ecology of Raptor Predation in Fall and Winter

CHAPTER 1

The Study Area and Technics

THE main objective of this study was to gather data which upon analysis might lead to understanding of raptor predation; that is, just how such predation operates, the mechanics of its complex and interrelated functioning. What does a population of hawks and owls take from an area of land; how and why do they take it; what is the effect of this predation upon known prey populations; how efficient is predation as a natural population control; what general conclusions concerning the role and mechanics of predation can be deduced from the facts accumulated? To what degree does it affect man's economic interests?

To accomplish this purpose, all major factors bearing on the nature of predation have to be measured simultaneously on a definite area of land and a great many data carefully recorded and analyzed, as a basis for the formation of conclusions. Thus the problem is approached by a land-area study of predator prey relationships.

For two years in Superior Township, Michigan, the year-round population densities, activities, movement, mortality, and productivity of the raptor and major prey populations were determined. The food habits of the raptor populations were also determined. The relationships existing between known quantities were then analyzed and interpreted to reveal the mechanics and the function of raptor predation, in terms of collective raptor and prey populations and their effects on each other.

Accumulation of isolated data over a period of years, though significant in many respects, does not, and cannot, show these complex relationships. The land changes, the cover changes, the populations of prey and predators vary, and, in fact, all ecological factors change. As

Heraclitus expressed it, "You can't step into the same river twice." All relevant elements have to be known at a given time and interpreted in relation to one another. Isolated data gathered over long intervals are excellent for purposes of comparison, but this information has yielded only partial understanding of predation. In this study complete data for a short period were considered more important than incomplete information over a long period. Superior Township, Washtenaw County, Michigan, was selected as the principal area for pursuing such a study (Plates 1 and 2), but data of a similar type were also gathered on a semi-wilderness area at Moose, Wyoming (Plates 3 and 4).

The effects of avian predation must be considered by any worker making a thorough biological study of a prey species. Past raptor predation studies have dealt with natural history observations, food-habit investigations broad in scope (McAtee, 1935), and studies of specific predators in respect to their food habits, as, for example, Errington and Hamerstrom's work on the Horned Owl (1940). Here qualitative food data were compiled for a single raptor species over an extended region. Still other studies analyzed the effects of predation on the population of a single prey species, such studies being typified by the works of Stoddard (1926) and Errington and Hamerstrom (1936), both studies dealing with the Bob-white. Tinbergen (1946), working in the Netherlands, determined in quantitative terms the mortality of prey species caused by a single rather specialized raptor, the old world Sparrow Hawk. Although he considered populations of both raptor and prey, the raptor population was not a collective one.

The role of raptors in controlling vertebrate population levels has been variously apprehended, and the literature on this important subject shows conflicting opinions as to both the function of predation and the mechanics of its operation.

The authors believe the major contributions of this book are:

1. A quantitative analysis of the mechanics and function of raptor predation.
2. A comprehensive study of raptor movements, ranges, territoriality, and inter- and intra-specific behavior as related to collective raptor populations.
3. Quantitative proof of the efficacy of raptor predation as a natural biological control or regulator, and clarification of the relationship of this phenomenon to other environmental forces that regulate the population levels of vertebrates.
4. A study of collective populations presenting a new approach to a further understanding of the forces or processes which regulate the densities of animal populations.

Literature concerning predation and the broader field of population-regulating processes is burdened with scientific terminology. Wherever possible, specialized terminology has been purposely omitted, with the intention of making this book more understandable to the layman and to a wider range of scientific workers.

GENERAL RESEARCH PROCEDURES

Predation involves an interrelation of animal life and its environment and therefore the first task was to select a suitable area where hawks and owls could be continuously studied, day after day, over a period of years. A survey of numerous areas in the vicinity of Ann Arbor, Michigan, indicated that the vegetational pattern, road system, land-use practices, topography, and convenient location of Superior Township rendered it suitable for this purpose. The small proportion (11 per cent) of woodland, uniformly scattered throughout extensive stretches of open country, and a network of section roads, making all parts of the area readily accessible, were the chief conditions considered in the final selection of this area. General reconnaissance the previous year indicated a winter raptor population and a spring nesting population that were sufficiently dense to provide study material.

Previous investigations of predation on a much smaller scale, but over a period of eleven years, indicated that a comprehensive, detailed ecological study of raptor predation could be conducted only on a large tract of land, one greatly exceeding the range covered in the daily movements of the individual birds. An area of 36 square miles was considered adequate and the value of this decision was rapidly borne out as the work progressed.

One of the greatest problems was to gather adequate data over such an extensive area. Without the section layout of roads, the homogeneous nature of the area, and the small, isolated character of the woodlots, the study could not have been conducted on such a scale (Map 1). These three characteristics of the area shaped study methods. The methods that were adopted had to meet the demands of both extensive and intensive field work. Sampling technics, though frequently employed, were used only when there was no alternative. Thus, from the standpoint of an areal approach, three main methods of investigation were used:

1. Studies embracing the entire Township
2. Sample studies of a representative eight-section block within the Township (Map 8)
3. A sample study of an 18-square-mile strip (Map 1).

Ecologically, the research naturally divided itself into:

1. Raptor movements
2. Raptor population studies
3. Prey population investigations
4. Raptor food habits.

The major research tool was planned systematized observation. Each investigator spent more than 800 days in the field observing hawks and owls, investigating prey abundance and studying raptor food habits or the many interrelationships of raptors to their environment.

Because of their relatively small numbers and extreme mobility, the hawk and owl populations were studied over the entire area. A sampling method was used in making a census of the wintering hawks, but a direct count was employed for owls.

Since the daily, weekly, and seasonal movements of the raptors were intricately involved in the determination of populations and in the predator-prey relations, raptor movements were studied over the entire 36-square-mile area.

Censuses of prey populations were in most cases confined to eight sections thought to be representative of the Township (Map 8). Quantitative methods were employed, and the sample data were applied proportionately to the Township. General observations of prey populations were made throughout the 36 square miles and compared with the sample findings.

Research on the raptor food habits and the predator-prey relations involved the entire area. The data from pellet analysis were correlated with the predation ecology.

As invariably happens in a field study, situations which warranted special attention arose. Such situations were classed as special projects, the intensity of study depending upon the degree to which they directly affected the general predation problem.

A cover map of the township was prepared. The value of this map as a tool for the recording, accumulation, and interpretation of data cannot be overemphasized. Every observation was recorded on field maps and dated. At the end of a day's work the data were transferred to other maps according to subject matter, such as hawk ranges, owl ranges, roosting areas, movement of individual raptors, activity and kill areas. Every raptor observation that appeared to have any significance was plotted. Gradually the data took shape. In addition to the maps, the same and additional data were kept on form sheets, where detailed notes were recorded. Just as certain characteristics of the area were fundamental to the execution of the study, so was the field map the key to the recording of data. The raptor population and movement studies involving thousands of individual observations would have been impossible on a quantitative level without an accurate cover map for field use.

In the course of making a cover map, which included the pacing of fence lines and the mapping of vegetation types, familiarity with the entire area was acquired and information that proved valuable in formulating plans was gathered. Specific data that applied directly to the predation problem were also gathered. The cover map (Map 1) includes, in addition to the 36 square miles of Superior Township, one square mile of land from the adjoining Ann Arbor Township and a one-quarter-mile strip of land wherever the hawk census route bordered the township. The section of land in Ann Arbor Township contained coniferous plantings and was included in the study for comparative purposes, since there were no conifers on the study area.

Predation studies made at Moose, Wyoming, (Plates 3 and 4) fall more appropriately under the spring and summer phase and thus study methods and a description of this area are presented in Chapter 8.

HAWK AND OWL WINTER CENSUS METHODS

It was essential to obtain accurate figures on the numbers of hawks and owls in the Township throughout each year. This required that seasonal or other fluctuations be taken into consideration and methods for counting both diurnal and nocturnal birds of prey were therefore devised.

Because the interspersion of vegetation types produced a rather homogeneous area, and because the grid of section-line roads provided opportunities for excellent visibility, it was decided that a strip census from a car might give an adequate sample of the hawk population during fall and winter. Owls could be counted in their daytime retreats throughout the township—a count that with repeated checks would approximate total population figures. Censuses of spring and summer populations were made by means of nest counts and range plottings.

Car Census

Experimental drives over the township facilitated selection of a satisfactory route of travel for use in making a census of the fall and winter hawk populations. The 40-mile route finally selected was considered representative of the Township, and wherever possible was limited to dirt roads (Map 1). Counting hawks within a quarter-mile of each side of the road provided a census of an area of approximately 18 square miles, or a coverage of half a township. Early experiments revealed that a quarter-mile distance was suitable for southern Michigan. Activity on a narrower strip might have been affected by the proximity of roads and farm buildings. Beyond this limit it was difficult to see small hawks, such as the Sparrow Hawk. Up to this range all species could be readily identified to genus and usually to species. This distance was also easily estimated from bordering quarter-section line fences. Reference to a cover map and practice in judging a constant distance were additional aids. An average speed of 13 miles per hour was the optimum rate of travel. This allowed sufficient time for complete scanning of the strip and for accurate identification.

Two observers with 20-15 vision made every census. Each was responsible for scanning the strip on his side of the road, but neither confined his observations to one side. One observer drove, and the other recorded data on form sheets and maps.

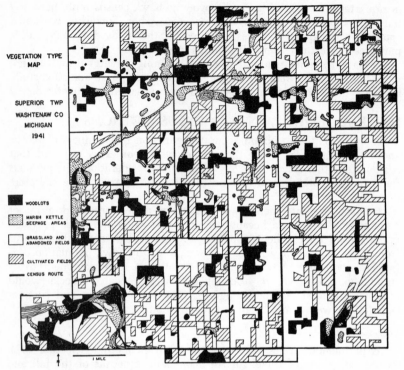

VEGETATION TYPE
MAP

SUPERIOR TWP
WASHTENAW CO
MICHIGAN
1941

◼ WOODLOTS

▨ MARSH KETTLE
SEEPAGE AREAS

☐ GRASSLAND AND
ABANDONED FIELDS

▨ CULTIVATED FIELDS

▬ CENSUS ROUTE

1 MILE

MAP 1. Vegetation type map, Superior Township

When necessary, very short stops were made for accurate identification. Binoculars (7 x 35) were carried and used. Identification to species usually was not difficult; all doubtful cases were in the genera *Buteo* and *Accipiter*. With 12-power glasses, experimented with in 1948, 100 per cent identification to species was possible.

Woodlots bordering the road were scanned by looking above and beyond them on approaching and leaving. As almost all woodlots were small, their interiors were viewed while observers were passing directly in front of them. The activity and the vegetation type associated with each raptor observation were recorded on a form sheet.

Activity was divided into flying and flight hunting, soaring, perching, courtship, and territorial behavior.

Hawk activities were correlated as closely as possible with vegetation types. In the case of a high-flying or soaring bird, the type recorded was that directly beneath the bird when first observed.

A hawk perched on the border of a woods but definitely hunting a field was recorded as hunting the grass type rather than the woods type.

Likewise, the activity of a hawk soaring over a woodlot during the nesting season and obviously not hunting the surrounding fields was associated with the woods containing the nest it was patrolling.

Most censuses were made from one to four p. m., sun time. This allowed a three-hour period relatively unaffected by seasonal changes in the length of day. No censuses were made prior to 8:00 a.m., as there was little early hawk activity on cold mornings.

Weather conditions prevailing during each census were recorded (Tables 63 and 64). The exact position of every hawk observation was plotted on a map, with information on species, sex, and stage of maturity indicated when possible. Individual characteristics, such as missing primaries or tail feathers, unusual color or markings, or the distinction of being a single representative of a species, were noted and recorded. Additional data were compiled when their desirability became apparent. For example, in 1948 the foot-candles of light measured against both sky and ground were recorded on each census, as well as the height at which hawks were seen perching or flying. The census was not only a method used to count hawks, but a systematized technic of studying a hawk population.

Censuses were continued beyond the winter period, but abandoned after the April 30 count, when development of foliage and confining nest activities made this census method unreliable. Eighteen car censuses were made from November 10, 1941, to April 30, 1942. Five winter censuses were made in 1947, 20 were made during the fall and winter periods of 1947-48, and four censuses (one fall and three winter) were made in 1948-49.

Foot Census

In 1941-42, a census on foot was undertaken to obtain relative abundance figures for hawks for comparison with car census results. Such a count, if eventually interpreted in terms of a known population, would give a relative abundance factor useful at future dates or on similar areas for computing approximate populations from rapid foot reconnaissances.

A method of correlating numbers and species observed during a known time and distance covered in the field was adopted as the most practical. The only hawks counted were those within a quarter-mile strip to either side of the observer. Usually two observers made simultaneous counts in widely separated parts of the township. The areas covered were arbitrarily selected and counts were made while doing other field work that did not interfere with taking observations.

Data similar in nature to those recorded during the car census were collected from October 23 to April 22 and divided into three seasonal counts—fall, winter, and spring (Tables 13 and 15). From the time the trees were in full leaf to the end of summer, the method was impracticable because of poor visibility and confining nest activities.

The fall count was made in the course of preparation of a cover map of the township, the winter count when observations on prey concentrations and raptor activities were being made and the spring count during search for nesting sites. Forty-seven counts, totalling one hundred hours, were made over a seven-months period. Each count was an hour or more in duration. A total of 121.5 miles was travelled, distances being computed by means of an accurate cover map, and 236 observations of hawks were recorded.

SEASONAL DIVISION OF STUDY

The study of hawk and owl populations during three years revealed that their activity can be divided significantly into four distinct phases: those of fall, winter, spring and summer which coincide roughly with those seasons of the year. The fall and winter phases form a natural group the spring and summer phases form a contrasting group. These periods of raptor activity, though merging one into the other, are well defined and the dates separating the periods can be very accurately ascertained. The importance of carefully delimiting these periods will become evident as the study progresses.

During winter, activity centered almost entirely around hunting. In contrast, summer activity was primarily related to nesting and family dependency.

The fall period is one of population instability and migration. It dates from the first week in September, the beginning of the first noticeable hawk migration, to the first week in December, the end of noticeable migration.

The winter period is characterized by the presence of a stable population with definite hunting ranges. During four years these conditions existed from the first week in December to the first week in March, which is the usual time for the return of migrating Red-shouldered Hawks.

The spring period is characterized by marked changes in the composition of the raptor population, with territorial selection and defense of nest sites by paired birds—this is accompanied by the breakup of winter ranges and a gradual departure of many winter residents. It dates from mid-February, time of nesting of the Great Horned Owl, to August 1, time of departure from the nest of the last young Marsh Hawks and Cooper's Hawks.

The summer period is characterized by a gradual cessation of direct nesting activities and is a period of hunting instruction and gradual separation of full-grown immature birds from parental dependency. It overlaps the spring phase, dating from about June 22 to September. June 22, the average date of departure of fledgling Red-shouldred Hawks from the nest, was selected because this species was the numerically predominant raptor. September brings the first indication of fall migration.

The latter part of the summer phase seemingly is a time of relative in-activity and seclusion.

THE STUDY AREA

To understand your next-door neighbor you must know something about his home and his office—the environment that to an important extent makes him the kind of man he is. Likewise, to understand hawks and owls and what they do, we must know something about the area in which they live and how the environment affects their activities. As this study of predation unfolds, we shall wish to refer to this chapter and to Appendix A to help visualize what the study area looks like, what plants and animals are to be found on it, how much it rains, how cold it becomes and what man does on the soil to make a living.

Superior Township, Washtenaw County, Michigan, lies north of the town of Ypsilanti and east of Ann Arbor.

It is situated in the glaciated plains of the central lowlands, where local differences in altitude seldom exceed 100 feet. The general eleva-tion of the upland varies from 800 to 1,000 feet above sea level. The topography has been modified by glacial deposition, supplemented by stream erosion. Swells, sags, kettles, eskers, and glacial outwash are conspicuous features of the area.

Weather conditions in the region are mild. Extremes of temperature and precipitation are rare. The mean annual temperature is 47.4°F. The mean annual precipitation, including an average annual snowfall of 37 inches, is 31.31 inches. Wind and evaporation are low; humidity is high.

A large proportion of winter days are overcast. The heaviest rainfalls occur during the months of May, June, and September. Snowstorms are light, 2 to 3 inches being the average, with a maximum of 14 inches in January, 1918.

HISTORY AND GENERAL LAND USE

The region was first settled in 1809 by French traders who located on the site of the present city of Ypsilanti. The first English-speaking families settled on the banks of the Huron River below Ypsilanti in 1823. By 1930, 83 per cent of the land was in farms, with an average area of 114.7 acres each. Small areas of marginal land unfit for cultivation occur on almost every farm.

Dairy farming now predominates and the greater part of the culti-vated land is devoted to raising forage crops. A sample of nine farms in Superior Township is fairly indicative of the local type of farming. The minimum size of a successful farm on the area is 80 acres, the maximum, with the exception of the Ford farms, is 360. The chief sources of farm income, listed in order of money value, are dairy products, small grain, butchering stock, forage, poultry, and cord wood.

Land use practices and changes were the principal forces altering the vegetation in this farming region. Vegetation changes were reflected by changes in density and composition of prey species, followed by adjustments in raptor populations. Thus the land and the use to which it is put must be considered in a study of predation phenomena.

ARRANGEMENT OF DATA

It should become apparent to the reader as he progresses in this book that the degree of accuracy in determining the raptor population on Superior Township has a very important bearing on some of the major conclusions drawn in the closing chapters.

To the research worker the experimental design, the research methods and the details of interpretation are as important as the general conclusions reached. He must know and is entitled to know how specific information is gathered and how major concepts are developed. In no other way can the soundness of the work be judged. On the other hand, the general reader is not concerned with details and their presentation tends to distract from the main line of thought. With this in mind essential but detailed information has been presented in the appendix. The research worker is urged to refer to this.

CHAPTER 2

Hawk Movements and Winter Ranges

L ITTLE is known about the movements of hawks as individuals, species, or populations yet neither hawk populations nor their effect on prey populations can be discussed satisfactorily until we catalogue the types of hawk movements and try to understand their nature. This will be the first task. We shall later show that raptor movements are the key whereby winter predation becomes a force controlling prey populations. Prior to treating movements specifically, it is desirable to look ahead.

In this and the following chapters we are dealing with numbers, maps, graphs, and tables. Each species is illustrated by a figure, the Red-tailed Hawk by one, the Great Horned Owl by another, meadow mice by still another. This process greatly facilitates comprehension, but it takes away the perceptual reality of the situation. The investigator has two different concepts—the mental, abstract, impartial one, constructed with the building stones of quantitative data, and the perceptual one, resulting from day-by-day, week-by-week, season-by-season observations of the dynamic activities of the animals being studied and of their environment. This latter is his awareness of an area of land more extensive than most large cities, a farming region of pastures and cropped fields; of cold winds blowing across wheat stubble of a morning sun warming the sides of shocked corn; of fields interspersed with woodlots—not woodlots as a vegetation type or as a percentage of the Township acreage, but individual woodlots—some seen with the late evening sun throwing long shadows on new, flaky snow, or another as a woodlot harboring a certain red oak whose crooked trunk serves as a protective perch for a Great Horned Owl. In a population table the owl is merely number 13. In the woodlot he is a live entity, scarcely visible, with the wet snow falling past him, a dominant influence upon animal communities about him. A grass field supporting 140 meadow mice per acre is etched with runways. It is a place where the mice scamper ahead and across one's path like dark shadows—where an American Rough-legged Hawk is regularly seen hovering—where telltale patches of fur in the grass or a bloodstain on the snow tell that he has been successful in hunting.

A Ring-necked Pheasant roost, merely a number on the map and a flush count on a table, is a panoramic view of whirring wings, of an early morning sky momentarily filled with birds, a sedge marsh covered with tracks, a roost deep in the snow, where a hen still sits, refusing to move on a cold morning. Thus each locale—the pastured fields, the plowed ones, the stream bottoms easily flooded, the woodlot with mature beeches and numerous hollow basswoods, or the one enclosing a buttonbush kettle where rabbits congregate in winter and wood frogs are vociferous in early spring—is a separate picture with distinct habitats containing individual birds and mammals. Memories of such things form a mosaic which to the observer is the Township. This mental image of an environment, with individuals or groups of animal life in their respective places, but intricately tied together by physical and biotic forces so sensitive that each small change may alter some details, cannot be completely transmitted, but must remain the possession of the observers alone. Such an objective understanding develops only when enough time is spent on an area so that almost every acre can be visualized and the observer has attuned himself to the land and the activities of the complex found. When such a condition is reached, a township has become as familiar as the well-known backyard. It has also become a living entity, in which man is the main influence in bringing about drastic changes in the economy of all animal species—some purposefully and others inadvertently.

A brief description of the seasonal activity of hawks and owls in Superior Township will be useful.

Each year, as fall approached, heralded by cool nights, coloring of leaves, harvesting of corn, and migrating of song birds, the immature hawks raised in the woodlots and fields of Superior Township grew restless. Movements occurred which resulted in population changes. Most of the immature, and many of the mature, hawks drifted out of the area and migrated southward. A few left nearly every day, until the majority were gone. Always some of the mature nesting birds—the Red-tailed Hawks and Cooper's Hawks, a few of the Sparrow Hawks, some Marsh Hawks, all of the Great Horned Owls, and apparently most of the Screech Owls—remained in the vicinity of their nest sites. Migrants arrived from northern regions, some stopping temporarily, others remaining to spend the winter. American Rough-legged Hawks stayed during early and late fall for varying periods. Some migrants remained to winter in numbers more or less in proportion to the food supply. In 1941-42, when the meadow mouse population was high, large numbers of American Rough-legged Hawks, immature Red-tailed Hawks (Plate 12), Marsh Hawks, Long-eared and Short-eared Owls wintered in the township. In 1946-47, 1947-48, and 1948-49, when less food was available, almost no immature hawks wintered and fewer mature migrants remained. During every winter the hawk population became stabilized by December, showing little change in composition or numbers until spring. This winter population of raptors adjusted itself spatially to take up all or most of the desirable hunting sites and, once established, the birds remained in definite areas which will be called ranges.

As spring approached, with occasional days of warm weather, the northward migration began, producing population changes that did not become stabilized until a nesting pattern was established.

Marsh Hawks and American Rough-legged Hawks, winter residents from farther south, were the first to return. Rough-legged Hawks paused on their northward flight for varying periods. Some Marsh Hawks passed through the area, while others became nesting residents. Red-tailed Hawks, Sparrow Hawks, and Cooper's Hawks returned—some, it seemed, to rejoin mates wintering in the township. The Red-shouldered Hawks swept into the township in late February and early March, sometimes *en masse*, at other times straggling in, a few at a time, but always coming in sufficient numbers to alter greatly the composition of the hawk population.

This is not the place for discussion and proof of this general account of major movements, but it is helpful for the present in giving a basis for an understanding that the investigators have obtained through laborious accumulation of field data.

DEFINITION OF HAWK RANGES

An important part of this study was accurate determination of the wintering hawk populations. Explanation of methods used requires an outline of the extent and nature of hawk movements. This in turn involves a thorough treatment of hawk ranges. Here we reach a dilemma, for, while movement cannot be discussed without considering the existence and nature of ranges, a thorough treatment of ranges requires an understanding of movements as well as of population characteristics. We shall start, therefore, with definitions of phenomena whose existence we shall later prove.

A winter hawk range is a rather limited area of land over which a hawk moves and hunts during a given period. It is an undefended area, thus differing from a territory as defined by Noble (1939), supported by Nice (1943) that is, "A territory is any defended area." Under certain conditions, defense of favorite hunting perches may be occasionally seen, but a hawk winter range as a whole is undefended. A winter or seasonal range can be subdivided into daily and weekly ranges. Some hawks maintained home ranges analogous to those described by Burt for small mammals (Burt, 1940). These were areas in which the birds not only wintered but over which they also moved while performing mating and nesting activities. Each home range existed for one or more years.

THE NATURE OF HAWK MOVEMENTS

Three types of movement were revealed by the studies. They are termed intra-range movement, drift, and seasonal migration. The extremes of these types merge: thus there is a point where extensive intra-range movement becomes drift, where continuous drift becomes migration. Nevertheless, all are commonly well defined.

Intra-range movement

Intra-range movement is movement within a rather definite and relatively small area or range. Range size varies with numerous environmental conditions, but its maximum is limited by the physical characteristics of the species. Such movement embraces the daily, weekly, seasonal, and even yearly movement of hawks in their numerous activities. In winter the principal activity is hunting. Intra-range movement greatly predominated over all other movements during the winter period.

Drift

Drift is movement beyond a generally established range. When daily intra-range movement is extensive and daily ranges are near the physical maximum for a species, drift is approached. It is extensive daily movement continued until an area which is capable of supporting a hawk is reached, thus permitting establishment of a range. Drift is therefore movement over an area much larger than a range and can be thought of as movement between ranges.

The stable nature of winter hawk ranges, some maintained for more than three months (Tables 2 and 3), as well as the numerical and compositional stability of the population as a whole (Chapter 3), show that there is little winter drift. Drift is most evident in late fall and early spring. In 1941-42 drift was best exemplified by a gradual increase of Marsh Hawks at a communal roost. During a three-month period the roosting population increased from 22 to 48 birds (Map 5). In 1947-48, when the meadow mouse populations were low, hawk movement was greater, daily, weekly and seasonal ranges generally larger, and drift was more apparent. Thus, a female Goshawk coming into the township on or about January 1, when a severe ice storm blanketed most of the state, was an example of drift; she remained in a local area for about two weeks, taking a toll of at least two pheasants from a large roost before leaving (Plate 47). American Rough-legged Hawks that drifted into the township about November 26, established temporary ranges but drifted out around the middle of December. Marsh Hawks completely disappeared between the middle of December and February 20 and began drifting in on February 20-26, following the first extensive thaw. Other examples could be cited, but the foregoing illustrate the existence and nature of drift.

Migration

Where drift is more or less continuous over long distances, it approaches migration. The term migration is used here, however, to mean seasonal movement of hawks in or out of the study area in sufficient numbers to be readily noticed. It is characterized by general mass movement over long distances during relatively short periods. Its daily manifestation may closely approximate drift, and ranges may even be established for short periods. In 1941 fall migration was first apparent on September 4 and continued until about December 7. Each year this migration was characterized by an influx of hawks into the area, correlated with the departure of many summer residents, the decline in Red-shouldered Hawk numbers being particularly significant. No mass flights, such as are found at focal points on migration routes, were observed. Hawk activity was not a succession of steady direct flights or high milling and circling, as it is at flight concentration points such as Cape May, New Jersey, or along the Appalachian ridges, but was manifested largely by the greater number of hawks present. It was further revealed by the appearance of large numbers of winter residents, such as the American Rough-legged Hawk. These birds arrived on October 25 and November 26 in 1941 and 1947, respectively. Because of the foliage, accurate hawk censuses could not be made prior to November 10; therefore the peak of migration could not be determined. Observation indicated, however, that in 1941 it might well have been during the middle weeks of November. Two fall censuses, during 1941, of 54 and 47 hawks show higher populations than those of any time during the winter, with its maximum count of 41 hawks. The highest fall counts of 19 American Rough-legged Hawks and 40 other Buteos, when compared with the highest winter counts of 15 and 27, respectively, show more Buteos present in fall than in winter. In 1947 the end of the fall migration and the beginning of winter were marked by the departure of Marsh Hawks and American Rough-legged Hawks around December 9-10. (The generic term *Buteo* is used throughout the text to mean Rough-legged Hawks, Red-tailed Hawks, Swainson's Hawks and Red-shouldered Hawks.)

Spring migration was more dramatic and more apparent than fall migration. In 1942, prior to February 27, only one Red-shouldered Hawk was known to be present. February 28, when four mature Red-shouldered Hawks were observed, marked the beginning of an influx or migration of this species. In subsequent days, paired Red-shouldered Hawks were seen defending territory over most of the woodlots. The majority of these spring and summer residents arrived on, or within a week of, the 28th. A nesting count (Chap. 10) indicated that some 46 Red-shouldered Hawks arrived in this migration wave. The foot census clearly indicates the population change. The analysis by species (Table 13, Chap. 3) shows that only one Red-shouldered Hawk out of 61 observations, or 1½ per cent, was recorded during winter and that the ratio increased to 30 out of 94 observations, 32 per cent, during spring. In subsequent years the Red-shouldered Hawk influx was somewhat later, less dramatic, and extended over a longer period. It became apparent about March 8, March 10, and March 6 in 1947, 1948, and 1949, respectively.

The foregoing examples illustrate migration in both its obvious and its barely perceptible form. Migration, not drift, was largely responsible for changes in the compositional and numerical status of hawk populations on the area. Drift was very slight during winter, but has been observed to be more pronounced in other regions under the stimulus of more drastic environmental changes. It only slightly altered the composition of the winter populations and had no noticeable or lasting effect on numerical stability. Intra-range movement is typical of hawks in winter in all regions of the country where we have observed them at that season. It is this type of movement with which we are most concerned.

METHODS OF DETERMINING WINTER RANGES

A range has already been defined as an area of land over which a hawk moves and hunts in a given period. Such a definition implies that hawks do not wander indiscriminately, but confine their activities within local, measureable areas. We shall now describe how the existence of winter ranges was detected and their characteristics were determined. Sight recordings of all hawks seen on the hawk censuses or at random were plotted on large detailed cover maps. Lines connecting the peripheral observations of any single individual indicate an observed range (Maps 2, 3, and 4). In 1941-42 a few individual hawks could be identified by unusual markings or color, missing feathers, by being the only known representative of a species on the area (such as the Red-shouldered Hawk), or by being isolated from other birds of the same species, as was the case with Sparrow Hawks and Cooper's Hawks. Isolation of these individual birds was determined by nearly simultaneous time recordings in widely separated places and each bird was thereafter observed again and again in the same area. Many individual birds were distinguished for a week or more and then intra-range movement caused range overlap, making further individual differentiation of the hawks impossible. In such cases, a range for the entire winter could not be obtained. Because of a smaller wintering population and less range overlap this difficulty was unimportant in 1948.

In making censuses, it soon was apparent that the same species of hawk and often the same recognizable individuals were seen repeatedly in the same localities (Table 1). On February 11, 12, and 13, 1942, hawk censuses were made at the same time each day and under nearly identical weather conditions (Table 63). The total counts were 32, 35, 35, respectively. On each of these three censuses, nine different hawks were seen three times in the same spots where they were first observed. Some were definitely the same individuals, and the others of the same species were thought to be so. Eighteen hawks (the same species each time) were seen repeatedly in the same spots on February 12 and 13, and some of these were identified as the same individuals. Twenty-two hawks were observed to be in the same spots on two of these censuses. Details are as follows:

$$
9 \left\{
\begin{array}{ll}
\text{6 Red-tailed Hawks} & \text{— 3 days} \\
\text{2 Marsh Hawks} & \text{— 3 days} \\
\text{1 Sparrow Hawk} & \text{— 3 days}
\end{array}
\right.
$$

$$
22 \left\{
\begin{array}{l}
18 \left\{
\begin{array}{ll}
\text{8 Red-tailed Hawks} & \text{— 2 consecutive days} \\
\text{3 Rough-legged Hawks} & \text{— 2 consecutive days} \\
\text{5 Marsh Hawks} & \text{— 2 consecutive days} \\
\text{2 Sparrow Hawks} & \text{— 2 consecutive days}
\end{array}
\right. \\[2em]
4 \left\{
\begin{array}{ll}
\text{2 Red-tailed Hawks} & \text{— 2 out of 3 consecutive days} \\
\text{1 Rough-legged Hawk} & \text{— 2 out of 3 consecutive days} \\
\text{1 Sparrow Hawk} & \text{— 2 out of 3 consecutive days}
\end{array}
\right.
\end{array}
\right.
$$

On the afternoon of February 24 and the forenoon of February 25 two more consecutive censuses were made. The total counts were 34 and 31 respectively. Twelve hawks, many recognized as the same in-

TABLE 1

OBSERVATIONS INDICATING STABLE NATURE OF WINTER HAWK POPULATION
(observations made on Car Censuses, 12/7/41–2/25/42)

Species	No. of observations of same individual in same area or spot	Number of days in period embracing sightings	Possible No. of observations
Red-tailed Hawk	5	36	10
Red-tailed Hawk	3	59	10
Red-tailed Hawk	7	73	10
Rough-legged Hawk	3	9	10
Rough-legged Hawk	5	36	10
Sparrow Hawk	4	36	10
Sparrow Hawk	4	24	10
Sparrow Hawk	4	53	10
Sparrow Hawk	6	68	10
Marsh Hawk	3	67	10
Cooper's Hawk	2	8	10

dividuals, and all recognized as the same species, were seen at the same spots on these censuses.

Individual birds of all species were kept under continuous observation from dawn to dusk and their ranges were plotted (Map 6). Such data supported fully the view that individual birds could be recognizable and identifiable at that season by their hunting ranges alone. Systematized census observations and the range plottings (Map 2) established the stability of the population and the existence of daily, weekly, and even seasonal ranges for each individual hawk. The same situation existed in 1947-48, when a much lower population inhabited the area (Maps 3 and 4, and Table 3).

WINTER RANGE PATTERNS

The general pattern of individual ranges was basically the same for all hawk species. For convenience, four range categories will be used:

1. Daily ranges (one day).
2. Weekly ranges (one to several weeks).
3. Seasonal ranges (approximately the fall or winter).
4. Home ranges (a year or more).

Daily ranges

Daily ranges were generally smaller than weekly ranges. Weekly ranges were contained within still larger seasonal ranges. The individual range pattern (Fig. 1) is diagrammatic, but illustrates the characteristic gradual expansion of hunting ranges. Although daily ranges in 1942 were relatively small, various stimuli would cause a hawk to fly one-half mile from one perch to another or to use one perch in the morning and move to another in the afternoon. In the course of a period of several weeks, hawks shifted back and forth between two or more dispersed hunting areas. A daily range was confined frequently to one such area. Several such areas might compose a weekly range. Occasionally the maximum daily range of any one hawk closely approximated a weekly range (Map 4). This was particularly true of hawk ranges in the winter of 1947-48, when the larger size of daily ranges was attributed to a lower and more scattered meadow mouse population, which induced longer flights on the part of the hawks.

Seasonal ranges

The seasonal range of any species were merely an extension of weekly ranges by the shifting from one daily or weekly range to another. The high hawk population existing in 1942 made it impossible to distinguish more than a few individuals; census observations, however, clearly indicated that the ranges of these were typical for all individuals in the population.

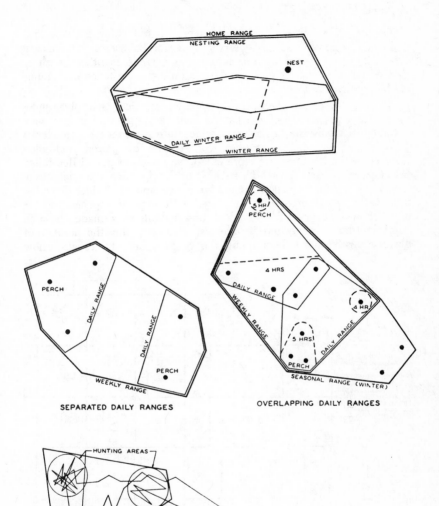

FIGURE 1. Generalized hawk ranges

WINTER RANGES

In 1941-42 a few seasonal ranges were plotted (Map 2). The area embraced is considered the observed range of the hawk, and in most cases closely approximates the actual range, while in others, such as that of the Cooper's Hawk, it indicates only the existence of a range in the area marked.

In 1948 the much lower hawk population enabled us to plot ranges for a larger proportion of the hawks. Data relating to all ranges were obtained, but only the 12 most nearly accurate seasonal ones are shown (Map 3). Four of these ranges were shared by paired Red-tailed Hawks, so that 16 hawks are represented. No ranges of paired Red-tailed Hawks were plotted in 1941-42. Because of the presence of a non-paired, immature population, such ranges were exceptional. Tables 2 and 3 indicate the range number as correlated with maps, the species and sex of the hawks, the period over which observations were made, the area embraced by a range and its greatest diameter, plus the number of observations used in determining the range. These tables also show

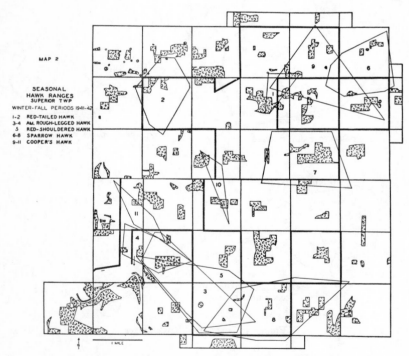

MAP 2. Seasonal hawk ranges, Superior Tp., 1941-42

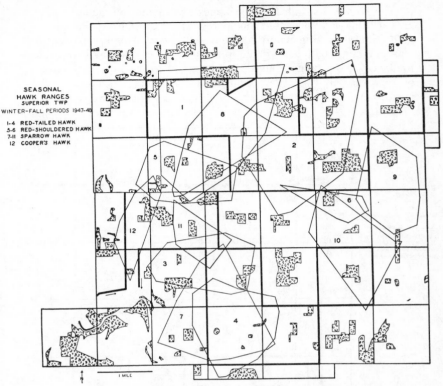

SEASONAL
HAWK RANGES
SUPERIOR TWP
WINTER-FALL PERIODS 1947-48

1-4   RED-TAILED HAWK
5-6   RED-SHOULDERED HAWK
7-11 SPARROW HAWK
12   COOPER'S HAWK

MAP 3. Seasonal hawk ranges, Superior Tp., 1947-48

that there is much variation in the area of ranges of individuals of the same species and that hawks of different species may have ranges of similar size. Since numerous factors determine the range of an individual hawk, we should expect such variation. It is evident, however, that range sizes are limited for all species and that some species characteristically range over larger areas than do others.

The American Rough-legged Hawk ranged over a larger area than any of the other Buteos, and its seasonal range was second in size only to that of the Marsh Hawk. Ranges 3 and 4 (Map 2) indicate variation in range size. These are accurate for the individuals, but are smaller than the average for the species on the area. Because of the difficulty of distinguishing individual Rough-legged Hawks, accurate range data for only two birds were obtained. These were relatively sedentary individuals located in areas of high meadow mouse density. It appeared that the usual seasonal range of birds of this species was four to six square miles in area.

The largest Red-tailed Hawk ranges were about four square miles in area (Table 3). The Red-shouldered Hawks exhibited the smallest daily ranges, were generally less active than the other Buteos, and had seasonal ranges with areas of 1.5 to 2 square miles. Sparrow Hawks often used small daily ranges, but all had larger seasonal ranges with an area of approximately two square miles each.

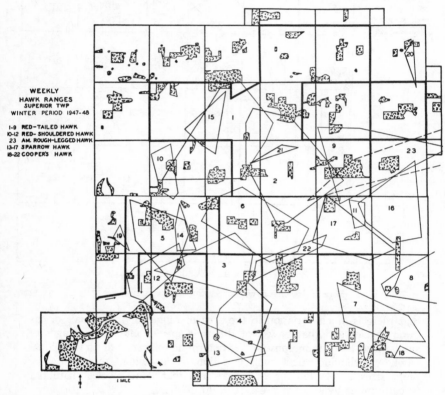

WEEKLY
HAWK RANGES
SUPERIOR TWP
WINTER PERIOD 1947-48

1-9   RED-TAILED HAWK
10-12 RED- SHOULDERED HAWK
23   AM. ROUGH-LEGGED HAWK
13-17 SPARROW HAWK
18-22 COOPER'S HAWK

Map 4. Weekly hawk ranges, Superior Tp., 1947-48

The seasonal ranges of Cooper's Hawks were known to include an area greater than is represented by Ranges 10 and 11 (Map 2). These appear small because of a scarcity of observations. Thus, in the case of Cooper's Hawks, range diameters are better indicators of range size than are range areas. An average winter range for this species would have a diameter of between 1.5 and 2 miles. Range 12 (Map 3) is the most accurately determined Cooper's Hawk range. For practical purposes no seasonal range of a hawk (Marsh Hawks excepted) exceeded

TABLE 2

Data Relating to Individual Hawk Ranges 1941-42
(single birds, no pairs)

Range No.	Observed area in sq. miles	Max. observed diam. in miles	Inclusive dates of observations	No. of days	No. of observations	Age and Sex
Red-tailed Hawk 1	.29	1.	11/8 –2/28	113	20	Immature
Red-tailed Hawk 2	1.15	1.5	12/7 –2/25	81	17	Immature
Rough-legged Hawk 3 ...	1.69	2.7	11/15–2/24	101	12	....
Rough-legged Hawk 4 ..	.68	1.5	12/7 –2/25	81	10	....
Red-shouldered Hawk 5 ..	1.94	2.5	12/14–2/25	74	18	Mature
Sparrow Hawk 6	1.16	1.6	11/8 –2/13	98	15	Male
Sparrow Hawk 7	1.64	2.0	12/14–2/28	77	13	Male
Sparrow Hawk 8	2.21	3.0	12/7 –2/28	84	15	Male
Cooper's Hawk 9	1.68	2.3	11/15–2/24	101	5	Mature female
Cooper's Hawk 10	.26	1.6	10/30–1/21	83	4	Mature female
Cooper's Hawk 11	.29	1.6	12/7 –1/29	54	4	Mature female

TABLE 3

Data Relating to Hawk Ranges 1947-48
(pairs and single birds)

Range No.	Observed area in sq. miles	Max. observed diam. in miles	Inclusive dates of observations	No. of days	No. of observations	Sex and Age
Red-tailed Hawk 1 (pair)	3.82	3.0	11/19–2/26	100	54	Mature
Red-tailed Hawk 2 (pair)	3.66	3.1	11/29–2/26	100	34	Mature
Red-tailed Hawk 3 (pair)	1.47	2.0	12/9 –2/20	74	53	Mature
Red-tailed Hawk 4 (pair)	1.82	1.7	11/19–2/26	100	50	Mature
Red-shouldered Hawk 5 (single)	1.50	1.8	11/29–2/23	88	21	Mature male
Red-shouldered Hawk 6 (single)	.49	1.6	11/26–2/26	94	21	Mature female
Sparrow Hawk 7 (single)	2.32	2.1	11/1 –2/26	118	28	Male
Sparrow Hawk 8 (single)	1.80	2.4	11/29–2/20	94	20	Male
Sparrow Hawk 9 (single)	1.45	1.9	11/26–2/7	75	14	Female
Sparrow Hawk 10 (single)	1.67	2.3	12/9 –2/9	63	8	Male
Sparrow Hawk 11 (single)	.65	1.3	11/23–2/26	97	11	Female
Cooper's Hawk 12 (single)	.74	2.0	12/6 –2/8	65	7	Mature female

a maximum diameter of three miles, and the smallest recorded range diameter was one mile. Ranges of all hawks were generally smaller in 1942 than in 1948. Because of the difficulty of recognizing individual Marsh Hawks, seasonal ranges for this species were not plotted. Individual flights, however, indicated that seasonal ranges of some Marsh Hawks were larger than those of any other species, sometimes covering half the Township. The ranges of a pair of Marsh Hawks and a single male were plotted over a period of 43 days, from February 26 to April 8, 1948 (Map 5, Sections 27 and 28). The ranges comprised areas of 0.55 and 0.63 square miles, respectively, with their largest diameters 1.06 and 1.56 miles. These ranges were smaller than the characteristic winter ranges.

Hawk flights

Hawk flights were found to be indicative of range sizes. Most single hawk flights in 1941-42 were very short, generally only a matter of a few hundred yards. Flight distances were recorded whenever a hawk was seen to fly more than one-quarter mile, the distance being measured in a straight line. In this year the Marsh Hawk was the only species whose flights measured more than a mile. The American Rough-legged Hawk flew for considerably more than one-quarter mile. Nine such observations averaged 0.64 of a mile, with three flights of one mile each. The flights of all other species were generally under one-quarter mile. The Red-tailed Hawk and the Sparrow Hawk occasionally flew one-half mile.

In 1947-48 flights were generally longer, but the majority were still under one-quarter mile. The Rough-legged, Red-tailed, Marsh, and Sparrow Hawks were seen to make flights of one mile or more. Thirteen of the longest Red-tailed Hawk flights averaged one-half mile and the longest Cooper's Hawk and Red-shouldered Hawk flights were one-half mile. The difference in abundance and distribution of meadow mice, as well as the difference in hawk densities during the two study years, affected the length of single flights.

Size of daily range

Daily ranges varied in size with different hawk species and different individuals. In general, the size of the daily range was determined by physical limitations of the species, by the available food, location of hunting perches, and whether a winter range was also maintained as a home range, with portions of it therefore defended by a pair of adults.

In 1942 daily ranges were small and often consisted of a prominent perch in good hunting territory and a surrounding radius of one-eighth mile or less. Such a confined daily range might continue for many days, as exemplified by that of the male Sparrow Hawk (Range 7, Map 2,

Plate 11) seen in the same spot on four out of ten winter censuses and observed day after day to hunt from the same perch during an entire day. In one month this bird accounted for nearly all of the 56 meadow mouse kills tallied on 0.25 acres. In spite of a generally limited daily range, the seasonal range of this hawk covered an area of 1.64 square miles (Table 2). Twenty observations of one Red-tailed Hawk (Range 1, Map 2) were made. This hawk was likewise sighted, day after day, hunting from the same perch over fields that had a minimum of 120 meadow mice per acre as determined by snap trapping. Even when going to roost, this hawk flew only 300 to 400 yards from his perch to a woodlot near by. In this case the daily range was small and the observed seasonal range of 0.29 square miles was also small. In 1948 the daily ranges of two pairs of Red-tailed Hawks Map 6) had a maximum diameter of 2.5 and 2.25 miles and an area nearly that of the seasonal ranges. These daily ranges were maintained for weeks at a time and thus were also weekly ranges (Map 4, Ranges 1 and 2).

A lone Red-tailed Hawk (Range 8, Map 4) confined much of his activity to less than 40 acres. This was not typical of the Red-tailed Hawks, nor did it typify the particular individual, since on subsequent days this bird's range was very much greater. It shows that there was variability in the daily range pattern of an individual even from day to day, but in general a routine whose basic pattern was similar for individuals and species alike was followed.

The daily range of one Red-shouldered Hawk did not exceed one-quarter mile in any direction from a central point, and was confined to an area of 160 acres. This daily range, which consisted of five favorite perches, was maintained for a period of many days and was observed in detail for four days thus in this case the daily and weekly ranges were practically the same.

In general, the daily ranges of the Red-tailed Hawks were about two miles in extent, those of the Red-shouldered Hawks one-half mile, and those of the Sparrow Hawks 1.5 to 2 miles. The range can best be described as a series of perches, some representing hunting perches, others merely temporary stops between hunting grounds. Perches on the periphery of the range of one bird were utilized by other birds of the same or different species with no visible conflict except in the cases of paired adults.

Daily ranges of the Marsh Hawk differed from those of other hawks. All the Marsh Hawks that hunted the Township in 1942 roosted in a community group in Section 25 (Map 5). This population was at its peak of 48 hawks during the last week of January and the first week in February. The Marsh Hawks fanned out each morning, flying varying distances to hunt, and returned at evening to roost. Their range patterns were a series of flights from the roost to intensively hunted areas. A daily range consisted of one or several such areas (Fig. 1), varying in size from 30 or 40 acres to a square mile or more.

Marsh Hawks were observed hunting in one or two fields all day and for consecutive days. The distance covered by a Marsh Hawk in any one day was generally great, even though the area hunted might be small. This tended to give the Marsh Hawk range a linear as well as an areal pattern. Over a period of time the Marsh Hawk ranges increased in total area, just as did the other hawk ranges.

Observations on the morning of February 25, 1942, illustrate this linear range pattern (Map 5). By 7:30 A.M. sixteen Marsh Hawks had been observed leaving the roost within an interval of a few minutes. Single hawks or small groups flew off in definite courses as though each were heading for a distant hunting ground. The main direction flight was N and NW over the Township, although a few birds went NE and W. When the last hawk had left, two observers drove over the Township, marking the location and checking the time of each Marsh Hawk observation. The distance of each hawk from the roost (measured in an air line) was correlated with the interval of time which had elapsed since dispersion.

MARSH HAWK MORNING DISPERSAL FROM ROOST

Elapsed Time Hrs. Mins.	Miles from Roost	Activity
8	1	Flying N over road
15	4	Hunting
34	5.5	Flying North
45	2.5	Perched
1 : 00	5.5	Hunting
2 : 10	5.0	Hunting
2 : 30	.25	Hunting
2 : 45	2.5	Hunting
2 : 55	3.5	Hunting

The lack of correlation corroborated other evidence that the hawks flew directly to hunting areas—some close and some distant—and then coursed over these areas. This activity pattern explained the presence of Marsh Hawks in the same spots on consecutive censuses (Page 19). Flying in a more or less straight line to these hunting ranges at a measured average speed of 25 miles per hour, the hawks were soon

	Per cent of area of total census strip	No. hawks	Per cent of total Marsh Hawk Numbers
1st Q	32.2	38	37.2
2nd Q	20.1	12	11.8
3rd Q	25.6	21	20.6
4th Q	22.1	31	30.4

widely dispersed. Throughout the winter, Marsh Hawks were seen in all sections of the Township, census observations indicating a rather even distribution in all quarters of the area.

The second quarter, farthest from the roost, had the lowest percentage of observed hawks in relation to the area covered by the census. The fourth quarter, containing the roost, had the highest per-

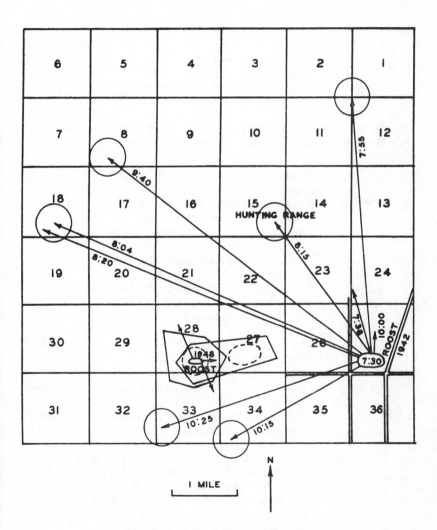

MAP 5. Marsh Hawk flight and dispersal

centage in that relation. It seems logical that the number of hawks hunting a given area should decrease with distance from the roost.

Activity Associated with Daily Ranges

In 1947-48 an effort was made to obtain more detailed information about daily hawk activity, since activity delimits a range. Accordingly, a number of birds of several species (Red-tailed Hawk, Red-shouldered Hawk, Marsh Hawk, Sparrow Hawk) were kept under constant observation from dawn to dusk. The daily movements of two pairs of Red-tailed Hawks are discussed here in detail and illustrated on Map 6.

January 20, 1948.

Daily Activity of a Pair of Red-tailed Hawks (Range 1, Map 4)

The following numbers correspond to numbers along the heavy lines (Map 6) showing the daily intra-range movement of a pair of Red-tailed Hawks.

Both birds were noted soon after they left the roosting site. Snow covered ground, sky clear, slight wind, temperature varied from 5° to 10°F.

1.	8:43- 8:45 A.M.	Hawks were first spotted more than one-half mile apart.
2.	9:05	Birds left perches simultaneously, one flying more than one-half mile to join mate.
3.	9:10	The pair left perch together, flew almost a mile in company, making several brief stops.
4.	9:12	Hawks took separate perches one-quarter mile apart and hunted.
5.	9:15	Both birds changed perches, separating them by one-half mile.
6.	9:35	Both flew to common perch, remained together less than a minute.
7.	9:36	The pair flew off simultaneously, covering one-half mile before perching together.
8.	9:38	Hawks perched side by side for 22 minutes.
9.	10:00	Hawks flew off together, but perched separately, one-quarter mile apart.
10.	10:10	They left their perches simultaneously and joined each other in flight. A male Sparrow Hawk dived at them in a playful manner.
11.	10:10 A.M.- 1:20 P.M.	Hawks perched separately, one lost to view. The bird that was lost to view joined mate from 10:40 to 1:20. They perched less than 50 feet apart (2 hours, 40 minutes). One bird did not move from its perch for two hours, the other made two strikes, but results could not be determined.

12. 1:20-   One hawk departed, followed 15 minutes later by its
     1:35 P.M.   mate. Both were lost to view until seen more than
                  one-half mile away, near original morning perch, on
                  eastern border of range.

13. 2:05   Both hawks perched along eastern border of range.
                  No defense by pair of hawks to east.

14. 2:40   Pair returned to other early morning perch, spend-
                  ing most of the afternoon hunting in the general
                  vicinity. During the course of the afternoon numerous
                  short flights were made and the birds frequently ex-
                  changed perches.

15. 3:12   Both hawks flew one-half mile to western border of
                  territory, then returned to starting point by different
                  routes.
                  Flight lasted 28 minutes. Appeared to be a purely
                  excursionary flight, as no attempt was made to hunt
                  or to defend western border of range.

16. 3:40-   From 3:40 to 5:26 the pair hunted in a very localized
     5:26   area. They hunted independently, but at no time were
                  they more than one-quarter mile apart. Perches were
                  exchanged frequently and all flights were short.
                  Twelve flights, none of more than 200 yards, were
                  recorded in one hour and 46 minutes.

17. 5:30   One hawk caught a meadow mouse running on top
                  of snow, 90 feet from the observers. Finished meal in
                  five minutes and then flew one-half mile to woodlot
                  to roost. Other hawk had preceded mate to roost by
                  eight minutes. Dark at 5:35 p. m.

                                          February 3, 1948.

Daily Activity of Pair of Adult Red-tailed Hawks (Range No. 2, Map 4)

The following numbers correspond to numbers along the light lines
(Map 6) showing the daily intra-range movement of a pair of Red-tailed
Hawks.

Both birds were seen soon after they left the roosting site in woodlot.
Snow covered ground, sky was overcast. Temperature varied from 15°
to 30°F. Generally the sexes could not be distinguished.

1. 8:25 A.M.   Male and female perched together on same tree.

2. 8:28-   One hawk left perch and flew one-eighth mile to a
     8:30   new perch, remaining there until 9:12. Other hawk
                  left and flew one-half mile in opposite direction to
                  perch, remaining there until 9:05.

3.   9:06-
     9:12

Both birds returned to early morning perch (8:25), remained together for two minutes, then flew one-half mile east, making several brief stops.

4.   9:19-
     9:20

One hawk took a high perch at edge of woods, remaining on same perch without moving until 11:24 (2 hours, 4 minutes). Other hawk took a low perch several hundred yards from mate. It changed hunting perches four times, made one attempt to catch a meadow mouse, then left area at 11:20, four minutes before its mate (2 hours, 1 minute).

5.   11:20-
     11:24

The first bird to leave flew one-half mile northwest to a hunting perch. The second bird flew more than one-quarter mile to perch, then made an unsuccessful strike at a meadow mouse.

6.   11:32

Hawks joined each other at common perch, then took off together at 11:40.

7.   11:40

Flying high, the birds went to northern extremity of their range. This flight was more than a mile in length.

8.   11:50-
     11:51

Birds perched in adjoining trees for one minute.

9.   11:51-
     11:58

One hawk made two short flights, taking perches used by pair of Red-tailed Hawks to the west. Pounced on meadow mouse at 12:00. Mate flew more than one-quarter mile to perch, made several short flights, and struck at meadow mouse at 12:20.

10.   12:20 P.M.

One bird flew one mile, making several stops, then circled female Cooper's Hawk with a Bob-white kill, but did not attempt to intrude. The other hawk flew more than one-half mile, making several stops, then perched from 12:48 to 1:04.

11.   1:04

Both birds made several short flights eastward, joining each other in same tree at 1:06. They left simultaneously at 1:30.

12.   1:31

Both flew one-half mile to same perch at edge of woodlot. One left almost imediately to fly south about one half mile. Left this perch, heading south and rising rapidly. A third Red-tailed Hawk in distance took warning and headed back out of territory. Following this sudden retreat of the intruder the hawk flew back to the first perch of the morning.

13.   1:53-
     1:54

One hawk flew one-quarter mile and remained hunting in vicinity until 3:06 (1 hour, 13 minutes). Other bird took perch its mate had vacated, then joined mate in hunting.

14. 2:00–
    3:06
Birds hunted in proximity to each other, frequently perching in same trees. One flushed a covey of 11 Bob-white and made a half-hearted attempt to catch one.

15. 3:05–
    3:07
One bird headed north at 3:07, flying more than one mile in direct flight to chase Red-tailed Hawk to west out of territory. Other bird struck at meadow mouse and may have been successful.

16. 3:30
After chasing intruder, the hawk made several short flights, then occupied perch used by pair of Red-tailed Hawks to west. Remained in the vicinity until 3:45. Mate continued hunting.

17. 4:00
Birds hunted together in same area.

18. 4:05
One hawk killed a meadow mouse. The other hunted edge of woodlot until 4:35.

19. 5:40
Hawks flew to roost in woodlot. Suspect they roosted together.

20. 5:42
Dark.

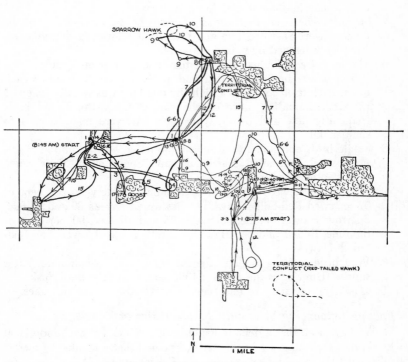

MAP 6. Daily activity of Red-tailed Hawks

The daily range (No. 2) consisted of two major hunting areas situated less than a mile apart. Approximately 7 hours and 40 minutes of nine hours were spent in hunting these two areas. The birds changed perches 52 times during the day, some perches being visited two or three times. The longest time spent at any perch without movement was 2 hours and 4 minutes. The shortest time was less than one minute.

Winter Daily Activity Pattern

From data on the daily activity of 5 Red-tailed Hawks, 1 Red-shouldered Hawk, 2 Sparrow Hawks the following general conclusions on activity can be drawn.

1. Individual birds and pairs followed definite activity patterns day after day.
2. Early morning "warm-up flights" of an hour or less were recorded for all species observed. The hawks visited peripheries of the range during "warm-up flights" and made no concerted effort to hunt.
3. A hunting range generally consisted of two major hunting areas, which were alternately hunted morning and afternoon, though not always in the same order. Most of the hunting hours of each day were spent in these major hunting areas.
4. Single perches were used for as long as two- and three-hour periods.
5. Perches were revisited two and three times during a day and frequently exchanged by paired birds.
6. The majority of flights were less than one-quarter mile each.
7. Paired hawks hunted together and at times were in view of each other. A shift of perch, or a flight by one bird was usually followed by a corresponding move on the part of the mate. Paired birds frequently flew and perched together, but were not observed to share a kill.
8. Single birds and immatures evinced no defense of a hunting range. but paired Red-tailed Hawks defended portions of their range against other Red-tailed Hawks.
9. High perches were preferred to low ones. Ninety-six per cent of 52 recorded Red-tailed Hawk perches were between 30 and 70 feet. Sparrow Hawks perched between 25 and 70 feet, 90 per cent of the time. Red-shouldered Hawks generally perched lower than Red-tailed Hawks.
10. Very little hunting is done the first one and one-half hours after sunrise. Hunting is the major activity and continues until dusk.
11. Hawks make far more unsuccessful strikes than successful ones.

Factors Influencing Size and Characteristics of Ranges

We have seen that a definite winter range existed for each individual hawk, that such ranges were limited in size and that a number of factors influenced the size and characteristics of these ranges. It was quite apparent that the abundance and availability of prey was a major factor influencing the size of ranges. Small weekly ranges, such as those of Red-tailed Hawk 1 and Sparrow Hawk 8 (Map 2), as well as that of Red-shouldered Hawk 6 (Map 3), were duplicated for numerous other hawks. These small ranges were always correlated with either very high or unusually vulnerable local meadow mouse populations. Likewise, the average daily ranges of hawks increased gradually in size from fall to spring, while meadow mouse populations went down.

In 1947-48 meadow mouse populations were much lower, and, although ranges were still limited and localized, they were generally larger than in 1941-42 (Tables 2 and 3, Maps 2 and 3). Single flights of practically all species were longer. The longest recorded flight of a Red-tailed Hawk was one and one-half miles. Hunting areas were poorer and not so numerous. To secure food, hawks had to fly farther. The seasonal Red-tailed Hawk Ranges 3 and 4 (Map 3) are much smaller than 1 and 2 of the same year. Ranges 3 and 4 were in the most extensive area of high meadow mouse density found in the township. Meadow mouse populations within Ranges 3 and 4 were decidedly higher than those within Ranges 1 and 2, and the meadow mouse habitat was less dispersed. Therefore, within certain bounds the size of ranges is inversely proportional to the food supply. We shall see later that the winter populations of raptors that are not normally year-round residents of a given area vary directly with the density of prey. When these raptor populations are low, ranges of some species, particularly those of the Marsh Hawk and American Rough-legged Hawk, may be so extensive as to approach the condition of continuous drift.

Environmental changes, such as flooding and freezing of large areas of prey habitat, plowing, burning, and grazing, affect ranges. When such changes occur, they cause prey populations to move, to concentrate, and to become more vulnerable generally. Hawks whose daily ranges are within sight of such environmental disturbances will move into these areas to hunt. Thus such disturbances may cause a temporary extension of an otherwise routine range and for a short period may cause the daily ranges of hawks present in the disturbed area to be very limited. Such movement of hawks in response to increased prey vulnerability brought about by environmental disturbances has been seen and recorded in the literature by numerous observers. A few examples, therefore, are sufficient to show the relation of such movement to ranges. For a two-week period in November, 1941, an area of 320 acres maintained a concentration of at least eight hawks. At times all of these birds were hunting over an area of 160 acres or less. This concentration was the result of both plowing and heavy grazing of fields whose meadow mouse populations were high. By December 7 meadow mouse density on this area had greatly decreased and the hawks dispersed. Similarly, in early December, 1947, two Red-tailed Hawks (Range 4) and a Sparrow

Hawk (Range 7, Map 3) took up daily hunting ranges near a freshly plowed field for a period of six days and then returned to former hunting areas. This movement was, however, within their seasonal ranges. A small drainage ditch adjacent to the field offered the only cover for the meadow mice whose habitat had been destroyed suddenly and completely. Consequently, within a short time the hawks had lowered the meadow mouse population to a point where it was more profitable to hunt elsewhere. This area was not hunted intensively again.

In Jackson Hole, Wyoming, Swainson's Hawks regularly concentrate in fields during plowing and even follow close behind the plow in order to pounce on meadow mice as their habitat is destroyed. In this same region, in the winter of 1947, American Rough-legged Hawks, previously widely distributed, concentrated in an area along the Gros Ventre River when flooding and freezing deprived meadow mice and other mice of habitat security. In another instance, a herd of elk so thoroughly trampled a grass-cattail marsh that protective meadow mouse cover was destroyed. For several weeks three American Rough-legged Hawks and three Short-eared Owls from a nearby roost, as well as coyotes and numerous Ravens, moved in to hunt. As the prey population decreased, the raptors again dispersed.

In all these instances only raptors from within a limited but variable distance moved in unless these concentrations occurred during migration. The distance the hawks moved in order to take advantage of increased prey vulnerability appeared to be influenced largely by the daily and weekly ranges of the birds and by their ability to sight other hawks already hunting the area. The latter factor is influenced by topography and by the hunting methods employed by the hawks. Soaring birds can be seen by other hawks for long distances. In Superior Township the level country and the dearth of high flying or soaring (Table 8, Chap. 3) limited this type of movement to a radius of approximately 1.5 miles from the disturbed area. Thus, hawks whose normal range peripheries were within 1.5 miles of such environmental disturbances might suddenly extend their ranges to hunt a relatively small area for a comparatively short time.

In Superior Township the Marsh Hawk and the American Rough-legged Hawk were found to be most given to range extension of this kind—due in part to their hunting technics, larger ranges, and greater flight activity. Fifty and 65 per cent of the American Rough-legged Hawks observed in 1942 and 1948, respectively (Table 9), were flying when first seen, often hovering to hunt. Marsh Hawks were found to be on the wing 50 per cent of the hunting day (Table 12). This leads to another important factor that was found to influence ranges.

The maximum hunting range of any species of hawk is limited by the physical characteristics of the species. For example, the Marsh Hawk is adapted for hunting in flight by its long wings, relatively light body, and powers of endurance. It has developed a low, coursing flight that takes it easily over long distances. The heavy, relatively shorter-winged Red-tailed Hawk must work hard to fly when there are few rising air currents, a characteristic of Michigan winter weather. Short flights and perch hunting, rather than soaring, are resorted to. The same is true of the American Rough-legged Hawk except that this species flies and hovers more than the Red-tailed Hawk if encouraged by breezes or rising warm air currents, such as exist even in winter on sunny days. The result is that under similar conditions the Rough-legged Hawk generally ranges over more territory than does the Red-tailed Hawk.

We have observed that the ranges of diurnal raptors of the same family or with similar hunting habits generally are directly proportional to the size of the bird. Thus, vultures and eagles range over more territory than Buteos, the American Rough-legged Hawks cover more ground than the Red-shouldered Hawks, and the Cooper's Hawks range farther than the Sharp-shins. Prairie Falcons and Duck Hawks have ranges that are larger than those of Sparrow Hawks or Pigeon Hawks. Falcon ranges, in general, are larger in relation to the size of the bird (Map 21).

In 1947-48 paired mature Red-tailed Hawks characterized the Red-tailed Hawk population. These mature birds exhibited defense of at least part of a winter range, this in turn being part of a home range containing a past and a potential future nest site. Such defense influenced the size and character of neighboring winter ranges.

The density of the hawk population affected the size and character of winter ranges, the ranges generally being smaller in 1941-42 when the township supported 108 wintering hawks than in 1947-48 when there were only 27 birds (Table 16). Ranges of the same and different species overlapped and were even one contained within another in 1941-42, although this is not evident from Map 2. In 1947-48 ranges of the same species overlapped on the peripheries, but tended to be isolated from one another. This was the most outstanding difference between winter seasonal ranges of the two years. Isolation of ranges of 1948 in contrast to the range overlapping of 1942 was the consequence of the presence of a smaller hawk population, a large number of paired birds, and the less dense and more dispersed meadow mouse population.

Interrelationship of Ranges

In 1942 the ranges of some hawks, both of the same and of different species, not only overlapped but lay completely within the range of other hawks. This was true of daily, weekly, and seasonal ranges. Two birds often hunted from the same perch side by side. Four or five hawks of the same or of different species hunted the same field at the same time. Often a hawk would fly to a perch already being used. Both birds might hunt for a short time and then move off to another perch.

Good hunting perches were interchangeable and used by all species whose ranges overlapped. A Red-tailed Hawk would use a perch one day, and a Red-shouldered Hawk the next, or a Rough-legged Hawk might hunt from it for several mornings, a Red-tail in the afternoon, and the next day a Marsh Hawk might course the same fields. At times all four species would hunt the same area at the same time. It appeared that the number hunting any such small area at the same time depended on the ability of each to secure the required food. Thus there existed a hunting tolerance which operated automatically to allow four or five birds or only one, depending upon availability of food, to hunt a field. Apparently no disputes arose. With the approach of a new hawk to a field where several already were hunting, frequently one hawk, the new arrival or one already there, would move off to another hunting area. It was as though each particular area had a definite carrying capacity which was recognized by the hawks. Thus within the limits of individual ranges of the birds the composition at any place would vary continually. Since the range of each bird in the area overlapped the ranges of several others, the result was a definite spatial relationship or density of hawks in relation to their immediate environment. The movement of hawks in response to movement of others could be thought of as unsocial behavior or a very slight indication of inter- and intra-specific strife. However, it much more satisfactorily illustrates, not strife, but a tolerance of other individuals, an automatic cooperative adjustment of every individual as part of a population.

This cooperative adjustment so thoroughly characterized the 1942 winter hawk population that only one instance of defense was observed. This occurred when two mature Red-tailed Hawks defended a small area of woodlot where they later nested.

In 1947-1948 ranges of the same species, though largely separated, still overlapped somewhat, and hunting perches were interchangeable. Conflict occurred within the overlapping areas when (in the case of pairs) one or both members of two pairs tried to hunt the same spot at the same time. Occasionally conflict between birds of different species occurred, but, more often than not, there was tolerance. Red-shouldered Hawks and Sparrow Hawks, as well as Red-tailed Hawks and Sparrow Hawks, hunted the same fields at the same time from perches situated close together. Conflict between different pairs of Red-tailed Hawks and less often between unpaired birds for hunting perches was observed. Nevertheless, it was still exceptional. Generally, when two to four birds came together in one small area of their ranges, an adjustment was promptly made by movement of some members of the group to other areas.

Duration of Ranges

Many of the winter ranges illustrated by Maps 2 and 3 existed as fall ranges. Some continued into spring. The winter ranges of fall migrants which wintered on the area and nested elsewhere were broken up as spring migrants arrived. Two pairs of Red-tailed Hawks nested and wintered for three consecutive years. One pair of Sparrow Hawks wintered and nested for two years. Pairs of other species were suspected of remaining in a home range for a year or more.

SIGNIFICANCE OF MOVEMENTS AND RANGES

The great mobility of raptors enables them to concentrate during migration in areas of high prey density. The establishment of definite, relatively small, stable winter ranges provides a continuous hunting pressure on prey species in proportion to their numbers. The importance of this hunting pressure will be analysed after presentation of other data equally necessary to understanding the mechanics of predation.

Following this discussion of the stability of winter raptor populations, we can undertake the problem of accurately determining the numbers of the hawks and owls and the composition of the raptor population.

CHAPTER 3

Fall and Winter Hawk
Populations

TO ARRIVE at reasonably correct conclusions concerning the management of wildlife, the basic population data must represent as closely as possible the actual conditions existing. The census is a tool that is developed and used for this purpose. Taking a census is usually a task that consumes time and energy and continues week after week and month after month and may span a period of many years. In this and in following chapters the census methods used will be described.

In making any census, numerous conditions and variables must be known and evaluated if accurate results are to be obtained. The more completely the habits of a single species are understood, the more accurate can a census of the species be made. This of course, is equally true of a mixed hawk population. The counts of hawks were obtained by making a census from a car, as described earlier. When interpreted in the light of known and measured influencing conditions, these counts give a total hawk population for both the census strip and the township. The car census was not used inflexibly to make a census of a mixed population but was the basic method, adjusted in the case of each species to take into account variations in activity, habitat association, and visibility. It should be borne in mind that, though a relatively simple census method adaptable to various conditions would be desirable, the main objective was to determine accurately a population on a definite area during a given period. The lack of such data has hitherto made quantitative evaluations of raptor predation impossible. An accurate determination of the raptor populations to express hunting pressure was essential to an analysis of predation. Special emphasis was attached to this phase of the problem, which is presented in detail. It is therefore advisable to look at some of the characteristics of the hawk populations and to discuss how and to what degree certain characteristics and conditions influenced the censuses. In evaluating the censuses, three categories of information must be considered.

First, the habits and characteristics of the hawk species and of the collective populations in which they were comprised.

Second, visibility factors and the observer's ability to see and identify accurately all the hawks on the strip or a constant proportion of them.

Third, recognition and measurement of conditions affecting the censuses and application of this information in computing winter populations.

CHARACTERISTICS OF THE HAWK POPULATION AS RELATED TO THE CENSUS

Although winter populations were computed for four years (Table 16), comparisons will be largely confined to the two principal study years, 1941-42 and 1947-48. In these years activity changes, as well as numerical and compositional changes in the hawk populations, made it desirable to divide the year into four periods—Fall, Winter, Spring and Summer. We are concerned for the present only with the fall and winter populations and with the spring population changes prior to the establishment of the nesting populations.

Species Composition of the Population

Six species of hawks, the Red-tailed Hawk, Red-shouldered Hawk, American Rough-legged Hawk, Sparrow Hawk, Cooper's Hawk, and Marsh Hawk, were represented in the fall and winter populations. This species composition remained relatively unchanged in the fall and winter of 1942 until the departure of the American Rough-legged Hawks, in April. In 1948, Marsh Hawks and American Rough-legged Hawks were not important constituents of the winter population. In this same year a Goshawk hunted the area in early January and then drifted on. No Sharp-shinned Hawks were recorded during either year, although this species was identified in neighboring areas. A few migrant Broad-winged Hawks appeared in spring, and Ospreys occasionally entered the area to fish the two lakes.

Although the composition of the population changed from season to season and also with the years, Buteos remained the numerically dominant element. Buteos composed an average of 50 per cent of the

winter populations during four winters. In 1942 the Buteo population
contained a predominance of immature birds, approximately 85 per cent
of the Red-tailed Hawks being birds of the year. This contrasted with
the 1948 Buteo population, with only one immature Red-tailed Hawk
present during fall, and a wintering population composed entirely of
mature birds. This predominance of mature Red-tailed Hawks was
characteristic of the other two winters.

In both years the American Rough-legged Hawks were most numerous in fall. The 1942 winter hawk population was characterized by a
large number of this species, which made up 20 per cent of the observed hawks. In 1948 American Rough-legged Hawks were absent
from the area during most of the winter and were only 2 per cent
of the total hawks observed (Table 4).

All Red-shouldered Hawks were mature birds. They were 2 and 8
per cent, respectively, of the hawks observed in the winters of 1942
and 1948.

In each fall Marsh Hawks were conspicious, being 19 and 18 per
cent, respectively, of the hawks observed (Table 4). In 1942 they were
39 per cent of the computed winter hawk population, but they were
not seen during the 1948 winter period until February 20 and the beginning of the spring migration.

The numbers of Cooper's Hawks and Sparrow Hawks fluctuated little
from winter to winter. The number of wintering Sparrow Hawks was the
same in 1942 and 1948, when they were 6 and 15 per cent, respectively,
of the hawks observed. Cooper's Hawks were only four and five per
cent of the hawks observed during both winters (Table 4), and were
10 and 22 per cent of the computed populations.

The hawk population in 1942 was substantially higher than in other
winters the average seen per census being 33.3, which may be compared with 10.8, 9.5 and 13 for the later years. These figures are indicative of the relative population densities of these years. Table 16 shows
the actual populations for four winter periods. The hawk population in
1942 was considered high in comparison not only with those of these
other years but in comparison with hawk populations observed in
other areas and other years in northern regions of the country (Table 14).

Association of the Hawk Population with Open Country

Although populations of six species of hawks were being studied,
ecologically all elements (except the Cooper's Hawks) functioned as a
single population. The Buteos, the Marsh Hawks, and the Sparrow
Hawks all hunted the open fields and fed primarily on meadow mice.
Cooper's Hawks, although observed hunting in the open, primarily
hunted birds in the woodlots. The degree to which the hawk population functioned as a unit was illustrated graphically by systematized
observational data recorded on both car and foot censuses. During the

TABLE 4

NUMERICAL AND PROPORTIONAL COMPOSITION OF HAWK POPULATION BY SPECIES, AS OBSERVED ON CAR CENSUSES, 1942 AND 1948

		Unidentified Buteos	Red-tailed Hawk	Red-shouldered Hawk	Rough-legged Hawk	Total Buteos	Sparrow Hawk	Cooper's Hawk	Marsh Hawk	Total Hawks
Fall Period	Number of individuals 1942	39	11	3	45	98	5	5	25	133
	1948	7	15	2	4	28	6	2	8	44
	Per cent of Total 1942	29	8	2	34	74	4	4	19	..
	1948	16	34	5	9	64	14	5	18	..
Winter Period	Number of individuals 1942	63	86	5	66	220	21	12	80	333
	1948	26	58	10	2	96	19	6	3	124
	Per cent of Total 1942	19	26	2	20	66	6	4	24	..
	1948	21	47	8	2	77	15	5	2	..
Spring Period	Number of individuals 1942	21	26	16	17	80	10	5	22	117
	1948	6	5	4	0	15	2	0	3	20
	Per cent of Total 1942	18	22	14	15	68	9	4	19	..
	1948	30	25	20	..	75	10	..	15	..
Combined Fall and Winter	Number of individuals 1942	102	97	8	111	318	26	17	105	466
	1948	33	73	12	6	124	25	8	11	168
	Per cent of Total 1942	22	21	2	24	68	6	4	23	..
	1948	20	44	7	4	74	15	5	7	..
All three Periods	Number of individuals 1942	123	123	24	128	398	36	22	127	583
	1948	39	78	16	6	139	27	8	14	188
	Per cent of Total 1942	21	21	4	22	68	6	4	22	..
	1948	21	42	9	3	74	14	4	7	..

combined fall and winter periods of 1942 and 1948, the associations of
466 and 168 hawks, seen on the car census, with types of vegetation were
noted. Of the hawks recorded, 92 and 95 per cent, respectively, were
observed hunting meadow mouse habitat, consisting of open grassland
(Table 5). Five and four per cent, respectively, were associated with
the woodlots. Eighty-seven per cent of the 142 hawks observed on the
1942 foot census were associated with the meadow mouse habitat in
open fields and seven per cent with woodlots. In addition, the majority
of the hawks associated with woodlots were not hunting them, but were
observed soaring or flying over them. Woodlots were systematically
combed by two or more observers to determine the use by hawks and
only Cooper's Hawks were found hunting in them. Other hawks used
the edges of woods as perches, but hunted the field habitats. The number of hawks hunting in the woodlots in 1948 did not increase, although
the meadow mouse population was decidedly lower, and the fox squirrel
population higher, than in 1942. The association of each species with
vegetation types (Table 6) again shows that all species, with the exception of the Cooper's Hawk, were hunting chiefly the open field habitats
during the hours when censuses were made and this likewise held true
for all hours of the day. Only in the most inclement weather did the
hawks seek daytime shelter in the interior of woodlots.

This association of the collective hawk populations with the field
habitats of meadow mice also is shown by the analysis of winter pellets,
in which 92.6 per cent of 8,146 food items and 91.1 per cent of 700
food items were meadow mice in 1942 and 1948, respectively.

The evidence clearly indicated that each year the hawk populations
functioned as single units, hunting the fields and preying largely on
meadow mice. Their almost complete association with open-country
habitats made the car census practical and accurate. Theoretically it was
possible to see almost every hawk on the census strip, since only the
Cooper's Hawks frequented the woods and this species was not numerous.

DISTRIBUTION OF HAWKS

During both winters, hawks concentrated in the better hunting habitats, though not always in localities of highest prey density. Concentrations, such as those illustrated on Map 11, are comparable and show
the hunting pressures in some local areas as contrasted to other locales.
Such hawk concentrations were distributed widely and rather evenly
in 1942, but were localized in 1948. In 1942 the distribution of hawks
on the census strip during a six-month period was such that counts
were directly proportional to the extent of land area on which they were
made. For instance, each quarter township supported approximately an
equal number of hawks (Table 7). Thus there was a numerically equal
distribution of hawks by quarter townships, but this distribution was

characterized by concentrations in both time and space. Meadow mice likewise were widely distributed and abundant throughout the study area. In 1948 both hawk and meadow mouse populations were distributed less evenly. The third quarter of the township contained the greatest concentration of hawks and likewise supported the highest meadow mouse densities. The first and fourth quarters supported correspondingly fewer hawks. Regardless, however, of the manner in which hawks were distributed, the census strip was large enough to yield a representative count (Map 1). Although this strip was only a 50 per cent sample, much more than 50 per cent of the hawks could be seen in the course of a long period, in consequence of intra-range movement of the birds.

TABLE 5

ASSOCIATION OF HAWKS WITH VEGETATION TYPES

(Observations made during car censuses of hawks)

November 10/41—April 30/42
November 19/47—March 24/48

Vegetation Types	Fall and Winter Periods				Spring Period			
	Individuals		Per cent of Total		Individuals		Per cent of Total	
	1942	1948	1942	1948	1942	1948	1942	1948
Grassland, abandoned and cultivated fields	430	159	92	95	94	13	80	65
Woodlots	25	7	5	4	13	7	11	35
Wet areas of marsh, kettle, seepage	11	2	2	1	10	0	9	0
Total No. of Hawks ...	466	168	..	..	117	20	..	..
No. of censuses	13	18	..	..	5	1	..	..

Relation of Population Stability and Hawk Movement to Census Figures

It is evident from the discussion of ranges (Chapter 3) that the winter hawk populations (within limits) were stationary populations (Maps 2 and 3). It was also true that for any given short period (for example, the three hours required to make a census) a population quite literally could be considered stationary. This resulted from the fact that the hawks remained on their ranges, and most flights were less than one-quarter mile. In 1942, meadow mouse populations were so dense and well distributed that long flights were unnecessary. Flights actually were more frequent, however, than in 1948, perhaps because of a need for more hawk movements in consequence of the hunting activity of a higher raptor population (Tables 8 and 9). In

TABLE 6

Seasonal Association of Hawks With Vegetation Types Expressed In Per Cent of Total Observations Made During 1942 and 1948 Car Censuses of Hawks

Vegetation type		Grassland, abandoned and cultivated fields		Woodlots		Wet areas of marsh, kettle, seepage	
Periods		Fall and Winter	Spring	Fall and Winter	Spring	Fall and Winter	Spring
Marsh	1942	100	73	0	0	0	27
Hawk	1948	100	100	0	0	0	0
Unidentified	1942	89	76	6	24	5	0
Buteos	1948	100	57	0	29	0	14
Rough-legged	1942	97	100	2	0	1	0
Hawk	1948	100	0	0	0	0	0
Red-tailed	1942	84	89	12	8	4	4
Hawk	1948	94	60	6	40	0	0
Red-shouldered	1942	75	62	13	25	13	13
Hawk	1948	100	25	0	75	0	0
Sparrow	1942	88	100	12	0	0	0
Hawk	1948	100	100	0	0	0	0
Cooper's	1942	71	40	29	40	0	20
Hawk	1948	75	0	25	0	0	0
Total	1942	90	83	7	14	3	4
Buteos	1948	96	50	3	44	1	6

1948 the longest flights were made in defense of winter hunting ranges or in flying from a hunting perch at one extremity of the range to a perch at the other extremity. Early morning "warm-up flights" were often equally long. The flights of Marsh Hawks to and from a communal roost are not considered, since they took place before and after the census period. Marsh Hawk flights within their hunting ranges were usually not extensive. For example, a pair of Marsh Hawks, observed from dawn to dusk, hunted almost entirely within a 40-acre field and made only an occasional swing away for a distance of less than one-half mile (Map 5, Chap. 2). The male hunted on the wing, frequently landing to watch for mice. For a 35-minute period in the morning he averaged one landing per minute; and in the afternoon he was observed to land 13 times in 12 minutes. Many flights were of longer duration, but not long in linear distance covered. This flight activity was duplicated by the female. Marsh Hawk flights of a mile or more were made while they were flying from one daily hunting spot to another. Of 12 flights recorded for a pair of Red-tailed Hawks during one hour and 46 minutes, none was more than 200 yards. During the combined fall and winter periods of 1941-42 and 1947-48, 42 and 69 per cent, respectively, of all hawks observed during the censuses were perching. The general procedure of most hawks was to spend from three to four hours a day in a local area and from one to two hours on a single perch. Thus, because flights were short and the birds remained perched for long periods, recounts were minimized or nonexistent. In one area where parts of the census route were separated by almost two and one-half hours in time, but by only a mile or so in distance, supplementary counts were made and compared with the regular census records for possible duplication. The totals obtained, as well as the species composition, did not indicate that recounts had been made. For practical purposes, therefore, the population could be considered stationary for a single census.

TABLE 7

COMPARISON OF NUMBERS AND PERCENTAGES OF HAWKS COUNTED DURING CAR
CENSUSES TO PER CENT OF AREA OF EACH QUARTER OF TOWNSHIP
INCLUDED IN CENSUS

Nov. 11/41–April 30/42
Nov. 19/47–March 24/48

Township quarters	Area of each quarter included in census	Per cent of area of total census strip	Total No. of hawks observed 1942	Total No. of hawks observed 1948	Per cent of total hawks seen on each quarter 1942	Per cent of total hawks seen on each quarter 1948
1st quarter	5.92 sq. mi.	32.2	199	34	36	18
2nd quarter	3.70 sq. mi.	20.1	104	43	19	23
3rd quarter	4.70 sq. mi.	25.6	140	83	25	44
4th quarter	4.07 sq. mi.	22.1	108	28	20	15
Total	18.39 sq. mi.		551	188		

In terms of two or more censuses, however, the population had to be considered mobile, as has been shown. Hawks moved within ranges which were definite spatial entities and existed for long periods. These ranges were relatively small and overlapping. Just as the individual hawk could be considered stationary for any short period, so could the ranges be considered stationary and fixed for long periods. The movement of any individual was necessarily movement within a fixed range (intra-range movement). Ranges generally extended farther than a cross section of the census strip and day-to-day movement within them was great enough to cause hawks to move in and out of the census strip. This made it possible for a single hawk to be on the census strip one day and outside it the next. Its perch or hunting area within the strip might be left unoccupied or, in consequence of overlapping of ranges and free interchange of hunting perches, might be occupied by another hawk. Movement of only a short distance by hawks near the boundary of the strip would make such changes. It can therefore be seen that though hawks can be considered as stationary for a single census, for the interpretation of several censuses, intra-range movement must be considered.

Wintering hawks remained within definite ranges for long periods, and it can be assumed that all or a large proportion of the population was stable. In fall a condition of partial stability existed simultaneously with drift and migration and was reflected in the censuses. In 1942 the average number of hawks per fall census was 44.3; for each winter census, 33.3. The fall counts of 1947, however, were not higher than the winter ones, as in 1941. The fall average of 8.8 hawks was actually lower than the average of 9.5 in the following winter. This may be explained as a consequence of the much lower meadow mouse population, which was not as effective in stopping and holding the migrating hawks. Differences in population composition indicated migration. For

TABLE 8

HAWK ACTIVITY AS OBSERVED ON CAR CENSUSES OF HAWKS

1942 AND 1948

Percent of Total

	Fall period		Winter period		Spring period		Fall and Winter		Combined	
	1942	1948	1942	1948	1942	1948	1942	1948	1942	1948
Perching	37	54	44	73	33	50	42	68	41	66
Flying	63	39	51	19	33	25	55	24	50	24
Soaring	..	7	5	8	28	25	3	8	8	10
Courtship	..	..	..	..	6	..	..	..	1	..
All flight activity	63	46	56	27	67	50	58	32	59	34

TABLE 9

FLIGHT ACTIVITY OF HAWK SPECIES FALL, SPRING, AND WINTER PERIODS

November 10/41–April 30/42

November 19/47–March 24/48

Determined from Car Censuses

Hawk Species	Total No. Flying		Total Observed		Per Cent Flying	
	1942	1948	1942	1948	1942	1948
Unidentified Buteos	51	6	123	39	42	15
American Rough-legged Hawk	83	3	128	6	65	50
Red-tailed Hawk	62	27	123	78	50	35
Red-shouldered Hawk	11	6	24	16	46	38
Total Buteos	207	42	398	139	52	30
Sparrow Hawk	11	3	36	27	31	11
Cooper's Hawk	11	5	22	8	50	63
Marsh Hawk	118	13	127	14	93	93
All species	347	63	583	188	60	34

example, presence of American Rough-legged Hawks and Marsh Hawks at the time of their migratory movement through the area contrasted with their almost total absence during the winter.

The fall, then, was characterized by a degree of population stability, expressed by individuals holding definite ranges, and also by a degree of change, expressed by movement of an itinerant population in and out of the area. The winter period began when there was no longer a shifting population, but individual birds settled in chosen hunting ranges. Both the hawk count and the Buteo count indicate that in 1941 this occurred about December 7. In 1947 movement stopped at approximately the same date. December 6 and 9 marked the dates of the last Marsh Hawk and American Rough-legged Hawk seen until the spring migration brought them back.

During the winter the Buteo and total hawk counts remained constant (within limits of error discussed later), indicating that no significant migration or drift occurred. A slight influx of Marsh Hawks during both years altered the composition of the winter population just prior to the mass spring migrations. It is possible that more drift than was recorded occurred, but, if so, it was so insignificant numerically as to be undetected on the census or through daily observations. The plotting of all winter hawk ranges in 1948 and the securing of a count of the winter hawk population further confirmed the stability of the winter population. Thus the hawk populations during the winters of 1941-42 and 1947-48 were stabilized, spatially fixed populations. The mixed composition of the population, its almost complete association with open habitat, the distribution of its various members and its numerical and compositional stability were all important conditions considered in taking and interpreting the hawk censuses.

VISIBILITY FACTORS AS RELATED TO THE CENSUS

It is evident that the accuracy of the censuses depended upon the ability of the observers to find and identify either all hawks on the strip or a constant proportion of them.

The numerical stability of the winter population having been established, the constancy of the total census counts would appear to indicate an ability to observe with a uniform degree of accuracy. Familiarity with the census route as well as experience in knowing where and how to scan reduced error, therefore the first few censuses of each year were not used. It is unlikely, yet theoretically possible, that 100 per cent of the hawks on the census strip could be seen on any one census. On the other hand, it is apparent that a hawk feeding on the ground, or perched at the limits of the strip in rainy or snowy weather, or sitting in the shaded edge of a woodlot could be overlooked. Since the ability to observe with a constant degree of accuracy depended largely on visibility factors, these will now be discussed.

Weather Conditions

Most of the hawk censuses purposely were made on clear, sunny, breezy days (Tables 63 and 64). Changes in temperature, humidity, and barometric pressure on these days had no noticeable effect on the number of hawks observed.

Severe weather conditions, including rain, snow, extreme cold and heavy overcast, however, affected hawk activity and decreased visibility, but even the most inclement weather did not affect hawk activity to the extent of preventing hawks from hunting. During periods of bad weather, flight activity decreased markedly. The census of January 20, 1942, gave a low count (22) which was partly the result of poor visibility. Weather during the census period was cloudy, cold, and raw. A count of 40 the following day, with increased sunlight, is in marked contrast. Variation in the number of hawks tallied could not be associated with hawks seeking shelter in woodlots, for intensive field work yielded no data that would correlate a low census count with a shift of hawks to the woodlots as a result of existing weather conditions. In 1942 censuses were made on February 11, 12, and 13 under similar favorable weather conditions (Table 63 and Fig. 2). The total hawk counts were 32, 35, 35, respectively, and the total Buteo counts 25, 21, 24, respectively. The recorded activity was also remarkably similar. Of the hawks seen on each census, 44, 46, and 46 per cent, respectively, were recorded as flying when first observed. Computing the township population from these three censuses, using the method shown later, we get a total winter population of 91 hawks, which may be compared with 96, the number computed from all the winter censuses. Similarly, in 1948, four consecutive censuses under nearly identical weather conditions (Table 64, Fig. 4) yielded, respectively, 8, 10, 9, 9, hawks observed; Buteo counts were 8, 8, 9, and 7, respectively. Visibility conditions were similar on these afternoons. The light conditions at the start of each census (1 P. M.) were as follows:

Date	Foot candles of light measured against the ground (snow)	Foot candles of light measured against the sky
1/16	450	700
1/17	450	700
1/18	900	500
1/19	600	550

In contrast, a day of poor visibility with snow on the ground was:

	9 A.M.	12 M.	4 P.M.
Ground	.8	300	70
Sky	25	400	100

A day without snow to reflect light was much darker, especially in regard to ground illumination. Light conditions for the census of December 6, 1948, corresponded to the figures above and the count of 7 was the second lowest of the season. Possibly hawks perched lower on exceedingly dark days in order to see their prey better, but no proof of this was obtained. If such were the case, low perching would affect counts on dark days. Clear, sunny days with slight wind movement induced flying and soaring, and hawks at the outer limits of the census strip were seen more readily.

We therefore can conclude that weather conditions affected visibility and hawk activity to such a degree that fewer hawks were observed on censuses during inclement weather. Since, however, most of the censuses were purposely taken in fair or excellent weather (Tables 63 and 64), such variations as occurred in the census figures cannot be fully explained on a basis of weather conditions.

Silhouetting Effect

The level, sparsely wooded landscape tended to silhouette both flying and perching hawks, a distinct advantage in finding them. The relatively large size of all species except the Sparrow Hawk increased the effectiveness of this factor. The absence during winter of other large birds, such as Crows and Vultures, which might have required stops for identification, was an advantage. Leafy vegetation reduced the silhouetting effect and the range of visibility, and, in consequence, no accurate censuses could be made in fall before the leaves had fallen nor in spring after the appearance of leaves.

Perching and Flight Activity

The perching and flight habits of the hawks made them easy to see and thus aided the making of accurate counts. The Buteos and Sparrow Hawks selected relatively high, conspicuous perches, from which they could scan a hunting spot (Plate 17). Some indication of the importance of this habit to the success of the censuses was obtained by estimating in 1947-48 the observed height of each hawk species above the ground level (Table 10). There was no indication that such data varied between the two study years. The average height of a hawk (flying or perched) above the ground, as obtained from 143 observations, was 47 feet. The average perching height was 31 feet, the maximum was 80 feet and the minimum three feet. The most frequently used perching heights were between 40 and 30 feet.

These figures reveal that in level, open country the perching height was such as to make the birds clearly and readily visible. The Red-shouldered Hawk, with an average perching height of 19 feet, which is next to the lowest, was the most difficult Buteo to find. Of the unidentified Buteos, only two out of 26 were observed flying. This illustrates the fact, quite apparent to the observers, that perching Buteos were much more difficult to identify than were those seen flying. During the fall and winter censuses of 1942 and 1948, 58 and 32 per cent, respectively, (Table 8) of all hawks observed were engaged in some form of flight activity when first seen and identified. Both the height of perching hawks above the ground and their flight activity were found to favor the accuracy of the censuses (Plate 17).

ERRORS AND THE DEGREE TO WHICH THEY AFFECTED CENSUSES

The degree to which the hawk counts err from the actual number present on the census strip is important. The consistency of the totals suggested that if some method of checking the accuracy of either the census counts or the counts of individual species were devised, one or more corrective factors could be found and applied to convert relative numbers into actual numbers. Long and careful observation made it

TABLE 10

HEIGHTS (IN FEET) THAT HAWKS WERE OBSERVED ABOVE GROUND LEVEL

	Red-tailed Hawk	Red-shouldered Hawk	American Rough-legged Hawk	Unidenti-fied Buteos	Sparrow Hawk	Cooper's Hawk	Marsh Hawk
Average height above the ground	61	33	70	41	30	26	7
Average perching height	36	19	43	39	29	29	7
Perching height most frequently used	40	20	40	30	30	..	..

possible to do this. Weather, activity, and distribution were found to have little effect on the counts. We believe that generally all the Buteos on the census strip were seen and counted and that even the maximum error for any census was negligible. It has been stated already that functionally the Cooper's Hawks had to be considered as a separate population. Their association with the wooded habitats where they could not be seen made counts of these hawks subject to much error. The small Sparrow Hawks, likewise, could not be considered as fully included in the census except by means of numerous counts. Unfortunately, these hawks hunted more by perching than by flying. Only 31 and 11 per cent in 1942 and 1948 respectively, (Table 9) were flying when first observed. During both years 93 per cent of the Marsh Hawks observed during the censuses were seen flying (Table 9 and Plate 31) and in this activity scarcely could be missed. Seven per cent, however, were observed perching on fence posts or on the ground. The chances of seeing even a small proportion of these hawks when they were perched in grass or crop fields were slight. If the number of Marsh Hawks perching on the ground was high, then the proportion of error in the Marsh Hawk counts would be correspondingly high. An average of only 8 Marsh Hawks was seen during the 1942 winter censuses, with a maximum of 12 and a minimum of 4 (Table 11), at a time when the roosting population was as high as 48 and the dispersion indicated that most of these birds hunted the Township (Map 5). This fact caused doubt as to the accuracy of the car census for these birds and it was considered necessary to determine the possible error in order to convert the census counts to total population figures for the winter of 1942.

Corrective Factor for Marsh Hawk Counts

Marsh Hawks hunted by coursing for long periods without landing. They also hunted by alternately flying and landing, sometimes alighting an average of once per minute. In addition they hunted from ground perches, which they occupied for periods varying from one second to 75 minutes. Marsh Hawks perched in grass and other herbaceous vegetation could not be seen, and a corrective factor was sought by determining the portion of their hunting time that was spent on the ground.

A pair of Marsh Hawks was followed and observed for an entire day, and others were observed for three-hour periods corresponding to the period when the afternoon census was taken. Weather on three out of four such days was typical of census days, but the fourth was stormy. The duration of ground perching periods of 60 seconds or more was noted. Table 12 shows that the Marsh Hawks spent an average of 57 per cent of their time perched on the ground, where they could not be easily seen. In inclement weather the proportion of time spent on the ground increased. For example, the female Marsh Hawk went to the ground 16 times for 60 seconds or longer during a three-hour period of light drizzle, rain and hail. She was settled in the grass for 152 out of the 180 minutes. The male was down 11 times for 60 seconds or longer, including 146 of the 180 minutes. Each bird was on the wing (therefore visible on a census) only for approximately one-half hour of the three-hour period.

Activity on the three days of good weather appeared typical of the hunting activity of the Marsh Hawks during the census periods of the two years. Therefore, if all of them spent approximately 57 per cent of their time perched on the ground, less than half the Marsh Hawks on the census strip were tallied. This would mean that for the winter period of 1942 the average of 8 Marsh Hawks per census, (a relative abundance figure) should be multiplied by 2.3 to give actual population figures for the strip. In other words, an average of 18.4 Marsh Hawks, rather than 8, were hunting the census strip. This would mean an average population of 37 Marsh Hawks in the township in winter and 38 in the fall. This population figure is in harmony with the number roosting on the

TABLE 11

CAR CENSUSES OF HAWKS

Superior Township

Maximum and Minimum Counts

1941-1942	Unidenti-fied Buteos	Red-tailed Hawk	Red-shouldered Hawk	Rough-legged Hawk	Total Buteos	Sparrow Hawk	Cooper's Hawk	Marsh Hawk	Total other Hawks	Grand Total
Fall max. ...	17	5	3	19	40	3	3	11	14	54
Fall min. ...	6	2	0	12	20	1	0	6	9	32
Winter max.	11	16	1	15	27	4	3	12	16	41
Winter min.	3	2	0	2	12	0	0	4	6	22
Spring max.	7	8	7	7	25	3	2	8	12	33
Spring min.	1	2	0	0	5	1	0	1	4	12
1947										
Winter max.	5	10	0	2	12	4	1	1	4	16
Winter min.	0	3	0	0	3	0	0	0	1	4
1947-48										
Fall max. ...	3	5	1	2	8	2	2	4	6	10
Fall min. ...	0	1	0	0	2	0	0	0	1	7
Winter max.	5	7	2	2	9	3	2	2	4	12
Winter min.	0	1	0	0	5	0	0	0	0	6

TABLE 12

MARSH HAWK GROUND PERCHING

Perching periods 60 seconds or longer

Date	Periods of ground perching		Total ground perching time (minutes)	Total time observed (minutes)	Maximum ground perching time
March 29	Male	4	62	180	29
1948	Female	3	129	180	75
April 7	Male	26	99	180	13
1948	Female	25	90	180	31
April 8	Male	17	107	180	25
1948	Female	16	131	180	34
Total		91	618	1080	
Per cent of time perching on ground		57			

area most of which were known to hunt the township.* Further evidence that this corrective factor is justified is the fact that a greater proportion of Marsh Hawks was seen on the foot censuses than on the car censuses because the perched hawks were often flushed. When the corrective factor is applied to the car census figure, we find that Marsh Hawks represented 33 per cent of the hawk population in fall and 39 per cent in winter. These figures compare favorably with the fact that Marsh Hawks seen on the foot censuses were 41 per cent of the hawks seen in both fall and winter (Table 13).

Total Count Method

It was evident that if the car census could be checked by a count of all hawks in the area made in a short period during winter stability of population, a correction factor could be obtained. This was impracticable in 1942, because of the large wintering population, the range pattern of the Marsh Hawks, and the overlapping of hawk ranges of the same species. The low wintering population of 1948, however, made such a count feasible.

Although the count was a one-day matter, with additional intensive daily observations over an ensuing 5-day period, the background of work and data that made it possible was obtained only by constant observation over the area for the previous three months. Prior to the count, practically all hawks had been located through censuses and random observations, and the exact ranges of some and the general

---

* Thirteen roost counts of Marsh Hawks between October and February of 1942 showed an average of 30 hawks with a minimum of 22 and a maximum of 48.

ranges of others determined. The distinguishing of individual hawks by range plottings, as well as by recognition of distinctive markings and sex, had given a nearly complete count of the Sparrow Hawks, Cooper's Hawks and Red-shouldered Hawks. Because of the pairing of Red-tailed Hawks and the overlapping of ranges, difficulty had been experienced in distinguishing all individuals of this species. The daily ranges and activity of several pairs (Map 6) had been determined. On February 3, while these pairs, with ranges already well defined, were kept under constant observation by two men, two additional observers drove over 86 miles from 8:00 a.m. until dark, identifying individual hawks previously plotted and definitely distinguishing them from others by time recordings and return observations.

Four pairs of Red-tailed Hawks with adjacent or overlapping ranges (1, 2, 6, 4; Map 4) were distinguished by this means. Others with known range limits were accounted for, and all areas not covered by the census were carefully checked from car or afoot for possible birds not previously seen. This single-day count of hawks over the entire area made it possible to distinguish 20 different individuals (Map 4): 11 Red-tailed Hawks, 2 Red-shouldered Hawks, 5 Sparrow Hawks and 2 Cooper's Hawks. The one Rough-legged Hawk that had occasionally appeared was not seen. The Red-shouldered Hawk (Range 12) and the Cooper's Hawks of ranges 19 and 22 were not observed, for no special effort was made to seek out birds whose presence and ranges already had been well established. These birds, however, were checked during ensuing days and continued daily observation through February 7 definitely distinguished the Red-tailed Hawk of Range 5 from surrounding birds and revealed the Cooper's Hawk of Range 18, which had not been seen on a census.

The population arrived at was:

13   Red-tailed Hawks—2 ranging on and off the area.
 3   Red-shouldered Hawks
 5   Sparrow Hawks
 5   Cooper's Hawks
———
Total, 26, of which 16 were Buteos.

In addition, a single Rough-legged Hawk occasionally ranged into the township. The afternoon census of February 8 and that of the morning of February 9 corresponded in total numbers with winter censuses taken prior to the total count, thus indicating that no population change occurred while it was being made. These censuses also indicated that the complete count was representative of the entire winter.

When the population figures obtained from the total count are compared with the winter censuses of 1947-48, it is found that twice the average number of Buteos (7.4) tallied on the winter census strip (representing a 50 per cent sample) would give 15 Buteos, or a number almost identical with the complete count of 16. When it is considered that two

TABLE 13

NUMBERS AND PERCENTAGES OF HAWKS SEEN ON FOOT CENSUSES

1942

	Marsh Hawk	Sparrow Hawk	Cooper's Hawk	Unidentified Accipiters	Rough-legged Hawk	Red-tailed Hawk	Red-shouldered Hawk	Unidentified Buteos	All Buteos
Fall Period									
No. of Individuals...(81)	33	3	2	1	10	6	2	24	42
Per cent of Total	40.7	3.7	2.6	1.2	12.3	7.3	2.6	29.6	51.8
Winter Period									
No. of Individuals...(61)	25	1	1	0	12	8	1	13	34
Per cent of Total	40.9	1.6	1.6	0	19.8	13.1	1.6	21.3	55.4
Spring Period									
No. of Individuals...(94)	17	3	4	0	9	15	30	16	70
Per cent of Total	18.1	3.2	4.3	0	9.6	15.9	31.9	17.0	74.5
Per cent for Combined Periods	32	2.8	3.2	—	13.9	12.1	12.0	22.6	60.6

of the Red-tailed Hawks included in the total count ranged in and out, one bird spending at least 50 per cent of its time off the study area, it is evident that the Buteo count obtained on the car census was accurate. Checking further, by applying the average for each species of Buteo as determined by the car census, we get: 4 unidentified Buteos, 9 Red-tailed Hawks, 2 Red-shouldered Hawks and no Rough-legged Hawks, a total of 15 Buteos. Thus for practical purposes the car census included nearly all the Buteos and no correction factor was necessary for this group.

Determination of Correction Factors for Census Counts of the Sparrow Hawk and Cooper's Hawk

A comparison of the total count of 5 Cooper's Hawks and 5 Sparrow Hawks with the 1948 winter averages of 1.5 and 0.5, respectively, per census indicates that, largely because of conditions already discussed, the counts of these birds were subject to error. The average of censuses would, if doubled, indicate 3 Sparrow Hawks and 1 Cooper's Hawk. In 1942, likewise, the plotting of census observations showed that there were 5 Sparrow Hawks and between 8 and 10 Cooper's Hawks in the township—at least 5 males and 3 females, but the winter census averages were only 2.1 and 1.2, respectively. Therefore, twice the average number of hawks tallied on the census strip does not give accurate populations for these two species. But if these total count figures are used as a basis for correcting the census counts, it is interesting to note that if the average number of Sparrow Hawks per census is multiplied by three instead of two in computing the population for the township the closest whole number approximates the actual populations for both years (2.1 x 3 = 6.3 and 1.5 x 3 = 4.5). Likewise three times the maximum number of Cooper's Hawks seen on any one census approaches very closely the actual populations during both years (3 x 3 = 9 and 2 x 3 = 6). Apparently only one-third of the Sparrow Hawks, on the average, was accounted for by the car census and only one-third of the Cooper's Hawks was seen when a maximum count of these birds was recorded. This correction factor can justifiably be applied to average and maximum counts obtained from a series of censuses.

INTERPRETATION OF CENSUS DATA

Major factors affecting the censuses have been discussed and correction factors determined. If we assume that the observers exercised a rather uniform ability from census to census, it is obvious that during each winter a remarkably constant number of hawks was tallied on individual censuses, indicating that a constant number of hawks were hunting the area comprising the census strip. Not only does this hold true for all hawks, but the tables disclose that it also applied to the numerically dominant Buteo population. The occasional low Buteo counts, such as those of January 20, 1942, and January 28, 1948, were followed immediately by counts indicating no change of population numbers. The stability of ranges further confirmed the view that occasional variations in counts could not be attributed to a population decrease. The Cooper's Hawk and Sparrow Hawk counts do not disclose any numerical changes in their populations, for a certain amount of fluctuation was to be expected in consequence of changes that, as previously mentioned, affected visibility. Although the Marsh Hawk counts fluctuated from census to census, both the maximum and the minimum winter counts were duplicated. This could indicate that, though the counts fluctuated, the winter population hunting on the township did not increase or decrease significantly as a result of migration or drift. Roost counts confirmed this. We come to the conclusion, therefore, that the constancy of the winter census counts reflects the numerical stability of the populations during the winters of 1942 and 1948.

In contrast to this numerical stability of the total counts, the species counts within small bounds, varied from census to census. We have here then a fluctuation of species numbers within the more or less constant totals. The maximum and minimum of these species counts (Table 11) are in every case, however, duplicated or closely duplicated by one or more other counts. This is characteristic of both years and is consistent with the fact that the population present did not change. Further study shows that among the numerically larger population groups, the fluctuation of the species counts takes on a rather definite pattern. The two principal Buteos, the Red-tailed Hawks and the American Rough-legged Hawks, tend to rise and fall inversely with each other. Likewise the count of unidentified Buteos, composed almost entirely of these two species, tends to rise with a low count in any one species or to drop with a high in one or both. Similarly, there is a marked tendency for the Marsh Hawk counts to rise or fall inversely with the Buteo counts. This is to some extent apparent in 1947-48, when the Marsh Hawk population was very low. An explanation of these fluctuating species counts within a rather constant total and of the compensating fluctuations of the larger population components is readily provided in terms of the known movement and activity of both the species and the population as a whole. A gradual drift or migration of hawks through the township in a rather constant species ratio might be a logical explanation for the fluctuating species counts comprising similar totals. There is, however, no basis in fact for such an explanation. The fluctuations can be explained on a basis of a stable population characterized by movement within fixed ranges, together with the daily spread of the Marsh Hawks over the township to their hunting ranges. If the hawk population is distributed over the township and beyond, with each individual hawk confined to a range and with intra-range movement extending farther than the cross section of the census strip, any one hawk could be on or off the census strip during any count, making it possible for varying numbers of any species to be recorded during various censuses. It is theoretically possible for almost all or almost none of the hawks to be on the census strip at one time. The constant total numbers and the constant Buteo counts indicate, however, that, though individual hawks moved on and off the strip, the census strip was sufficiently large and representative to offset such movement. In other words, the movement of one individual off the census strip was generally compensated for by movement of another one on, and *vice versa*. This compensatory movement was largely due to the spatial adjustments made by hawks in minimizing competitive hunting pressure. The amount of hunting that any small area could support for a definite period and still reward the predators, varied, however, with environmental conditions. Thus environmental changes caused temporary hawk concentrations. In other words, at times a majority of the hawks in the area would be on the strip, and at other times the reverse would occur. The low total hawk and Buteo counts of 22 and 14, respectively, on January 20, 1942, followed the next day by 40 and 25, the low Buteo count of 13 on January 29, 1942, and the low total hawk count of six on January 28, 1948, followed by a relatively high count the next day, are all attributable to movement of hawks off the strip. The higher counts of 41 on December 7, 1941, and 40 on January 20, 1942, may be attributed to movement of hawks into the census strip. Thorough investigation of other possible influencing factors already discussed, such as weather conditions, individual hawk activity, and visibility, led only to the conclusion that they could not explain those counts that varied most from the average. Intra-range movement that temporarily placed more or less than the average number of hawks on the strip at one time does, however, explain these counts and was often observed. The fluctuations in the numbers of each species making up the constant Buteo total can also be attributed to the compensatory character of intra-range movement. This movement, characterized in 1941-42 by the overlapping of ranges of the same and different species, with free exchange of hunting perches, caused such fluctuations within the Buteo group as a change from a count of 16 Red-tailed Hawks and 3 Rough-legged Hawks on February 11 to a count of 8 Red-tailed Hawks and 12 Rough-legged Hawks on February 25. A rise or fall of counts of either species would have been compensated for by a corresponding fall or rise in the other to an even more striking degree if all the individual hawks could have been identified. Of interest in this respect is the count on January 21, when the unidentified Buteo, the Red-tailed Hawk, and the Rough-legged Hawk counts were 8, 8, and 9, respectively. At no time in 1941-42 were all the Buteos identified specifically. For example, the Red-tailed Hawk count on December 7 was 0, but the Rough-legged Hawk count was 15, a striking difference. The count of unidentified Buteos, however, was 11, all of which may have been Red-tailed Hawks. A week later, the Red-tailed Hawk count rose to 10, and the Rough-legged Hawk count

The Land

PLATE I. Michigan woodlots cover 11 per cent of the study area, furnish winter cover and roosts for hawks and owls and support prey species, such as fox squirrels, rabbits, white footed deer mice and small birds.

PLATE II. Typical winter pheasant roost in southern Michigan. Flooding caused the birds to vacate this roost until the water froze. Similar flooding and freezing occurred in meadow mouse habitat and made mice more vulnerable to raptors.

PLATE III. The western area differed from the Michigan area in topography, altitude, climatic conditions, fauna, flora, and the dispersion of vegetation types. From front to back may be seen the broad topographic and vegetation divisions—Black-tail Butte, sagebrush bench, river bottom, sagebrush bench, on far side of river, and lodgepole pine moraine. Peaks of the Teton range appear in the distance.

PLATE IV. Predation studies were conducted on a semi-wilderness area at Moose, Wyoming, and the findings compared with those resulting from the work in agricultural southern Michigan. The foreground is habitat of Blue Grouse, Ruffed Grouse, and snowshoe rabbit; the background, habitat of Sage Grouse, ground squirrels and white-footed mice.

PLATE V. High rodent populations affect vegetation adversely. Crab apple tree girdled by meadow mice. Some shrubs and saplings were completely stripped of bark to the extremity of every limb. Melting of snow blankets exposed meadow mice that had destroyed their food, vegetative cover, and their security.

PLATE VI. Girdling of soapberry by meadow mice. Heavy populations destroy the habitat necessary for themselves as well as for other forms of life.

67

dropped to 6. The unidentified Buteos were still 11. The complete Buteo count for each day was 27. This time it was the low Rough-legged Hawk count that could be explained by reference to the count of unidentified Buteos. Half of this number would bring the Rough-leg figure up to 11 and all of them would increase it only to 17, two above the number identified on the previous census. This, plus intra-range movement, fully explains the constant Buteo totals accompanied by even the greatest fluctuations in counts of the component species.

The Marsh Hawk counts, with their tendency to rise or fall inversely to the Buteo population (Fig. 2), likewise can be explained in terms of compensatory movement. The Marsh Hawk counts (1942) fluctuate more than those of any other species and, although their "ground perching" habits may partially account for the fluctuation, they do not explain the striking inverse rise and fall of Marsh Hawk and Buteo counts. This resulted from the fact that the township supported a constant number of hawks. Each morning and evening the Marsh Hawks moved diagonally across the township, to and from hunting ranges. In consequence of numerous environmental factors, these ranges varied in both size and duration. The result was much less stabilization of ranges among the Marsh Hawks than was characteristic of the Buteos. The Marsh Hawks each morning headed for definite hunting tracts. They were moving, however, into an area containing a stable population of other hawks moving only within their ranges. The Marsh Hawks could and did hunt the same fields as the Buteos, but they tended to shift in response to hunting pressure; that is, to move to a temporarily unhunted field, thus filling niches left by the Buteos. Hence, if a low number of Buteos were on the census strip, this half of the township offered more unoccupied hunting area for the Marsh Hawks than did the other half. Hunting grounds on the strip, vacated by the Buteos for a field outside the strip, were filled by the more mobile Marsh Hawks. Here again is a compensatory movement, an adjustment of individuals within a population to the movements of others. This compensatory movement, the numerical stability of the total population and the existence of definite ranges whose bounds were set by intra-range movement characterized the winter population and made possible accurate total population counts. These same phenomena existed in 1947-48, but the lower population and reduced range overlap of birds of the same species made them less obvious. Figures 4 and 5 show a slight fluctuation of the 1948 Buteo population, inverse to that of the combined population of other hawks, as in 1942. Here also is illustrated (Figures 2 and 4) the similarity in counts on consecutive days characterized by more or less identical weather conditions. The same is true of the counts made in the afternoons and mornings of consecutive days (Figures 3 and 5). In all cases the rise or drop in count of one group (Buteos, Marsh Hawk, or other hawks) is

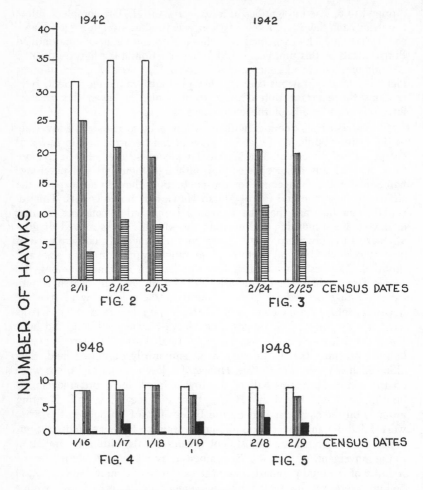

FIGURE 2. Census graphs, 1942
FIGURE 3. Census graphs, 1942
FIGURE 4. Census graphs, 1948
FIGURE 5. Census graphs, 1948

TABLE 14

CROSS COUNTRY CAR CENSUS OF HAWKS
36-mile strips one-half mile wide
Winter 1947

TRAVELING EAST

Date	Location	Bald Eagle	Golden Eagle	Red-tailed Hawk	Am. Rough-legged Hawk	Ferruginous Rough-leg	Unident. Buteo	Red-shouldered Hawk	Sparrow Hawk	Marsh Hawk	Cooper's Hawk	Total
Jan. 30	Jackson, Wyo	3			4							7
Jan. 30	Pinedale, Wyo		1		1							2
Jan. 30	Rock Springs, Wyo											0
Jan. 30	Piker Springs, Wyo						1					1
Jan. 30	Laramie, Wyo											0
Jan. 31	Cheyenne, Wyo		2		2	1						5
Jan. 31	Pine Bluffs, Wyo				2	2						4
Jan. 31	Potter, Nebr		1		5	10				6		22
Feb. 1	Maxwell, Nebr				1				2			3
Feb. 1	Kearney, Nebr				1		2		2	1		6
Feb. 1	Columbus, Nebr			1			2					3
Feb. 2	Denison, Iowa			2			1		2			5
Feb. 2	Ames, Iowa											0
Feb. 2	Ames, Iowa			1								1
Feb. 3	Cedar Rapids, Iowa			1								1
Feb. 3	Clinton, Ill											0
Feb. 3	Aurora, Ill										1	1
Total		3	4	5	16	13	6	0	6	7	1	61

TRAVELING WEST

Location	Total	Bald Eagle	Golden Eagle	Red-tailed Hawk	Am. Rough-legged Hawk	Ferru-ginous Rough-leg	Un-ident. Buteo	Red-shoul-dered Hawk	Spar-row Hawk	Marsh Hawk	Coo-per's Hawk	Date
Jackson, Wyo.	6	2			4							Mar. 5
Pinedale, Wyo.	3		2		1							Mar. 5
Rock Springs, Wyo.	0											Mar. 5
Piker Springs, Wyo.	1				1							Mar. 5
Laramie, Wyo.	0											Mar. 5
Cheyenne, Wyo.	2		1						1			Mar. 3
Pine Bluffs, Wyo.	3		2				1					Mar. 3
Potter, Nebr.	16	2	4		4	3	2			1		Mar. 3
Maxwell, Nebr.	6		1		2		1		1	1		Mar. 3
Kearney, Nebr.	5				2				1	2		Mar. 2
Columbus, Nebr.	4				1				1	2		Mar. 2
Denison, Iowa	4			2	1				1			Mar. 2
Ames, Iowa	0											Mar. 2
Ames, Iowa	0											Mar. 2
Cedar Rapids, Iowa	5				2		3					Mar. 1
Clinton, Ill.	3			1					2			Mar. 1
Aurora, Ill.	3				1		1		1			Mar. 1
Total	61	4	10	3	19	3	8	0	8	6	0	

ADDITIONAL EAST WEST CENSUSES

Location	Total	Bald Eagle	Golden Eagle	Red-tailed Hawk	Am. Rough-legged Hawk	Ferru-ginous Rough-leg	Un-ident. Buteo	Red-shoul-dered Hawk	Spar-row Hawk	Marsh Hawk	Coo-per's Hawk	Date
Rock Springs, Wyo.	3				2	1						Mar. 5
Table Rock, Wyo.	0											Mar. 5
Medicine Bow, Wyo.	8		3		5							Mar. 5
Roscoe, Nebr.	10		1		3		2			4		Mar. 3
Chapell, Nebr.	5				2					3		Mar. 3
Columbus, Nebr.	7						1		3	3		Mar. 2
Denison, Iowa	6			4	1		1					Mar. 2
Cedar Rapids, Iowa	7						4	1	1	1		Mar. 1
Cedar Rapids, Iowa	7			3	1			2		1		Mar. 1
Total	114	4	14	10	33	4	16	3	12	18	0	

compensated for by a corresponding drop or rise in another component of the population.

The changes in the number of hawks, their close correlation with the meadow mouse populations, and the numerical stability of the hawk populations, indicate that the township had a definite hawk carrying capacity for each year. This tendency for an area of land to support a stable winter hawk population is a characteristic of other large land areas. It is clearly shown by cross-country censuses of hawks (Table 14). Hawks on 17 areas of land (each containing 18 square miles), corresponding to the census strip in Superior Township, were counted by car between Jackson, Wyoming, and Aurora, Illinois, and these counts were repeated a month later, when the observers were travelling from east to west. The 17 censuses from west to east, representing a relative abundance count on 612 square miles, gave a total of 61 hawks. The count a month later gave exactly the same number (61). The various strips differed greatly in their carrying capacity for raptors, yet strikingly enough, after a month's interval, each of the 17 strips yielded an almost identical count. High counts were associated with good rodent cover and low numbers with overgrazed or waste land.

COMPUTATION OF TOTAL TOWNSHIP POPULATIONS

The various factors affecting the census counts having been discussed and the degree of error in the relative abundance counts having been measured, it is possible to compute the hawk populations in the township by using the data presented. It has already been shown that the 18-square-mile census strip was representative and can therefore be used as a 50 per cent sample. The following formulas are based on population data previously presented. Fractions are carried to the nearest whole figure. The 1941 fall population is thus computed:

Fall and Winter Populations as Determined from Census Figures

Twice the average number of unidentified Buteos, plus twice the average number of Red-tailed Hawks, plus twice the average number of Red-shouldered Hawks, plus twice the average number of American Rough-legged Hawks, plus three times the average number of Sparrow Hawks, plus three times the maximum number of Cooper's Hawks, plus 2 (2.3 times the average number of Marsh Hawks) equals the population.

The population in the fall of 1941 was: 26 unidentified Buteos, 7 Red-tailed Hawks, 2 Red-shouldered Hawks, 30 American Rough-legged Hawks, 5 Sparrow Hawks, 9 Cooper's Hawks, and 38 Marsh Hawks, making a total of 117 hawks.

Using the same method, the population in the winter of 1941-42 is computed as 13 unidentified Buteos, 17 adult Red-tailed Hawks, 1 Red-shouldered Hawk, 13 American Rough-legged Hawks, 6 Sparrow Hawks, 9 Cooper's Hawks, and 37 Marsh Hawks; a total of 96 hawks.

If the corrective factors for the Sparrow Hawks and Cooper's Hawks are omitted, and the total number of these species distinguished by means of random observation, repeated censuses, and range plottings is substituted, we get a similar total: 13 unidentified Buteos, 17 Red-tailed Hawks, 1 Red-shouldered Hawk, 13 American Rough-legged Hawks, 5 Sparrow Hawks, 10 Cooper's Hawks, 37 Marsh Hawks; a total of 96 hawks.

Applying the corrective factors to the fall and winter census figures of 1947-48 gives: 3 unidentified Buteos, 6 Red-tailed Hawks, 1 Red-shouldered Hawk, 2 American Rough-legged Hawks, 4 Sparrow Hawks, 6 Cooper's Hawks, and 7 Marsh Hawks; a fall total of 29 hawks; 4 unidentified Buteos, 9 Red-tailed Hawks, 2 Red-shouldered Hawks, no American Rough-legged Hawks, 5 Sparrow Hawks, 6 Cooper's Hawks, and 1 Marsh Hawk; a winter population of 27 hawks. When this method is used, the winter number is 27 hawks, which may be compared with 26 obtained by a complete count. No Marsh Hawks were present when the total count was made.

Hawk Populations of the Spring Transition Periods

From the arrival date of the Red-shouldered Hawks up to and beyond the first nesting of these birds and of the Red-tailed Hawks is a transition period during which some winter residents remain, others leave the area, and new arrivals continue to come in. Car censuses made in spring, although helpful in revealing the remnants of the winter population and the population changes taking place, did not yield reliable data for determining the population. This was largely because major hawk activity switched from open-field hunting to prenesting and nesting activities in the woodlots. This is partially indicated by the rise in proportion of hawks seen in woodlots on both car and foot censuses (Tables 5 and 15). In the spring of 1942 this transition period became evident in late February. At first there was no indication of any mass movement of hawks, only the appearance of new forces in action, forces about to dominate the activity of the hawk population. With the exception of a few birds, the Red-tailed Hawk population during the transition period was still predominantly immature. The mass arrival of Red-shouldered Hawks on February 28 marked the beginning of a sudden population change that gave impetus to the break-up of winter ranges and started a gradual departure of many winter residents. By March 19, the composition, distribution, and activity of the hawk population had changed completely. The winter concentrations and ranges had broken up; some winter residents had left, others still remained, and migrants had arrived and taken up territories. Soaring, territorial defense, courtship and nesting activities, though fluctuating in intensity with weather conditions, were exhibited more and more as spring advanced. Although many wintering immatures had left, there were still a number of these birds as late as April 14. A census made on this date yielded significant information about the composition of the population during the transition period. Twenty-seven hawks, 15 of which were Buteos, were counted. It is of particular significance that of these 15 Buteos, seven were identified as immatures, four others were American Rough-legged Hawks, and of the remaining four, three were very likely immature birds but could not be so classed with certainty. Thus few, if any, of the Buteos could be classed as nesters or potential nesters.

Eight of the 27 were Marsh Hawks, a species not yet nesting, as the earliest laying by this species in the two years was April 26.

TABLE 15

ASSOCIATION OF HAWKS WITH VEGETATION TYPES

Foot Census

October, 1941–April, 1942

VEGETATION TYPES	FALL		WINTER		FALL AND WINTER		SPRING	
	Individuals	Per Cent of Total	Individuals	Per Cent of Total	Individuals	Per Cent of Total	Individuals	Per Cent of Total
Grassland, abandoned and cultivated fields	65	80.2	58	95.1	145	88.5	42	44.7
Woodlots	9	11.1	1	1.6	10	6	40	42.6
Wet areas of marsh kettle and seepage	7	8.6	2	3.3	9	5.5	12	12.7
Total No. of hawks	81		61		164		94	

Three were Sparrow Hawks and one was a Cooper's Hawk. Thus 23 out of 27 hawks recorded were known to be either immatures or non-nesters. Three more were very likely immatures and only one was a bird that could possibly have had a nest with eggs—a mature Red-shouldered Hawk. Therefore the transition population was composed of the immature winter residents which could be counted, plus a large population of mature birds that because of their nesting activity could not be counted from a car. The fact that only one mature Red-shouldered Hawk was tallied on the census when there were 19 pairs nesting indicates that the car census in mid-April missed nesting birds almost completely, although still functioning well for the non-nesting population.

By April 30 the wintering immatures as well as the American Rough-legged Hawks had left, completing the transition from a wintering to a nesting population. This transition period is then characterized by a population density higher than that of either the winter or the nesting period. This fact has great significance in evaluating the dynamics and function of raptor predation and will be discussed in later chapters.

In computing the population of the transition period, the remnants of the winter population as well as the nesting population must be considered. By the middle of March, 1942, it was evident that most of the breeding residents were present, so from this date the area will be considered as supporting the number of hawks found to be nesting on it. A complete nesting count of all hawks revealed that 38 pairs nested in 1942. In addition, holdover individuals from the winter population hunted the area until approximately April 30. This element was well included in censuses taken by the car method, as their hunting activity and habitat association remained characteristic of the wintering population. Using average spring census figures, we can determine this non-nesting population just as we determined the non-nesting winter population. It consisted of 8 unidentified Buteos, 10 Red-tailed Hawks, and 7 American Rough-legged Hawks, a total of 25.

Addition of these figures to the known nesting population gives an average hawk population of 101—38 nesting Red-shouldered Hawks, 4 nesting and 10 non-nesting Red-tailed Hawks, 7 non-nesting American Rough-legged Hawks, 4 nesting Sparrow Hawks, 16 nesting Cooper's Hawks, 14 nesting Marsh Hawks and 8 non-nesting unidentified Buteos.

There was, however, one unmated Sparrow Hawk, and more Marsh Hawks hunted the area than the pairs that had selected nest sites would indicate, since the winter roost of these birds did not break up until April 21. The spring census figures indicate 20 Marsh Hawks [2 (4.4 x 2.3) = 20.2] in the township. Therefore, adding one Sparrow Hawk and using a count of 20 instead of 14 for the Marsh Hawk gives a still more exact population figure. Thus there was an average population of 108 hawks in the spring transition period from February 29 to April 30. There were migrants that passed through, sometimes remaining a day or two, but their effect on the prey populations was insignificant.

In 1948 the transition from the wintering to the nesting population began with the return of a few Marsh Hawks on February 20. February 26, when a greater influx occurred, will be used, however, as the terminating date of the winter period. Actual population figures were not altered noticeably until March 10, when the Red-shouldered Hawk influx began. This was not a mass return, as in 1942, but occurred as a scattered migration continuing over approximately a week. There was no wintering population of immature Buteos to be displaced gradually, but there was a wintering population of mature Red-tailed Hawks, most of which remained to nest. If the nesting population, with the exception of a few late arrivals, is considered to have been present on or soon after March 15, the average number of hawks during the 1948 transition period, March 15 to April 25, can be determined. This latter date marks the departure of the American Rough-legged Hawks. From the nesting count (Chap. 10) it was known that 41 pairs of hawks, including 16 pairs of Red-shouldered Hawks, 9 of Marsh Hawks, 7 of Cooper's Hawks, 5 of Red-tailed Hawks and 4 pairs of Sparrow Hawks, nested. In addition there were non-nesting single and immature birds. These included one Sparrow Hawk, two immature Red-shouldered Hawks, five immature Red-tailed Hawks and two immature Cooper's Hawks. These were determined from random counts and spring range plottings. From March 31 to April 25 two American Rough-legged Hawks established definite hunting ranges, while others passed through in migration. The prey taken by the hawks passing through was negligible, but the two American Rough-legged Hawks that hunted from March 10 to April 25 must be considered. This gives an average of 94 hawks during the 1948 period of transition from winter to spring.

Winter Hawk Populations of Southern Michigan

Table 16 summarizes the number of hawks hunting during the fall, winter, and transition periods of two years, the winter period of 1947 and the fall and winter periods of 1948-49. It is evident that the winter populations fluctuate from year to year and that the 1941-42 fall and winter populations were high compared with those of other years. They were also higher than any of the winter populations determined across the country (Table 14). The highest census count of 22, near Potter, Nebraska, would give a total population for a township area of $16 \times 2 + 2$ (6 Marsh Hawks x 2.3) = 59.6. Hawk populations comparable in density to those of Superior Township in the fall of 1941 have been observed in other areas during the fall migration, but no winter population of the same magnitude has been observed by the authors in northern regions of the country. The average winter hawk population in Superior Township during four years was 48 hawks or 1.33 per square mile. If this figure is applied to the hardwood forest region of southern Michigan (18 counties, containing 18,064 square miles), with vegetation and land use similar to those of Superior Township, an average winter hawk population of 24,000 in that region is indicated. About half of these will be Buteos. Widespread observations in this region, as well as the cross-country hawk census, tend to justify this extension of the data. Though obviously liable to error, this computation shows that hawks and their predation are not confined to small areas or to specialized habitats. Predatory pressure of the intensity described for Superior Township for all seasons (see Chap. 10) is characteristic of much larger land areas. This fact is important in connection with the function of raptor predation, described and analyzed later.

On the average the number of hawks per square mile in Superior Township was as follows:

TABLE 16

YEAR ROUND HAWK AND OWL POPULATIONS ON SUPERIOR TOWNSHIP
ILLUSTRATING THE CONTINUAL APPLICATION OF RAPTOR PRESSURE

Species	1941-42					1946-47	1947-48					1948-49		
	Fall	Winter	Spring Trans.	Spring	Summer	Winter	Fall	Winter	Spring Trans.	Spring	Summer	Fall	Winter	Spring
Unident. Buteos	26	13	8	0	0	4	3	0	0	0	0	7	0	0
Red-tailed Hawk	7	17	14	5	6	11	6	12	15	15	19	14	14	17
Red-shouldered Hawk	2	1	38	40	77	0	1	3	34	34	60	1	2	31
Am. Rough-legged Hawk	30	13	7	0	0	2	2	0	2	0	0	1	4	0
Sparrow Hawk	5	5	5	7	15	5	4	5	9	9	19	11	6	10
Cooper's Hawk	9	10	16	16	31	5	6	6	16	16	30	6	6	14
Marsh Hawk	38	37	20	14	30	2	7	1	18	18	18	0	9	16
Total Hawks	117	96	108	82	159	29	29	27	94	92	146	40	41	88
Total Buteos	65	44	67	45	82	17	12	15	51	49	79	23	20	48
Great Horned Owl	11	11	11	11	17		13	13	16	16	19	14	14	14
Long-eared Owl	7	7	2	2	6		0	0	1	1	1	1	1	0
Screech Owl	14	14	26	26	57		22	19	30	30	70	22	22	28
Short-eared Owl	22	31	10	0	0		0	0	4	0	0	0	0	0
Barn Owl	0	0	2	2	2		0	0	2	1	0	0	0	0
Total Owls	54	63	51	41	82		35	32	53	48	90	37	37	42
Total Bird Predators	171	159	159	123	241		64	59	147	140	236	77	78	130

	Fall	*Winter*	*Spring Transition*
1941-42	3.25	2.67	3.0
1947-48	.78	.75	2.6

The foot censuses in 1941-42 revealed a fall average of 3.3 hawks per square mile, a winter average of 3.8 and a spring average of 4.7. These figures are generally higher than those obtained from the car censuses, the cause undoubtedly being recounts and greater difficulty in judging the quarter-mile distance. Foot counts conducted with these possibilities for error in mind could produce population estimates sufficiently accurate for practical purposes. Under environmental conditions comparable to those of Superior Township, three or four consecutive winter car censuses can give accurate population figures on the Buteos, but Cooper's Hawks will tend to be low. For practical purposes, an accurate total of a hawk population can be found by making three or four consecutive or closely spaced car censuses and supplementing these with random observations on Cooper's and Sparrow Hawks.

As the hawks were only the diurnal raptors, it is necessary for the purposes of this study to determine the nocturnal raptor populations.

CHAPTER 4

Fall and Winter Owl Populations

O WLS, as well as hawks, are birds of prey. They are, however, not closely related to the hawks but are more closely related to the Nighthawks and Whip-poor-wills. Functionally, hawks and owls are much alike. Through the ages they have evolved similar adaptations to fit them for a life of preying on other animal forms—such adaptations as the heavy, curved beak, powerful grasping talons, keen eyesight, and strong flight muscles. Ecologically, owls complement hawks. They hunt primarily at night, and thus their food is largely crepuscular or nocturnal animals. Though less conspicuous than hawks, owls are present in all types of country that support their prey. Just what species and how many were present must be known before their hunting pressure on the prey populations in Superior Township can be determined.

Only four of the six species of owls present in the Ann Arbor region wintered in the study area. These were the Great Horned Owl, the Long-eared Owl, the Screech Owl, and the Short-eared Owl. Species not present as winter residents were the Barred and Barn Owls.

A census of the owls of the entire area was made by direct daylight counts. Woodlots were searched for Horned Owls, Long-eared Owls, and Screech Owls; the heavy grasslands and sedge swamps for Short-eared Owls. Systematic searches, began in October, were continued throughout the winters. Finding pellets, hearing calls, and observing the activity of other birds all helped to locate owls. Once an individual or roost was located, frequent visits were made to determine numbers, movement, and the length of time that the owls remained.

THE GREAT HORNED OWL

Census Method

In Michigan the Horned Owl prefers extensive woodland areas or large woodlots of mature trees. This is generally true elsewhere, though in unpopulated desert areas they hunt over and even nest in sagebrush and treeless expanses. Small isolated woodlots of 10 acres or less were usually unattractive either for wintering or nesting, although one pair of owls did winter in a small woodlot (#5) in 1942 (Map 7), and in 1948 a single bird wintered in another small woodlot (#1) and later nested there. Only a part of the wooded area was suitable for wintering and this preference for large woodlots simplified the task of locating and observing Horned Owls. In fall, and also in early spring, the calls of flocks of Crows and Blue Jays often helped. In winter and particularly immediately following the hunting season, when the owls were especially wary, we worked together to locate and count them. One of us, stationed at one end of a woodlot, remained motionless, while the other meandered through from the other end, directing his course toward likely roosting spots. If Horned Owls were present, they usually flew ahead of the moving man and were generally seen by one of us on the first run through the woods. In large woodlots where field signs indicated the presence of owls, but where they were not observed on the first run, both observers would wander back through the woods. This procedure invariably gave evidence of the presence of Horned Owls and usually an accurate count. If only one owl were seen, the procedure was repeated on following days and in adjacent woodlots until there was no doubt as to whether a single individual or a pair was present.

In the absence of coniferous trees, the Horned Owls roosted in the small white oaks and beeches that held their leaves throughout the winter. Such trees were easily found in a woodlot and regularly inspected for roosting owls, and the ground beneath was scanned for fresh pellets.

Movement

Movement of the owls had to be closely followed, for only by knowing their movements could the owl population be determined accurately and such related questions answered as: Are the Horned Owls permanent winter residents on the area, or do they drift through; how long do the various species actually hunt the area; is there much fluctuation in numbers over a period of time; and, do they maintain definite winter ranges?

Frequent counts indicated that the same Horned Owls recorded in late fall wintered in the woodlots where they were first observed. Every

Horned Owl originally located in the fall of 1941 was accounted for throughout the winter, and remained to nest or take up a territory in the same area. There was no increase or decrease in this winter resident population, although the possibility exists that other Horned Owls stopped for a short time as they moved through. If this occurred, they were in such small numbers and remained for such short periods that they were undetected and could not have exerted an appreciable influence. In the spring of 1948 there was evidence of some movement during the mating season. One owl moved into Section 2 to join a lone resident, and two owls, later established as non-nesters, moved into Section 7, on the outer edge of the Township. Both of these movements may have been local.

Even local movement among the resident population, however, was slight. There was a strong tendency for the Horned Owls to confine their movements to one or two woodlots. A pair might remain for several weeks or a month in the same woodlot, move to an adjacent one for a time, and then return to the first. Although there was no positive way to identify individual birds, the regular use of favorite perches, with accumulation of pellets, as well as certain individual characteristics, such as wariness and flight behavior, indicated that the same birds were present and that movement was local in nature. For example, some individuals habitually remained perched until very closely approached, whereas others moved at the first appearance of an observer. The exact spot to which an owl would fly on being disturbed could be predicted, so uniform were its flight reactions. Such observations, frequently repeated with other owls, indicated that the same individuals were under observation. The relatively small size of the woodlots and their isolation, combined with the habits described, reduced to a minimum the possibility of confusing observations (Map 1). Baumgartner (1939) believed that adult Horned Owls may range over several miles during the postnesting season, but do not normally desert the region in which they are accustomed to nest. Observations indicated this to be true in Superior Township. The Horned Owls located in late fall generally roosted within an area of half a square mile or less during the winter, being found regularly at favorite roosting sites. There was no evidence of drift in or out of the area nor were there any data to indicate that Horned Owls when once established are wide-ranging wanderers.

Winter Range

Observations indicated that the owls were paired but not closely associated during fall and early winter. On December 8 a pair was observed perched in the same tree and the sexes definitely were distinguished. This was the first certain record of pairing, although, previous to this, birds of undetermined sex had been seen together in several woodlots. In

1942, all Horned Owls except one (#8) were paired. Members of a pair generally roosted in close proximity to each other and, as far as could be determined, hunted the same general locality. At the approach of the nesting season, they selected and defended a nesting territory almost identical with the winter hunting and roosting area. The one exception (#8) did not obtain a mate. In 1948 there were three unmated owls; two of them, numbers 3 and 10 (Map 7), were both at locations where in 1942 there had been pairs of non-nesters. This might indicate that these were old, sterile birds, unable to nest in 1942 and single by 1948. Number 9, on the other hand, was an individual that evidently had not yet paired. This was an immature bird that had been banded in 1947 at a nest four miles away, outside the Township. It had moved into this area and, finding a woods unclaimed by other Horned Owls, had set up a winter range where it remained as a non-nester during the spring. The fact that the winter roosting areas and the nesting areas were in every case identical, and the fact that no two pairs wintered in close proximity to each other (Map 7) strongly indicates a defense of a winter home range similar to that described for the Red-tailed Hawk.

The paired owls became more social as the winter progressed, until by February they were nearly always seen in pairs, and their activities were obviously more closely related and dependent upon each other. This tendency of the Horned Owl to winter in the vicinity of the nest site is supported by Baumgartner (1939), who stated, "In most cases resident birds are found near the nest site at all seasons of the year with the possible exception of a few months in late summer and fall." Errington (1940) stated, "In Wisconsin it was observed that many but not all Horned Owls wintered in the vicinity of future nesting sites." Our nesting range data (Chap. 11) indicated that the adult owls likewise tend to spend the summer in the vicinity of the nest site. Thus, if the adult Horned Owls under normal conditions do any wandering, it very likely occurs during a short period in the fall and in most instances it is probably not extensive. This, however, does not preclude extensive movement or migration by Horned Owls in response to unfavorable environmental conditions. Banding returns are beginning to show that such movements occur. It may be that more banding and trapping will show that extensive movement of Horned Owls is confined chiefly to juveniles because most of the suitable wintering sites on the areas observed were occupied, thus compelling the offspring to move out, or to occupy unfavorable sites.

Home Ranges

Examples of Horned Owls wintering and nesting in the same vicinity year after year (Chap. 10) are evidence that one or both of the adult birds maintain a home range (an environmental niche or complex, meeting the biological requirements of the species throughout life) and that the young birds usually move off and eventually establish their own home ranges or replace a lost mate in a range already established.

Pair No. 4 (Map 7) illustrates the permanent type of occupancy characteristic of these owls. In 1941 and 1942 a pair wintered and nested in their home area. Nesting was unsuccessful. No additional observations were made until 1946-47, when a pair of owls wintered and nested in the same woodlot. Nesting again was unsuccessful. In the fall of 1947 a pair roosted in the same woodlot, but during the winter moved one and one-half miles to another woodlot, where they nested. They were again unsuccessful in that attempt, and by fall of 1948 had returned to the original woodlot. This may have been the same pair, or at least one member of the original pair may have been involved, although such a conclusion cannot be proved. The home range or area was, however, occupied continuously.

TABLE 17

OCCUPANCY OF A HOME RANGE BY HORNED OWLS
Superior Township and Section 36, Ann Arbor Township

Home Range No.	Period of Occupancy			Type of Occupancy
	1942	1948	1949	
1		X		Single
2	X	X	X	Pair
3	X	X	X	Pair in 1942; Single in 1948; Pair in 1949
4	X	X	X	Pair in 1942 and 1948; single in 1949—Also occupied in 1947
5	X	X	X	Pair
6	X			Pair
7		X	X	Pair
8	X	X	X	Pair—Logging forced birds to move to new woodlot in 1949
9		X	X	Single both years
10	X	X	X	Pair in 1942; Single in 1948 and 1949
11			X	Pair
Total	13	14	15	

Data gathered during the winter and spring of 1949 supplied further evidence that a home range tends to be occupied throughout the seasons, year after year. Six home ranges out of seven that existed in 1942 were occupied in 1948 and 1949, and eight ranges out of nine occupied in 1948 were still occupied throughout the winter and spring of 1949 (Chap. 3). The portion of a home range regularly utilized or ranged over during any one year will vary (Map 7), but the owl or owls are confined to an ecologic niche that has definite spatial limitations. It may be that in Superior Township there were only a limited number of such niches and that with slight variations these tended to be occupied year after year. Undoubtedly certain combinations of woodlots, fields, and swamps, with their respective prey populations, were favored; the continuity of occupancy (Table 17) strongly suggests, however, that the same individual owls were involved. In other words, a certain pair of owls maintained a home range until one died, whereupon the survivor attracted a new mate. In time, both the original birds may be replaced by new ones, but generally there is one bird associated with the home range to hold it, attract a mate, and continue the occupancy.

Winter Hunting Ranges

Sight records of individual owls were few compared with sight records of hawks. Therefore, the wintering areas (determined from sight records) only approximate the winter hunting ranges. A winter hunting range was generally smaller than the related home range, and this would naturally be expected, since the area ranged over during the lifetime of an individual or pair (migration or movement in search of a mate excluded) would exceed the area ranged over during a season. It is improbable that there was any considerable overlapping of hunting ranges, since the roosting sites of a pair were limited to one or two adjacent woodlots and the roosts of different pairs were separated widely. The greatest distance between temporary roosts within the same winter range was one and one-quarter miles. Usually the maximum distance between major roosts of a pair was one-quarter of a mile. Baumgartner (1939) found that, throughout most of the year, Horned Owls restrict their hunting to relatively small areas and that in no case did the feeding range extend more than one-quarter mile from the nest. The winter hunting ranges in Superior Township exceeded this. Some were observed a half-mile from their usual roost, and some moved as much as one mile to other woods. The radius of a winter range, however, seldom exceeded a half-mile. Thus it is possible that ranges may overlap, but in only a few cases. For instance, the two pairs of owls closest together would have overlapped ranges if each cruised five-eighths of a mile toward the other.

Populations and Distribution

Eleven Great Horned Owls, confirmed by numerous counts, wintered in the township in 1941-42. This is a density of one bird to approximately every 2,000 acres or a pair to every 6.2 square miles. In 1948, 13 Horned Owls, two more than in 1942, wintered; and in 1949, fourteen wintered. This population stability is due to the fact that in southern Michigan these birds are permanent residents and thus only a forced migration or an exceptionally heavy adult mortality would appreciably lower the population density. This stability would also indicate that the local carrying capacity is between 11 and 14 owls and that under present conditions any substantial population increase is not probable.

Baumgartner (1939) observed denser nesting populations, averaging from one to three pairs per square mile, and winter populations apparently at least equally dense, since he found that resident birds generally remained in the nesting vicinity throughout most of the year. Fitch (1947), in studying 2,000 acres of rolling California foothills, made winter counts of 17, 15, and 16 Horned Owls. He concluded that there was at least one owl per 100 acres. This would be a winter density 20 times as great as that encountered in Superior Township in 1942. This high density may represent a concentration resulting from a drift from mountain areas of deep snow. A drift of a portion of the Horned Owl population out of the Jackson Hole region of Wyoming was observed in 1946 and R. L. Patterson observed a similiar movement into the sagebrush desert adjacent to the Wind River Mountains in 1948 and 1949.

Carrying Capacity

From these data it would appear that the study area was not optimum Great Horned Owl range. Lack of extensive wooded areas, rather than food supply, apparently was the factor limiting the Horned Owl population. There was room for many more Horned Owls if spatial needs alone were the limiting factor, but the dispersion of the birds seemed to indicate that only certain woodlots or combinations of woodlots generally were habitable. There was no indication that available food was a major factor in determining the carrying capacity.

The intra-specific intolerance of the owls undoubtedly played a role in determining population density, distribution, and thus the carrying capacity. Each pair of owls defended an entire woodlot while adjacent woodlots in the vicinity of a pair of owls were left vacant in 1942, 1948, and 1949. This strongly indicates that the owls would not tolerate near neighbors under existing woodlot conditions.

The dependence of the owls on the size, type, and arrangement of woods is illustrated by a comparison of the north half of Superior Township with the south half, and the east half with the west half.

	1942	1948	1949		1942	1948	1949
North ½ (Owls)	8	8	9	East ½ (Owls)	5	6	5
South ½ (Owls)	3	5	5	West ½ (Owls)	6	7	9
Total	11	13	14	Total	11	13	14

The north half of the township contained the greater number of wintering Horned Owls because there were more large woodlots of mature timber there than in the southern half. The relatively equal east and west distribution was due to a more nearly equal distribution of suitable woodlots in these two halves of the township. The cover map shows a rather similar distribution of large woodlots in each quarter of the township. This at first glance would seem to be contradictory. A number of the larger woodlots in the southern half are composed, however, of second-growth trees, with only scattered mature trees. These did not offer suitable cover or protection for Horned Owls.

Thus it seems that the wintering population density was primarily dependent on the pattern of suitable woodlots. Intra-specific intolerance prevented occupation of some otherwise suitable woodlots by additional pairs. These two interrelated conditions largely determined the carrying capacity irrespective of food supply, which, with the exception of relatively small areas, was everywhere sufficient for the population. This does not mean that food supply was not or could not be influential. It seems very probable that the large mature woodlots which furnished protection and nesting sites also furnished more extensive and perhaps better hunting grounds, and were thus intimately related to food availability. Undoubtedly other conditions influenced owl density and distribution. The close proximity of huntable areas to the roosting woodlot may well be significant, since the hunting range of these owls is small. Near every woodlot supporting wintering owls were fairly large areas of uncultivated land. A glance at the cover map will show that most of the Horned Owls were located where marsh, swamp, and kettle types were most numerous. Another factor is the presence of suitable hollow stubs or stick nests. Since the Horned Owl does not build or even repair a nest and since it seems to select a home range and maintain it throughout the year, the selected area must have one or more suitable nest sites. In other words, the wintering areas as a part of the home range must meet many or all of the biological requirements of the species.

Thus it is apparent that any one of a number of conditions could limit the density of Horned Owls and determine their distribution, but on the study area the presence of large woodlots seemed to be controlling.

THE LONG-EARED OWL

The Long-eared Owl population was determined by a careful search of favorable roosting sites and by pellet groups. Coniferous woods seem to be very important to this bird, and because of lack of this type of cover roosting owls were few (Plate 37). Those that were present roosted in tangled thickets of grape vines. Such retreats are, however, used only if conifers are not available. The restriction of Long-eared Owl roosts to places providing such a narrow range of conditions was a great aid in making counts of individuals of this species.

Immediately adjacent to Superior Township, in Section 36 of Ann Arbor Township, a roost of Long-eared Owls was found in a coniferous woods. Because it was observed that some of these birds were hunting on the study area, special attention was paid to them regularly. Although they were not discovered until February 27, hundreds of pellets in 12 large accumulations, corresponding to the number of owls observed, indicated clearly that they had been using the area during most of the winter.

TABLE 18

LONG-EARED OWL ROOST—SECTION 36
ANN ARBOR TOWNSHIP

Date—1942	No. Owls Observed
February 27	12
March 10	12
March 15	9
March 20	3
March 29	4

Observation at the roost in Section 36 and four other roosts not on the study area indicated that the Long-eared Owl, if undisturbed, returns to the same roosting tree, frequently the same perch, every morning throughout the winter. In 1940 a Long-eared Owl at Mud Lake Swamp, near Whitmore Lake, Michigan, was observed on the same perch nine times in as many visits. The same year two in a coniferous plantation were observed roosting in trees 30 feet apart. They exhibited some shifting of the roost, but never were found more than 15 or 20 feet from the major pellet accumulations. These observations agree with those of Cahn and Kemp (1930), who found that the number of Long-eared Owls at a certain roost varied from four to seven, the latter number being counted on 44 occasions. As a result of this roosting habit, Long-eared Owl pellets are not scattered, and a fairly accurate estimate of the number of owls regularly using a roost can be made by counting the larger pellet accumulations. Moreover, allowing one pellet per owl per day will result in a fairly accurate estimate of the number of owl days per pellet accumulation. On this basis, the twelve owls were estimated to have arrived in late December or early January. After February 27, direct counts were made at the roost.

The number of Long-eared Owls on Section 36 showed, in 1941-42, a gradual reduction between March 10 and 29. Thereafter, none was observed. The one pair that nested in the township started laying approximately April 12. The progress of pre-nesting dissipation of a communal roost is indicated by the data presented in Table 18.

Population

Four Long-eared Owls wintered and hunted in the township, but there was no way of telling how many owls from the roost in Section 36 hunted in the study area. It seems fair to assume, however, that an equal number went out in each direction from the roost. On this basis, it was concluded that three Long-eared Owls from this roost regularly hunted the study area in 1941-42. The three estimated owls and the four known residents made a total of seven, the number of Long-eared Owls that regularly obtained winter food on the study area.

The roosting sites in conifers were not the hunting territories used by these owls. Some owls from Section 36 were observed hunting open fields and that this was a normal habit is indicated by the fact that meadow mice made up 88.6 per cent of their food.

Therefore it seems that Long-eared Owls roost communally during the winter, usually in coniferous cover, and fan out from their roosts in various directions to hunt open fields. This conclusion is substantiated further by the fact that Long-eared Owls were observed roosting undisturbed within the territory of Great Horned Owls. Had the hunting ranges for individuals of both of these species been in woodlots, it is highly probable that some of the Long-eared Owls would have fallen prey to the Horned Owls, but this evidently did not occur. For instance, the roost in Section 36 was within several hundred yards of a pair of Horned Owls, and two Long-eared Owls in Section 14 of Superior Township wintered in the same woodlot with a pair of Horned Owls without being molested by the larger birds.

Other Long-eared Owl concentrations located this same year on areas nearby were as follows:

WINTERING CONCENTRATIONS OF LONG-EARED OWLS

Roost Site	Average No. of Owls
Pine Plantation 4 miles W of study area	8
Spruce Stand 5 miles SW of study area	12
Spruce Stand 12 miles W of study area	2

These same coniferous stands did not support similarly large concentrations during the winter of 1947-48; only a single bird was found on the entire study area, and that one was in the coniferous stand in Section 36. This change in population density was directly related to the change in density of meadow mice between the two years.

Although some of the Long-eared Owls in 1941-42 may have been summer residents, observation led to the belief that most of them migrated into the area in late fall or early winter. There they found mice abundant, and, like the Marsh Hawks and American Rough-legged Hawks, they remained, held by the plentiful food supply. In contrast, no influx of these owls was observed in 1947-48, and there was no indication that an owl wintered within the township. Similar conditions, also correlated with a low population of meadow mice, existed in 1949.

THE SCREECH OWL

Census Method

Screech Owls are the most difficult owls to count, as they almost invariably spend daylight hours in hollows and leave little sign. They avail themselves of a great variety of hollows, such as bird houses, barn eaves, and hollow trees, not only in woodlots, but in fields, along fences, and in farmyards. A census of this little owl was made by inspecting all likely looking hollows if nearby pellets, owl feathers, or kills indicated that a Screech Owl was in the vicinity. When such hollows were found to contain either an owl, fresh pellets, or Screech Owl feathers, the area was recorded as harboring an owl. In addition, other hollow trees in the vicinity were examined, even though no sign was apparent. Every site in which Screech Owls were found had from a few to numerous pellets, either directly below it or very close by. Most of the nests found in spring were located in these wintering areas. The Screech Owl, like the other owls, showed distinct roosting preferences. Hollows that were poorly protected from snow and rain were not used. Those used regularly by fox squirrels were avoided; and old apple orchards, perhaps because of the numerous fine hollows and because of a generally available supply of meadow mice, were favored. Danger from the Great Horned Owl was also less in such places. The Screech Owl is fairly vulnerable to the Horned Owl because both often hunt the same habitat (Table 25). Low hollows seemed to be preferred to very high ones. This belief is supported by observations made previously in various localities. Of more than 40 nesting and roosting hollows observed in eastern United States, none was more than 40 feet above the ground, and the average was between 15 and 20 feet. Hollows with small to medium-sized openings (three to five inches) are preferred to those with large entrances.

Winter Roosts

Unless disturbed, a Screech Owl tends to use only one or two hollows as winter roosts. It frequently perches at the entrance, where it can be easily seen. The least disturbance, however, will cause it to slide down out of sight, for all the world like a piece of decayed wood falling into a hollow. In 1949 one owl was observed 21 times out of 23 visits, while another was observed 30 times in 37 visits. Fresh pellets beneath the roosts indicate the presence of an owl. Only a portion of the total pellets cast by a Screech Owl is found at such roosts, and thus the period of occupancy cannot be computed from the number of pellets. A knowledge of roosting preference and the owl's habits aided greatly in finding the Screech Owls.

Occupancy of a Home Range

Bent (1938) stated that Screech Owls are supposed to be permanent residents throughout their range, but probably some migration from the northern portion of their summer range takes place. Some and probably most of the Screech Owls in the township were year-round residents. It is possible that the immature birds and some adults migrated, but no information on this subject was obtained. Data accumulated during three winter periods (Table 19) show that the same area, frequently the same woodlot and even the same roosting hollow, is utilized winter after winter. Only banding can definitely determine whether the same individuals are involved, but the data presented in Table 19 and the nesting data of Chap. 10 strongly indicate that the Screech Owl is a permanent resident that occupies a home range in the same manner as the Horned Owl.

Pairing

There are no reliable data to indicate whether Screech Owls remain paired throughout the year. The fact that their mating calls were heard in February would indicate perhaps a closer association at this time, even though nesting activity did not begin for several months. Whether this pairing was of residents or of migrants could not always be determined, but it took place in areas where wintering Screech Owls were known to be present. Observations in various localities over a period

TABLE 19

OCCUPANCY OF A HOME RANGE BY SCREECH OWLS

Superior Township

Home Range No.	Period of Occupancy			Type of Occupancy
	1942	1948	1949	
1		X		
2	X	X	X	Same hollow occupied
3		X		
4		X	X	Same orchard occupied
5		X	X	Same hollow occupied
6		X	X	Same woodlot
7	X			
8	X	X	X	Same hollow occupied 2 years
9			X	
10	X	X	X	Same woodlot occupied
11			X	Observed in same hollow 21 times out of 23 visits during winter period
12			X	
13	X			
14	X			
15		X	X	
16	X			
17	X	X	X	Observed in same hollow 30 times out of 37 visits in winter 1949. Same woodlot—2 years in same hollow
TOTAL	8	10	11	

of 15 years indicate that during the winter these owls often are loosely paired, like Horned Owls, and that both birds, though not always together, usually remain in the same vicinity. Though no intensive work was done on their individual ranges, members of a pair were found to confine themselves to an area comprising a woodlot or roost site and adjoining fields.

Populations

In 1942 and 1948 eight and ten regularly-used Screech Owl hollows, respectively, were found and at some of these there was evidence of occupancy by two birds. Interpretation of the data on number of roosting hollows, actual sight counts, and later pairing and nesting strongly indicates that the winter population of Screech Owls was at least fourteen in 1942 and nineteen in 1948. This is as close to a complete count as could be obtained, but it is recognized that it may be less accurate than any of the other raptor population counts. The wintering population of both years was appreciably lower than the observed nesting population (Table 16). If, as is probable, the Screech Owl, like the Horned Owl, has a winter population closely approximating the nesting population, the estimate of overwintering Screech Owls is low.

THE BARRED OWL

The Barred Owl is a common winter and summer resident in the region of Ann Arbor, Michigan, but none wintered or nested in Superior Township in either 1941-42 or 1947-48. In the spring of 1942 a Barred Owl roosted in the pine plantation of Section 36, Ann Arbor Township, but did not nest there. Another was seen in the spring of 1948, but searches failed to reveal a pair or a nest.

Barred Owls prefer large woods and require stands that are sufficiently mature to provide numerous nesting and roosting hollows. Superior Township woodlots were notably lacking in mature basswoods, whereas 10 per cent of the trees in some Lodi and Scio Township woodlots are basswood. As a result of attack by the heart-rot fungus, mature basswoods are usually hollow, only the outer shell of the tree remaining sound. The lack of these conditions, plus the presence of Horned Owls in all suitable woodlots, were major factors in determining the absence of Barred Owls in Superior Township. The nesting study in Lodi, Scio and Freedom Townships, west of Ann Arbor (Chap. 10), where the woodlots were extensive and large hollow basswoods abundant, revealed three pairs of nesting Barred Owls on a 36-square-mile area in 1947 and two pairs in 1948. We may conclude, therefore, that the Barred Owl is not an important predatory bird in areas like Superior Township.

THE BARN OWL

No Barn Owls were observed in Superior Township during the winter

of 1941-42. Four of these birds, however, were observed roosting near
the Long-eared Owls in Section 36, Ann Arbor Township. They used
the roost for only a few days.

In a pine plantation several miles distant, a Barn Owl was observed
on November 19. Pellets indicated that the owl had been there at least
a week or 10 days. On February 22, five owls were counted in this roost,
and pellet remains indicated that at least one or more owls had used the
site as a permanent winter roost. Barn Owls were frequently heard at
night around city buildings in Ann Arbor.

No Barn Owls, castings, nor signs were seen in the study area during
the winter of 1947-48. On April 24 a single bird was flushed in a swamp
woods in Section 10 and another single was seen on April 30 in a woodlot
of Section 14. These birds probably were migrants.

Hollow trees as well as silos and barns that might furnish suitable
roosts for Barn Owls were examined, and farmers were questioned, but
no evidence was found to indicate that any Barn Owls wintered in either
year. This owl (Plate 35) does, however, winter regularly in the region
of Ann Arbor, and the wintering Barn Owls were higher in numbers in
1941-42 than in 1947-48, probably indicating a response to the difference
in supply of meadow mice in the two years.

THE SHORT-EARED OWL

During the winter of 1941-42 the Short-eared Owl (Plate 10) was
the most abundant resident owl. It was first observed on October 1 and
last seen on April 11, residing in and hunting the area for a period of
about six months. Not one Short-eared Owl wintered in 1947-48, although
four birds came into the area on February 27, at the close of the winter
period, and left on or about May 9. No Short-eared Owls remained to
nest (Chap. 10). The Short-eared Owl is a migratory bird and, although
some winter in their nesting areas, most of them move south unless an
abundant winter food supply is available to hold them. The migratory
habits of this species enable it, like the Long-eared Owl and the American Rough-legged Hawk, to take advantage of high populations of
meadow mice. If, in the course of migration, dense populations of
meadow mice are encountered, these owls are likely to stop and remain
as wintering populations, the size of the concentration depending on the
amount of available food and the number of migrant owls that enter
the area. The high wintering population in 1942 and the complete absence of these owls in 1948 are believed to have been correlated directly
with densities of meadow mice in the two years.

Census Method

Like the Marsh Hawks, the Short-eared Owls formed communal roosts
on the ground. They were counted when flushed at the roosts. Soon
after the first owls were observed, systematic searches revealed the roost

areas. Although grass and marshland composed 54 per cent of the township, less than three per cent of that grass and marshland had cover suitable to serve as roosting sites. Previous experience indicated that, almost without exception, Short-eared Owls roost in timothy, fox-tail grass, brome-grass, or other light-colored vegetation that blends with their plumage, in preference to darker cover types. Vegetation at the first roost further supported this fact, and with this clue in mind all fields of tall light-colored grass, as well as the grass-sedge kettles and marsh areas were searched. In early winter two large roosts were present; later they divided to make four.

The large roost #2, map 7, in Section 14, discovered on November 1, acted as a distributing center and thus a detailed account of this roost and of the activity of the owls is necessary for an understanding of the Short-eared Owl population and movements. This roost was among sedge and fox-tail grass in a small depression in a large expanse of open land in the immediate vicinity of high meadow mouse populations.

Roosting Habits

The owls generally left the roost about dusk, several rising at a time and quartering the fields near by before disappearing in the distance. They were never observed hunting during the day, and it was difficult to flush them before 4 p. m. The Short-eared Owl is noted for its diurnal hunting, being especially active in daylight hours during the nesting season, but those observed during the winter of 1941-42 remained on the roost throughout the day. It is probable that the abundance of meadow mice permitted the owls to obtain sufficient food between dusk and dawn, and hence there was no need for daylight hunting. The large number of hawks with which they would have had to compete may also have discouraged diurnal hunting.

When flushed in the morning or at mid-day, a few would rise at a time, circle about overhead, perch on fence-posts near by, and finally leave the area. In the late afternoon all would flush simultaneously, "bark" several times, and quickly disappear. Accurate counts could be made at this time. The owls did not return to the roost the same day they were flushed but scattered in all directions, soaring to heights of 300 or 400 feet (Plate 30). On one occasion there were five American Rough-legged Hawks soaring directly above the roost when the owls were flushed. The hawks did not attempt to dive on the owls from their vantage point, but allowed them to fly about unmolested.

The owls returned to their roost before daylight, but whether they returned as one flock, as the Marsh Hawks did, or drifted in singly was not observed.

The owls roosted close together, but shifted from one spot to another within the general roosting site. Some birds were only two or three feet apart, whereas others roosted 20 to 50 feet from the main group. Some roosted in heavy cover; others perched on the snow in small openings. Density of cover did not seem to be as important as its color, since dense spiraea was not utilized, and, likewise, in other roosts, the light-colored and often less dense cover was selected in preference to heavy dark-colored vegetation. There seemed to be a tendency for the owls to return to the same roosting spot each morning, or at least to select a spot that had been used previously. Pellet accumulations gave evidence of this. When the snow melted, the pellets in the original roosting spots were examined and counted. Many roost sites contained 2, 3, 4, and 5 pellets; the largest accumulations were 16, 14, 12, and 9. Pellets dropped after observations began were more scattered, there being only one to three in a group, indicating that the birds were reacting to the disturbance by shifting roosts.

Feeding and Pellet Ejection as a Clue to Movement

Frequently the Short-eared Owl casts a pellet in the late afternoon or evening, before leaving the roost. This is not a general rule, and the pellets at a roost do not represent all those ejected.

On January 15, after a fresh snow, 15 Short-eared Owls were flushed in late afternoon. A thorough search was made for pellets and only seven fresh ones were found, indicating that by late afternoon only half the birds had cast. Similar results were obtained on other occasions, indicating that the birds often cast in late afternoon and that each owl would eject no more than one pellet at the roost per day and might not cast until after it had left the roost. Guerin (1928) found that there were at least two daily ejections of pellets by the Barn Owl—one in the hunting area, and one at the roost. Chitty (1938) concluded that the length of the interval between the start of a meal and the production of the pellet in the Short-eared Owl makes it seem likely that, as in the Barn and Tawny Owls, there are two daily ejections of pellets. The fact that the owls did not eject all their day-time pellets at the roost and that in addition they may normally eject a nocturnal pellet toward the close of the night's hunting, would indicate that when characteristic pellets were found distant from a roost and in mouse habitats, it could safely be assumed that the owls hunted that vicinity. Pellets on hunting areas could be differentiated from those at a roost by the lack of excrement, so characteristic of even a single night roost. Such evidence indicated that the owls frequently hunted one and one-half miles from the major roost, and may have hunted as far as three and one-half miles away. Thus, from pellet distribution and direct observation of the owls, the hunting pattern was pieced together. Like the Marsh Hawks, the owls fanned out from their roosts to hunt. Many probably hunted definite areas, returning to them again and again. Their hunting was by coursing or by still-hunting from low perches on the ground, on fence posts, or on haystacks. It resembled in many respects the Marsh Hawk's hunting technic. Their low coursing flight, resulting in a sudden appearance, startled mice into moving and made them visible to the owls. At 5 p.m. an owl caught a mouse in this manner 100 feet from the observer, who was concealed in a haystack and on numerous occasions this hunting technic was observed at lesser vantage points. The areas that the owls hunted, as determined from pellet distribution, showed that they ranged over most of the township, though for a period of a week, or sometimes longer, each bird hunted only a relatively small area.

Winter Population

Roost No. 1 (Map 7) was very close to that of the Marsh Hawks; in fact, some owls occasionally roosted among the hawks, apparently returning before the hawks left in the morning and frequently remaining until after they returned in the evening. This roost was small, and there was no great fluctuation in numbers. Due to the patchy clumps of fox-tail that served as individual roosts, these birds tended to be scattered over many acres and, as a result, flush counts were difficult to make. The maximum number of owls on 12 counts was five, the minimum one, and the average 2.25. Because of the difficulty of flushing all birds under prevailing conditions, it seems likely that some birds were missed. It is assumed, therefore, that an average of three owls used the roost regularly during a 63-day period in 1941-42. They were not found there after the end of January. Deep snow that reduced cover may have been the cause.

On November 1, Roost 2 contained nine owls. The size of the roost gradually increased until a maximum of 28 was attained on January 27. During the last of January and the first week of February, the population dropped sharply. It did not thereafter increase above five. On February 13 there were no owls at the roost. The first deep snow of the year, which fell a few days earlier, may have made the roost undesirable, as it covered much of the concealing vegetation. That same day, five Short-eared Owls were located in Section 28, Map 7 (Roost 3). They were seen in the same spot the following day, but were not observed there after that time. There was no evidence of the roost having been used before, as only a few pellets were located. It is likewise probable that these birds were part of Roost 2, which had broken up. Pursuing this theory, possible roosting sites were rechecked, and on March 13 Roost 4 in Section 12 was located (Map 7). The greatest number flushed at this roost was five, and none was observed after March 14. Pellet evidence indicated that at least 14 individuals had used the roost for a minimum period of 12 to 15 days. This was determined by finding 14 separate roosting spots, each containing 12 to 15 pellets. In addition, there were many scattered pellets, two or three at a spot. These birds were believed to have come from Roost 2, Section 14. Roost 2 broke up on February 11 and there was definite evidence of use of Roost 4 about

February 26. It is quite possible that the birds remained scattered in small groups for several weeks after deserting Roost 2 and only gradually found new sites, such as Roost 4, where large numbers could congregate. All evidence indicated that by late March most of the Short-eared Owls had left.

Briefly, the roosting situation was this: two winter roosts, 1 and 2, were formed in November. Roost 1 remained fairly constant in numbers and broke up about February 1. Roost 2 fluctuated greatly in numbers; it consisted of 9 owls in early November, increased gradually to 16, 17, 22, and 26, and reached a peak of 28 before breaking up about February 1. This roost was located in and near areas of high meadow mouse density. It appeared that birds from all over the study area, with the exception of those at Roost 1, gradually congregated here to hunt and eventually formed a single roost. As migrants arrived, they hunted and roosted in scattered groups, but gradually congregated. Some of these groups of two and three owls were observed on Sections 28, 29 and 32 in early winter but none became a permanent roost. No such small groups were seen at the height of the roosting concentration, indicating that most of these isolated units had joined the major group at Roost 2. It is believed that snow conditions caused the initial break-up of both permanent roosts, and, as a result, some of the owls left entirely, while the remainder formed small roosts, only two of which were located—Roosts 3 and 4. It is perhaps significant that both Roosts 1 and 2 disintegrated at about the same time. On January 27, just before the break-up, the maximum number of owls was observed—three at Roost 1 and 28 at Roost 2, a total of 31.

It was desirable to determine the number of owls hunting throughout late fall and winter, and an average of the 27 roost counts does not give an adequate result. The fluctuation in numbers at the roosts was due partly to the fact that all birds did not always return to the same roost and to the fact that Roost 2 gradually increased in size by attracting owls already present and new migrants.

An estimate of the fall population was made by combining the highest fall counts at Roosts 1 and 2 and adding the scattered individuals that were observed. This indicated a population of 22 in November and December.

On January 27, 31 owls were seen at the two major roosts and the periodic roost counts showed that this is close to the number present from January 1 to February 1.

There was gradual reduction in numbers from February 1 to April 11, the date on which the last owl was observed. Combining the two highest counts at Roosts 2 and 4, made on March 13 and 14, gives a total of 10, which represents an average number from February 1 to April 11. Scattered individuals were observed during this period, indicating that the owls dispersed after Roost 2 broke up, and then perhaps gradually emigrated.

THE INTERRELATION OF OWL AND HAWK POPULATIONS

The fall and winter owl populations may be summed up as follows: There were two population constituents. One, composed largely of year-round residents, included Horned Owls, Screech Owls, and a few Long-eared Owls; the other consisted largely of migrant Long-eared and Short-eared Owls. The resident population remained stable, while the migrants fluctuated. The residents hunted definite ranges, while the migrants roosted communally and scattered widely to hunt. There was evidence that individual Long-eared and Short-eared Owls, like the Marsh Hawks, hunted in definite areas, rather than by moving indiscriminately from place to place. The Barred and Barn Owls were both year-round residents in areas near by, but were absent from the study area during winter. The owl range and movement, then, resembled those of the hawks—one group more or less fixed in position, with definite ranges, and another group working from common roosting sites to widely dispersed hunting ranges.

The number of owls hunting during fall and winter was:

| | 1942 | | 1948 | |
	Fall	Winter	Fall	Winter
Great Horned Owls	11	11	13	13
Long-eared Owls	7	7	0	0
Screech Owls	14	14	22	19
Short-eared Owls	22	31	0	0
Total	54	63	35	32

These figures are very close to the actual population that was present, but they may be somewhat erroneous in two respects. The Screech Owl counts cannot be considered complete, since it is possible that the Screech Owl population of 1941-42 was as great as that of 1947-48. The 1948 decrease in this species from fall to winter was due to known mortality. The relative instability of the Long-eared and Short-eared Owl populations makes error possible in determining the average fall and winter density.

The difference in the population densities of the two years is largely due to the absence of the Long-eared and Short-eared Owl migrants in 1947-48. In 1942 an average of 54 and 63 owls hunted during fall and winter respectively—in 1948, 35 and 32, respectively. Year-round residents present in winter in 1942 and 1948 were 34 and 32, respectively.

Before considering the prey populations and the relationship of predators to prey, it seems desirable to emphasize the interrelationships of the hawks and owls and to bring out their combined patterns of ranges and movement.

Understanding individual bird ranges and the movement of birds within the raptor population is an important feature of this study. As observations accumulated, a point was reached where the populations on the area were known and the individuals distinguished from one another so that almost every observation could be related to a definite bird (or roost, in the case of some owls) and could be correlated with observations of individuals near by. In other words, once all the pieces of the complex puzzle were "tagged" and accounted for, the job of putting them together was facilitated greatly.

During the fall of 1941 an average of 117 hawks and 54 owls, or 171 raptors, hunted the township (Table 16). During the winter there was a slightly lower population of 159 raptors, averaging 96 hawks and 63 owls. The hawks, except for the Marsh Hawk, maintained definite locations, each hunting a limited area or range. Every morning a wave of Marsh Hawks spread out from their communal roosts, to which they regularly returned at night. They tended to hunt in the same general areas, but within these ranges they filled in the hunting spots not in immediate use by the Buteos or other hawks. As the Buteos went to roost in the woodlots and the Marsh Hawks settled down for the night, the night shift of owls took up the ceaseless hunting, exerting most of their pressure upon the meadow mouse population. Short-eared Owls rose from their roosts to spread by night, just as the Marsh Hawks had done during the day. The Screech Owls and Horned Owls hunted small, limited ranges, including fields as well as woods, but they tended to prefer the woodlots. The Short-eared and Long-eared Owls combed the extensive stretches of open country. The owls not only functionally replaced the hawks, but in some respects did so species by species. The Short-eared Owls were ecological equivalents of the Marsh Hawks, and both they and Long-eared Owls hunted the same prey as the Marsh Hawks, in much the same way and in similar habitats. The place of the Cooper's Hawk and the Red-tailed Hawk was taken at night by the Horned Owls.

In areas where both the Red-shouldered Hawk and the Barred Owl were found, they were co-actors in the same drama—one by day, the other by night, hunting the same habitats in pursuit of similar prey. This was even more true of these two birds in spring and summer than in winter. The Barn Owl and the Rough-legged Hawk similarly could be compared, and the Sparrow Hawk's nocturnal complement was the Screech Owl. The outstanding feature is the dovetailing of individuals and species to form a pattern of hunting that largely eliminated direct competition. This is particularly striking since most of the raptors were feeding largely upon the same food supply (Table 33). It was a population of predators composed of different species, each adapted to hunt either a different habitat, or the same habitat in a different manner or at a different time. Direct competition was eliminated further by the fact

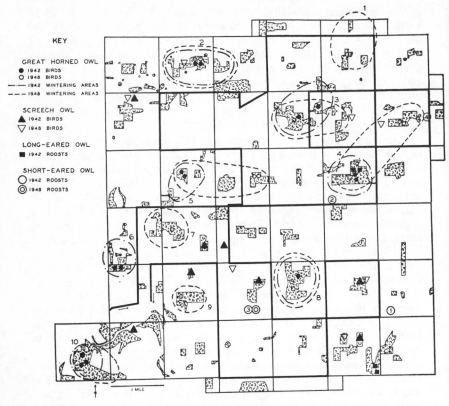

KEY

GREAT HORNED OWL
● 1942 BIRDS
○ 1948 BIRDS
—·— 1942 WINTERING AREAS
— — 1948 WINTERING AREAS

SCREECH OWL
▲ 1942 BIRDS
▽ 1948 BIRDS

LONG-EARED OWL
■ 1942 ROOSTS

SHORT-EARED OWL
Ⓞ 1942 ROOSTS
◎ 1948 ROOSTS

MAP 7. Winter distribution of owls

that hawks and owls of the same species did not have overlapping ranges during the winter of 1947-48 when the staple food supply (meadow mouse) was low. In 1941-42, when overlapping of the ranges of the same and different species was common, a compensating movement of individuals in the population occurred, so that direct competition did not anywhere become intense. The raptor population was a team working in harmony for the benefit of each individual and at the expense of the prey populations. The efficiency of the raptor population as a whole thus was increased, and the number of birds the area could support was greater than it would have been had strife and competition among individuals and species been the rule.

In the winter of 1946-47, the average hawk population was 27 (Table 16), a population level considered more usual among the hawk winter residents in the Ann Arbor area. This population was duplicated numerically in the winter of 1947-48, and the fall population of 28 hawks showed no noticeable increase. The migrant owl populations during both fall and winter were low. The total raptor populations, 63 and 59, respectively, for the fall and winter periods were just a little more than a third of the 1941-42 populations. The pattern furnished by the Marsh Hawks and Short-eared and Long-eared Owls was lacking in the winter, but this same pattern on a much reduced scale was formed by the migrants during the spring transition period (Map 7). We have observed the hunting pattern of these Michigan winter populations in hawk and owl populations in various parts of the country. It appears that these relations are a characteristic of raptor populations.

CHAPTER 5

Winter Prey Populations

BECAUSE predation is essentially a relation of population numbers, counts of both predator and prey are necessary to the understanding of this phenomenon.

The hawk and owl populations have been determined (Chap. 3 and 4), and their food habits will be revealed by analysis of pellets (Chap. 6). Before we can interpret these food habits, it is necessary to know the approximate amount and kind of prey that was available and this information must be eventually correlated with the prey actually taken, as indicated by the raptor diet. It is essential, therefore, to express the available prey in terms of approximate populations. It obviously is impossible to work with all the varied prey populations over such an extensive area. Therefore, intensive studies of the prey populations generally were confined to eight typical sections—two areas of four square miles each, designated Areas A and B (Map 8). The eight-square-mile area was considered to be large enough and the vegetation and its distribution sufficiently similar to those of the township as a whole so that prey population data obtained would be representative. These studies were continually supplemented by specific observations and counts throughout the township to see if prey conditions on the samples were generally representative.

An average of 24 man-hours a week was spent in making, in the two areas, censuses of prey populations—the meadow mouse, Ring-necked Pheasant, Bob-white, cottontail rabbit, and fox squirrel. All

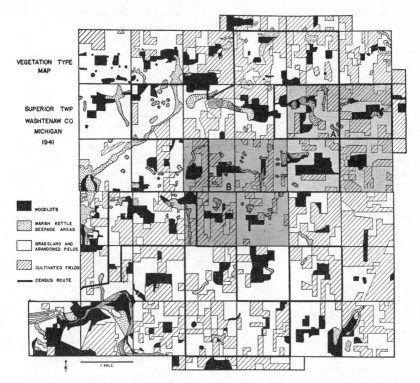

MAP 8. Sampling areas for prey, Superior Tp.

censuses were made after the hunting season. The degree of accuracy attained varied with the species; censuses of some populations, such as pheasants and Bob-white, were quite accurate, while those of meadow mouse, fox squirrel, and cottontail rabbit were approximate, and those of white-footed mice and small birds were estimates only. As will be shown later (Table 34), however, the knowledge of the relative densities of the various prey populations existing over this large area, expressed in round numbers, is all that is necessary to determine the dynamics of predation.

MEADOW MOUSE POPULATIONS

Much emphasis was placed on the study of meadow mouse populations, as it was evident that they formed the chief food of most of the raptors and that their activities and population densities were reflected directly in the activities and population densities of these birds.

Two major questions required answers: First, what is the meadow mouse density in local areas where raptors are hunting? Second, what is the average fall meadow mouse population for the entire area? The conditions and population densities under which meadow mice live are so varied that generalization as to home range, daily activity, size of winter litters, and population densities cannot be drawn accurately and applied to specific land areas in the present state of our knowledge. But in spite of the recognized chance for error, we believe that the data obtained were sufficient to make the estimates of meadow mouse numbers reasonably accurate.

General Method

In the fall of 1941, the population of meadow mice was at a high level. This afforded an unusual opportunity to study predator-prey relations. The population of raptors was also high. A survey of the township was conducted to find and evaluate meadow mouse habitats and to determine what relation, if any, existed between these and the areas of high raptor density (Map 11). In conjunction with this, an intensive meadow mouse survey was conducted on Areas A and B to evaluate and map all habitats of this animal on the basis of relative population density. This, the first step toward making a population estimate, revealed the following facts.

1. The meadow mouse population was high, but not of mass outbreak proportions.
2. High densities were localized but were scattered fairly uniformly throughout the area.
3. High densities were associated with dense grass or weed cover.
4. Overgrazed pastures and crop land supported relatively low populations.
5. A relation existed between meadow mouse density and raptor density.
6. Runways, dung piles, girdled trees, and scurrying meadow mice were indicators of areas of high density.

Intensive Survey

Quadrats were laid out and burrows, dung piles, and runways counted in an attempt to obtain for meadow mice a rating index based on quantative data. The method proved too time-consuming for our purposes and was abandoned in favor of an observational rating system.

The method of rating meadow mouse habitats on Areas A and B consisted of walking back and forth through a field to observe signs of these mice and scoring the area high, medium, low, or very low.

High density areas were recognized readily. They were characterized by a heavy grass or weed cover consisting of one or more of the following plants: blue grass, orchard grass, timothy, ragweed, and goldenrod. Runways and holes were abundant. Dung piles an inch or more across and one-third to one-fourth inch high lay at frequent intervals along the runways, being especially dense near the burrows. A short walk through any area of high density exposed to view several meadow mice, sometimes as many as a dozen. With a light film of snow on the ground, two observers tallied 64 meadow mice on a five-acre plot in three-quarters of an hour by walking seven feet apart and covering the area in three runs. Environmental destruction varied from the partial consumption of the non-woody vegetation to the girdling of small trees and bushes, and, in extreme cases, almost the complete destruction of

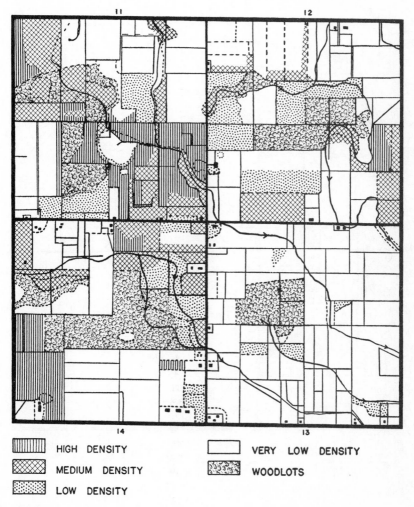

HIGH DENSITY		VERY LOW DENSITY	
MEDIUM DENSITY		WOODLOTS	
LOW DENSITY			

MAP 9. Density & distribution of meadow mice, Area A

all vegetation within small areas. This was observed only in late winter (Plates 5 and 6).

Areas of medium and low density were more difficult to distinguish. Areas of medium density were considered to be those where the meadow mouse environment was favorable, but where runways were not exposed and signs of the animals not glaringly evident. Areas of low density were those in which little choice meadow mouse cover existed, but where there were occasional signs of the presence of such mice. These were

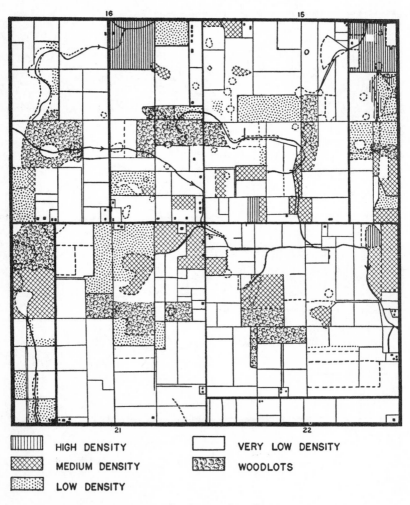

HIGH DENSITY   VERY LOW DENSITY

MEDIUM DENSITY   WOODLOTS

LOW DENSITY

Map 10. Density & distribution of meadow mice, Area B

represented largely by heavily grazed fields. Clear-cropped fields and those freshly plowed or sown to winter wheat were found to be poor meadow mouse habitat and were rated as areas of very low density.

The density maps 9 and 10 show the general extent and dispersal of meadow mouse environment and indicate high, medium, low, and very low densities. Each field or area on the map was rated individually. Wherever the cover was spotty, consisting of areas of high, medium, and low density in the same field, an average rating was given. Thus, areas of medium and low density may contain small isolated units of high density. The areas of high density were largely homogeneous, although they, too, frequently possessed small isolated areas of medium and low density, which do not show on the maps.

Areas A and B (Maps 9 and 10) contained 303 acres of habitat of high density, 455 acres of medium density, 406 acres of low density, and 3,486 acres of very low density. Woodlots, which were not rated, comprised 470 acres. Thus, 5.9 per cent of Areas A and B supported high meadow mouse densities, and 8.9, 7.9, and 68.1 per cent of the area supported, respectively, medium, low and very low densities. Assuming the acreage percentages to be reasonably representative, the above percentages can be used to determine the total acreage of each of the four rated habitats on the 36-square-mile area (Table 20).

Trapping

To obtain indices of relative abundance for the high and medium meadow mouse densities, representative areas were trapped. Four trap lines were run four consecutive nights. Snap traps were baited with a mixture of peanut butter and oatmeal.

Traps were set in runways separated by approximately half a home range, or 33 feet, (half a home range computed from Blair's figures (1940).) Eighty-eight and 83 per cent of the animals taken on sites 1 and 2 were meadow mice, while on the sites of medium density (3 and 4) only 39 and 50 per cent were of that group. A high percentage of meadow mice to other mice characterized all densely populated areas.

The number of individuals caught in four nights did not decline progressively, and a similar condition prevailed during the six nights of trapping on Sites 1 and 2. The trap-night indices for meadow mice (the number of mice caught divided by the product of the number of traps set and the number of nights traps are in service) (Grinnell, 1914) (Table 21), based on four nights of trapping, indicate that the average meadow mouse densities on the high areas were approximately three times as great as those on medium areas, but the total small-mammal densities, as expressed in trap-night indices (Table 21), indicate that the populations on the areas of high density were greater than those on medium areas by only about two-fifths. These data show relative densities, but do not reveal how many meadow mice per acre were available to raptors.

To obtain the greatest value in studying the predator-prey relations, it is desirable to express the meadow mouse population in areas of high density in terms of numbers of individuals per unit of area. Until more is known about meadow mouse populations over extensive areas and at peak densities, attempts at expressing them in densities per acre will be cruder than indices of relative abundance. At the present state of our knowledge, however, a trap-night index, such as .492 or .410, cannot be visualized readily and has full significance only when compared with other indices and when these, in turn, are compared with indices for some area where both total population and trap-night index have been determined.

With this in mind, a trapping experiment was conducted in 1948 to determine trap-night indices on an area of high desity. The total population inhabiting this same area was found by exhaustive snap trapping. Thus the trap-night indices could be interpreted in terms of numbers of meadow mice per acre.

TABLE 20

MEADOW MOUSE HABITAT RATING EXPRESSED IN DENSITY PER ACRE
1941-42

Rating	Meadow mice per acre corresponding to rating	Acreage for eight sections	Per cent acreage for eight sections	Acreage for 36 sections	Estimated Population for Township
High	139	303	5.9	1,364	189,596
Medium	50	455	8.9	2,048	102,400
Low	2	406	7.9	1,827	3,654
Very low	0.5	3,486	68.1	15,687	7,844
Woodlots not rated	...	470	9.2	2,115	......
Total meadow mouse population					303,494

A 40-acre field of nearly pure timothy, completely isolated on three sides by plowed fields and on the fourth side by both a macadam road and a plowed field, was selected for the trapping experiment. This field showed signs of a high meadow mouse population and would have been rated "high" in 1942. October plowing undoubtedly forced meadow mice into the timothy field from the adjoining plowed fields, but after this initial movement in consequence of the isolated situation of the field, there was apparently little or no movement in or out of it. Early and persistent snow prevented trapping operations until February 24.

A five-acre quadrat, bordered by plowed fields on three sides and by the remaining 35 acres of timothy on the fourth, was marked off for exhaustive snap trapping. Prior to saturating it with traps, three lines of 37 traps each were run for four consecutive nights. The trap lines were placed 66 feet apart and the distance between consecutive traps in each line was 33 feet, or one-half a home range (Blair, 1940). The trap-night indices thus obtained were .222, .263, and .291. The plot was then saturated nightly with 333 snap traps for 10 additional nights.

A catch of 465 meadow mice, 16 shrews, and 1 white-footed mouse was taken from the five acres in 3,774 trap nights. It required 14 trapping nights and 8.1 traps per meadow mouse taken to depopulate the unit. No traps were set within 33 feet or half a home range of the western border of the five-acre plot. Thus the total population could be computed for exactly five acres. Two and three meadow mice, respectively, were taken off the quadrat on the last two days, and the area was considered to be trapped out, although there were indications that there still remained a few meadow mice that refused to enter the traps. The meadow mouse population was thus computed to be 93 per acre.

Movement

Toward the close of exhaustive trapping, control lines were run to determine whether movement of mice into the trapped area was occurring. Lines set in the plowed fields caught no meadow mice and two examinations following light snow flurries gave further evidence that none was present except in the portions bordering immediately on the timothy. No signs of movement across the plowed fields were observed, and movement into the quadrat could have occurred only from the side bordered by timothy. Two parallel trap lines set 66 feet apart in the timothy and situated 132 feet or two home ranges distant from the quadrat caught 48 and 47 meadow mice in four nights of trapping. These yield trap-night indices of .263 and .263 respectively, which are comparable to those recorded on the adjoining five-acre plot. From these data it is evident that, toward the end of the period of exhaustive trapping, the adjoining timothy had a population density similar to that existing originally on the five-acre plot. If there had been any considerable drift into the quadrat, it would be reasonable to assume that it would have lowered the trap-night indices in the adjoining timothy, the area of origin. Likewise, the total catch of the eight trap lines used in depopulating the site during the ten-night period (with the exception of some random traps) was, from the timothy border to the plowed field, 45, 39, 25, 31, 40, 51, 32, 16 meadow mice. The figures give no evidence of a drift into the quadrat from the remainder of the field. It was expected that the line adjacent to the timothy would yield a high total and that the line paralleling the plowed field would be low. To check further the possibility of drift into the five-acre tract, the trap lines were reset following a ten-day interval. Individual traps were staggered with regard to their former positions. Ten meadow mice were caught the first night, ten the second night, and none the third. Some of these, and possibly all, were remnants of the original population. Neverthless, had all of them moved into the trapped area, it would have been at an average rate of only two meadow mice per day. A similar rate of movement during the earlier trapping operations would not have significantly influenced the results. Thus it seems safe to conclude that the computed meadow mouse population of 93 per acre is reasonably accurate.

TABLE 21

TRAP-NIGHT INDICES ON FOUR LOCATIONS

Fall—1941

Rating	Habitat	Number of Traps for (4 nights)	Total of small mammals	Total meadow mice	4 night Trap-night index for meadow mice	4 night Trap-night index for total of small mammals
1—High	Heavy grass *(Poa)*, dense goldenrod, matted dry grass, scattered hawthorn	120	59	52	.433	.492
2—High	Matted grass *(Poa)*, dense goldenrod	112	46	38	.340	.410
3—Med.	Matted grass *(Poa)*, goldenrod, scattered hawthorn	76	29	11	.145	.382
4—Med.	Wheat stubble with dense clover cover ..	152	40	20	.131	.263

Computation of Meadow Mouse Populations

A .259 average of the trap-night indices of .222, .263, and .291 obtained on the five-acre plot may be considered indicative of a population density of approximately 93 meadow mice per acre. Other areas of high density were comparable to the experimental area and thus also supported meadow mouse populations of approximately 90 per acre.

Since the areas rated high in 1941 yielded trap-night indices of .433 and .340, we can assume that they supported correspondingly higher meadow mouse populations. An approximate population figure for these sites can be obtained by using the average trap-night index of both years in a proportion where the one population is already known. Thus:

$$.259 : 93 : : .386 : X$$
$$X = 139$$

The areas of high density in 1941 (Table 21) thus supported approximately 139 meadow mice per acre.

By using this population figure in a similar proportion the average population density of the sites of medium density (Table 21) can be computed. Thus:

$$.386 \; : \; 139 \; : \; : \; .138 \; : \; X$$
$$X = 50$$

In 1948 there was no serious destruction ("eating out") of vegetation such as that recorded in habitats of high density in the winter of 1942. This fact plus other observation further indicated that the areas rated as high density in 1948, though comparably rated with high density areas of 1942, supported lower numbers of meadow mice per acre. Certainly the total small-mammal indices of .492 and .410 (Table 21) indicate a much higher total population of small mammals in 1942.

The point to be emphasized is that areas rated as having high meadow mouse density supported populations approximating 140 and 90 meadow mice per acre in 1941 and 1948, respectively. Extremely unfavorable weather conditions interfered with trapping activities on the habitats rated as having low and very low densities. Such data as were obtained indicated, however, an approximate density of two of these animals per acre on areas rated low and one per two acres on habitats rated very low. Because of the general distribution of habitats of high density (Map 11), the eight sections rated in 1942 were thought to be reasonably representative. On this basis the trapping and rating data for 1941 and 1948 will serve to estimate roughly the meadow mouse populations. By projecting the acreage for eight sections, the acreage of the variously rated meadow mouse habitats is obtained (Table 20).

Using these acreages and the population per acre figures derived from trapping, a meadow mouse population of approximately 303,500 was computed for the township in the fall and early winter of 1941. In round figures, then, there were approximately 303,000 meadow mice in the early winter of 1941-42.

Ninety-six per cent of these were concentrated in 14.8 per cent of the township on scattered areas that were fairly evenly distributed over the 36 square miles. It can be further estimated that in the fall of 1941, at the peak of the hawk population, there were present at least 2,589 meadow mice for every hawk, or 1,771 for every raptor.

In the fall and winter of 1947-48, the meadow mouse population on local areas of high density was comparable to the density of 93 per acre determined on the five-acre trapping site. Such areas were, however, relatively few in number and unevenly distributed as compared with 1941-42. They were the areas most consistently hunted by the largest number of raptors. The total acreage of high meadow mouse densities was somewhat less than a square mile. On the remaining area the density was low, with patches of medium density. An estimate placed the meadow mouse population at about 75,000, or approximately one-fourth of that in 1941. Thus, during the winter, there were, in theory, approximately 2,777 meadow mice available to every hawk present, or 1,271 to every raptor. The ratio of meadow mice to raptors was approximately the same in both years.

There have been few reliable estimates of meadow mouse populations in mass outbreaks or plague areas and no accurate counts. Elton (1942) offered one example of what he considered a good estimate of such populations in plague proportions—more than 8,000 meadow mice per acre. Piper (1909) estimated 8,000 to 12,000 mountain voles per acre in the lower Humboldt Valley of Nevada, 1906. Thus, neither the estimated population of 303,000 meadow mice in the fall of 1941 nor 75,000 in 1948 approaches plague proportions.

Hamilton (1940) cited meadow mouse populations of 200 and 300 per acre in favorite localities, these populations determined by "reasonably accurate census methods." He also reported that meadow mice reached densities of from 160 to 230 per acre during the 1935-36 mouse year in New York State (Hamilton, 1937).

The areas of high density, with 140 meadow mice per acre in what could well be termed a mouse year (1941-42), are comparable to the populations given by Hamilton.

POPULATIONS OF WHITE-FOOTED MICE
(INCLUDING DEER MICE)

No specific trapping of white-footed mice was conducted to determine populations. Some indication of their numbers can be obtained from the data of Burt (1940) and Stickel (1946). Burt, live-trapping in the George Reserve, Michigan, found populations of 10.5 and 10.87 white-footed mice per acre in November, 8.15 in December, and 7.25 in February. Stickel found September populations of between six and seven adults per acre on the Patuxent Research Refuge, Maryland. If Burt's November figure of 11 white-footed mice per acre is assumed to be reasonably representative of the population density of the woods in the study area in the fall of the year, a total population of 26,500 wood mice can be estimated on the basis of 2,405 acres of woodland. Burt (1940) stated: "This species is found chiefly, but not exclusively, in areas which are wooded or covered with heavy brush." Trapping returns indicated that 26 per cent of the white-footed mice trapped in 1942 were common white-footed mice; these were caught in open habitat. Seventy-four per cent were prairie deer mice. If an average home range of 225 feet, derived from Burt (1940), is used to compute the acreage of the trap lines, using a buffer strip of one-half this distance, it appears that on the meadow mouse habitats of high density (Table 21) there was approximately one white-footed mouse per acre, and on the meadow mouse habitats of medium density an average of 2.5 white-footed mice per acre. This is computed on a basis of four night catches, as it is felt the high meadow mouse populations prevented trapping out the white-footed mice in three nights. These figures, projected on a basis of acreage (Table 20), would give an approximate white-footed mouse population of 6,500 for the open land areas in the township in 1942. Approximately 4,800 of these were prairie deer mice. Trap lines set in various open habitats in 1948 caught only two white-footed mice. This was so low that no open land population was estimated. The combined white-footed mouse population, as estimated for the township, approached 33,000 individuals in 1942 and 27,000 in 1948. These estimates are not submitted as accurate population figures, for it is realized that the most accurate population data taken during a given year and at a given location cannot be applied to other years and other locations. Such estimates are valuable, however, in showing the relative numbers of white-footed mice as compared to meadow mice possibly available to raptors. The meadow mouse population of 75,000 in 1948 would be nearly three times that of the white-footed mouse, and the meadow mouse population of 303,000 in 1941 would be about ten times as great. The white-footed mouse population was confined largely to the woods (11 per cent of the total land area), and because of its habitat and generally nocturnal habits was available to only a small percentage of the raptors. Prairie deer mice were at a higher population density in 1942 than in 1948, as evidenced from trapping results. Twenty-eight were caught in 460 trap-nights in 1942, and only one in 3,774 trap-nights in 1948. Their greater abundance is also revealed by their presence in the 1942 diet (Tables 27 and 28) of raptors that hunted open land.

GAME POPULATIONS

Periodic censuses of the game populations on Areas A and B (Map 8) were made to determine the winter density of pheasants, Bob-whites, rabbits, and fox squirrels—all potentially important prey species. The area was large enough so that no appreciable change occurred as a result of movement on or off it during the winter. Although some movements of pheasants and fox squirrels occurred in the fall and spring, these movements, as far as could be ascertained, were compensatory.

It was not feasible in this study to determine the winter mortality of all game species, nor was it practicable, except for the Cooper's Hawk, to investigate the extent of predation through direct field observation. Data on game predation were largely determined from pellet analysis (Chap. 6), which was supplemented by observational data of game kills and periodic counts of game populations. Errington (1936) determined the extent of predation on Bob-whites from fluctuations in census counts and from direct field observation. He concluded that predation was confined to the vulnerable surplus, or, in other words, the overpopulation of wintering Bob-whites. In our case winter predation on pheasants was so small that it was impossible to express such a loss in numerical differences between population counts in winter and in spring. Winter mortality of Bob-whites, however, was determined in this way, but the full part that predation played in the loss could not be ascertained. The prime purpose of the game censuses was not to determine mortality or predation losses, but to find the average number of individuals of game species present and potentially available to raptors.

RING-NECKED PHEASANT AND BOB-WHITE POPULATIONS

Censuses of Bob-whites were taken by making flush counts of coveys, following Errington's technic (1936), and those of Ring-necked Pheasants by roost counts (Wight, 1931). Areas A and B were covered within a ten-day period with the aid of a bird dog. This rapid census, which served to locate all the pheasant roosts and Bob-white coveys, was followed up by check counts twice a week from January 10 to March 15. Each area was covered completely every three to four weeks. Pheasant counts were averaged to determine a roost population, or the best count was taken. Tracking observations (Errington, 1936) supplemented the flush counts and were particularly valuable in locating small groups and scattered individuals (Plate 41). As there was no measurable decline in numbers, the pheasant population arrived at in January was considered as existing throughout the winter period.

Conditions Affecting Pheasant and Bob-white Censuses

As Errington (1936) has pointed out, it is easier to take reliable censuses of populations securely situated than of those insecurely situated. The relative stability of the pheasant flocks enabled accurate population counts to be made throughout both winters.

The taking of censuses was facilitated by the habit, exhibited by pheasants in this region, of congregating and roosting in swamps, kettles, and other areas of dense herbaceous cover (Plate 2), by their reluctance to leave the roost on very cold days until mid-day, and by their relatively small radii of movement.

These advantages were offset at times by the temporary splitting of roosting flocks in response to environmental changes caused by flooding and freezing of swamp roosts (Plate 2), and by the tendency of cocks to segregate at times and again to join the hens. As low temperatures tended to bring all individuals of a flock together into a single roost, the most nearly accurate counts were made at such times.

In 1942 Bob-whites, when once located, were easily counted because coveys were widely separated, winter territories were small, and coveys were stable. The insecurity of the 1948 Bob-white coveys had an opposite effect. The stability and isolation of coveys made possible the determination of winter Bob-white mortality based on the numerical differences between the first and last flush counts (Errington, 1936).

The daily movement of pheasant flocks was largely dependent on the arrangement of food and roosting cover. This movement near the borders of sections included in the census had to be considered in arriving at a population figure for Areas A and B. Some flocks ranged one-quarter mile, others occasionally one-half mile, but most of them moved only a few hundred yards from roost to corn field and seldom wandered farther unless disturbed. The birds were generally on the roost by dusk and off soon after dawn, except on very cold days (0° F.) when they were, on occasion, flushed as late as 11 a.m. or even 2:30 p.m. After the morning feeding, flocks frequently would seek protected brushy fence rows and remain relatively inactive for several hours. Activity again would increase with late afternoon feeding and, as dusk approached, the birds generally would fly back to the roost as one flock.

Seasonal movement into winter habitats occurred throughout the late fall, after which flocks were relatively stable until the advent of mild weather, when sexual stimulation initiated scattering.

Size and Composition of Pheasant Flocks

Flocks varied in size from four to 54 members. The temporary merging of two large ones formed a flock of 105 birds. They roosted as a unit in a large sedge swamp and fed on the extremely abundant and readily available fruits of hawthorn. Marginal areas, where roosting cover was adequate but food was scarce, supported one or two individuals. Often these were cocks isolated from the larger congregation of hens (Plate 41). The winter sex ratio, as determined from counts, was one cock to four hens in both years.

Bob-white Coveys

In early January, 1942, the largest of seven Bob-white coveys consisted of 25 birds, the smallest of 10. There was only one record of two coveys merging temporarily. Coveys were well situated in regard to food and cover, and four suffered no winter mortality. All the winter losses were in coveys 2, 3, and 7. The relative insecurity of these coveys appeared to be directly associated with the presence of large pheasant flocks. Roosting and escape cover were not limiting factors, and the quantity of food was sufficient, although there was decided competition between quail and meadow mice for ragweed seeds. We believe that the relations between the Bob-whites and the pheasants made the habitat of coveys 2, 3, and 7 less secure than that of other coveys on the area. This effect was attributed to direct competition for food and other relations between pheasants and Bob-whites that adversely affected Bob-white activity, and to the fact that large pheasant concentrations invariably attracted house cats, stray dogs, and raptors, and the predators then found Bob-whites easier prey than pheasants.

In the fall of 1948 Bob-white populations were extremely low. Areas A and B each supported only one covey of Bob-whites, and these coveys, as evidenced by their movement and losses, were insecurely situated. By mid-winter only one covey of twelve birds could be found on the eight sections.

Computation of Pheasant and Bob-white Populations

Table 22 shows the number of pheasants at each roost as found by a series of counts throughout the winter periods of 1942 and 1948. Table 23 shows the maximum counts of Bob-whites in early January and mid-March, the difference being the winter loss. The pheasant winter populations were computed separately for Areas A and B and then for the combined sample areas, adjustments being made in both instances for movements of flocks situated just inside or outside borders. For example, in 1942, of the 35 birds in Roost 5 (Table 22) four-fifths were assigned to Area A and one-fifth to Area B, on a basis of daily movements in feeding. On an average, about half the birds of Roost 9 (Table 22) fed outside Area B. Therefore only half of them were counted in computing the population per area. On the other hand, the 19 birds of Roost 3, though situated near a border and occasionally moving beyond it, were nevertheless contained within the sample sections and therefore were all tallied in the areal count.

In 1942 Area A supported 100 wintering pheasants and Area B, 126. Since winter mortality, as well as movement on and off the areas, was negligible, these 226 pheasants for practical purposes, were, the combined wintering population of eight sections. In round numbers, this is

TABLE 22

WINTER PHEASANT FLOCK COUNTS

January 1 to March 15, 1942 and 1948

Area A—4 sections				Area B—4 sections			
Roost No.		No. Pheasants		Roost No.		No. Pheasants	
1942	1948	1942	1948	1942	1948	1942	1948
1	1	15	12	5	8	35(1/5)=7	6
2	2	12	5	6	9	12	40(2/3)=27
3	3	19	6	7	10	12	6
4	4	21	6	8	11	6	27
5	5	35(4/5)=28	25	9	12	54(1/2)=27	35
..	6	..	4	10	13	51	26
..	7	..	34	11	..	6	..
Total in roosts ..		95	92			121	127
Singles		5	2			5	1
Total birds including single		100	94			126	128
Sex Ratio 1942		1948					
4 ♀ to 1 ♂ —		4 ♀ to 1 ♂					

28 birds per section, or one per 23 acres. In 1948, Area A supported 94 and Area B, 128 pheasants, 222 birds in all. This also is 28 birds per section, or one per 23 acres.

In January, 1942, Area A supported four coveys of Bob-whites, containing 69 birds. By March 15, there were 56 birds, indicating a winter loss of 19 per cent (Table 23). Area B contained three coveys, with 47 birds in January and only 40 by March 15, a winter loss of 15 per cent. Thus there was a mortality of 20 birds, or 17 per cent, for the combined areas.

TABLE 23

WINTER BOB-WHITE COVEY COUNTS

January 1 to March 15, 1942

Area A				Area B		
Covey	No. Bob-whites Jan. 1	Bob-whites March 15	Winter loss	Covey	No. Bob-whites Jan. 1	Bob-whites March 15	Winter loss
1	14	14	0	5	10	10	0
2	25	18	7	6	21	21	0
3	16	10	6	7	16	9	7
					—	—	—
4	14	14	0		47	40	7
	—	—	—				
	69	56	13				

January ratio, birds to acres = 1:37.1 .... 1:54.5
March ratio, birds to acres = 1:45.7 .... 1:64.0

1948

One covey of 20 Bob-whites   One covey of 12 + Bob-whites
January ratio, birds to acres = 1:128 .... 1:213
March ratio, birds to acres — No Bob-white could be located on the 5,120 acres

The early wintering population of 116 Bob-whites on Areas A and B was reduced to 96 by early spring. Thus the mean of the winter population was 106 Bob-whites, or an average density of 13 per section or one per 48 acres. In 1948 these eight sections supported only two Bob-white coveys, one of 20 birds and one of 12. No data were available to determine why the wintering Bob-white population was so low, but similar conditions existed throughout the township. One covey disappeared in February; possibly it removed completely from the study area or was eliminated by unusually severe ice conditions. We suspect the latter. The other covey suffered from severe weather conditions and predation. A female Cooper's Hawk, a house cat, and a red fox all took toll, the Cooper's Hawk hunting the covey quite regularly and accounting for at least two members. This covey also could not be found after February. The Bob-white mortality, which may have approached 100 per cent, was notably higher than in the winter of 1942. The maximum winter population of Areas A and B was 32 birds, an average of four birds per section, or one per 160 acres. Spring observations on paired birds in the township also indicated an extremely low over-wintering Bob-white population.

In 1942 the combined pheasant and Bob-white population (using the mean winter Bob-white population) was 332 birds (Table 24). In 1948 it was 254 (taking the maximum winter Bob-white population).

It is of interest that in 1942 the winter density of Bob-whites and pheasants together on Areas A and B was nearly uniform, being 1:15.7 acres on A and 1:15.1 acres on B. The 1948 game bird population was 23.5 per cent less than in 1942. This lower winter population was due entirely to the decrease in Bob-whites, since the pheasant population practically was unchanged.

In 1948 the average ratio of pheasants to acres on Areas A and B (1:25.6 and 1:20.3) differed little from the corresponding data of 1942 (1:27.2 and 1:20.0), but the low Bob-white density (correspondingly lower in Area A) reduced the total game bird density of Area A to 1:22.5 acres and that of Area B to 1:18.3 acres. Nevertheless, both areas supported comparable densities for these two species together in 1942 and 1948, and both had negligible pheasant losses and comparable Bob-white losses. This would indicate similar carrying capacity on both areas for Bob-whites and pheasants together.

Pheasant and Bob-white Distribution

Tables 22 and 23 give the numerical composition of pheasant roosts and Bob-white coveys. With the exception of a blank area in Section 13, the distribution of both pheasants and Bob-whites was fairly uniform throughout Areas A and B. There was, however, a good deal of variation in the number of pheasants using a roost. Variation in the size of flocks was related directly to suitable combinations of available food (chiefly

TABLE 24

BOB-WHITE AND PHEASANT POPULATION DENSITIES ON AREAS A AND B
1942-1948

Land Area	Pheasants		Bob-whites		Pheasants and Bob-whites	
	1942	1948	1942	1948	1942	1948
Areas A and B	226	222	106*	32	332	254
Av. No. per section Area A	25	23.5	16	5	..	..
Av. No. per section Area B	31.5	32	11	3	..	..
Av. No. per section Areas A and B	28.25	27.75	13.25	4	41.5	31.8
Av. acreage per bird Areas A and B	1:22.7	1:23.1	1:48.3	1:160.0	1:15.4	1:20.2

* = The average number of wintering Bob-whites

standing corn) and roosting cover. In 1942, clean farming was practiced on Section 13. No corn was present and, because of lack of protective travel lanes, other food, for the most part, was unavailable. Good roost cover was lacking. In 1948 this section contained four fields of standing corn, and roosting cover in the form of a fallow field which in 1942 had been overgrazed heavily. It supported three pheasants in 1942; 32 in 1948.

Large areas that did not support pheasants apparently also were blank or marginal areas for Bob-whites. The grouping of quail into coveys and of pheasants into roosting flocks produced a distribution of bird concentrations which affected the situation with respect to predators. They tended to attract diurnal raptors but served as protection against nocturnal raptors, especially the Horned Owl.

Computed Pheasant and Bob-white Populations

The winter pheasant and Bob-white populations were 163 and 169 for Areas A and B, respectively, in 1942; and 114 and 140, respectively, in 1948. In other words, the densities of game bird populations on these two areas in 1942 and 1948 were comparable. The general distribution could also be considered comparable, Area A having five and seven pheasant roosts in 1942 and 1948, respectively, and Area B having seven and six roosts. Similarly, Area A supported four coveys of Bob-whites in 1942, as compared with three for Area B, and each had one covey in 1948. In both years the size, number, and distribution of roosts throughout the Township, determined by less intensive work, was similar, area for area, to that on the eight-section study area. This was likewise true of Bob-white cover, roosts, and covey size. The eight-section area was, therefore, considered representative of the larger area. By projecting the population data, the approximate population of winter game birds in the Township was computed to be 1,017 pheasants and 477 Bob-whites in 1942, and 1,000 pheasants and 144 Bob-whites in 1948.

Relation of Pheasant and Bob-white Density to Raptor Density

In 1942 there were approximately 1,500 game birds theoretically available to 159 raptors, or a ratio of one raptor to every 9.4 game birds. Since a large proportion of the raptor population either could not take these species, or would not except under very unusual conditions, the ratio of one Cooper's Hawk or Horned Owl to 71.4 game birds has greater significance. In 1948 approximately 1,150 of these species were theoretically available to 59 raptors, or a ratio of one raptor to every 19.2 game birds, or one Cooper's Hawk or Horned Owl to every 60.5 game birds. It is immediately evident that, although the raptor-game ratio (including all raptors) varied from 1:9.4 in 1942 to 1:19.2 in 1948, the ratio of the Horned Owls and Cooper's Hawks (both of which can and did prey regularly on game) to the game birds present showed less variation for the same periods (1:71.4 and 1:60.5).

Mortality

Pheasant mortality from all causes was so small during both winters that it could not be measured by differences in population counts. Bob-white mortality averaged 17.24 per cent in 1942 and was close to 100 per cent in 1948. In spite of intensive efforts to locate game kills, only two hen pheasant kills (both in October) were found on Areas A and B in 1942. Neither of these could be assigned definitely to raptors. In 1948, ten pheasant kills, six of which were found during late February beneath a Horned Owl winter perch, were discovered. Two were accounted for by a migrating Goshawk (Plate 47), and two more were taken by a single female Cooper's Hawk that regularly hunted a large pheasant flock. In 1942, although there was a mortality of 20 Bob-whites, only four kills of this species were found. Horned Owl pellets contained evidence of one (Table 25) and Cooper's Hawks accounted for three others. We suspect that the Cooper's Hawk was responsible for more of the Bob-white mortality than direct field evidence indicated. It is doubtful, however, if raptors were responsible for even half of the winter mortality of Bob-whites in 1942.

In 1948 four Bob-whites were known to have been killed by Cooper's Hawks on Areas A and B, and it is quite likely that some others met the same fate.

The positive field evidence of predation on Bob-whites and pheasants is meager. The negative evidence, however, is significant. Past experience indicates that, had there been much predation on game, the time spent in the field would have resulted in the discovery of more kills. Such predators as Cooper's Hawks and Great Horned Owls, though present, did not prey significantly on game birds. Losses of pheasants and Bob-whites by predation were relatively greater in 1948 than in 1942, although the ratio of these game birds to their two most destructive bird predators in the two years was the reverse. The greater predation losses of 1948 were influenced by the lower meadow mouse population of that year.

RABBIT POPULATIONS

No rapid field technic for making an accurate census of cottontail rabbits with only a moderate expenditure of time and man-power has been devised, but a combination of several technics that yielded reasonably reliable estimates was employed. For this study, trapping and marking were too slow and time-consuming to be practicable, and it was questionable whether data obtained on small areas would be superior to those obtained by less precise methods over larger areas. Furthermore, the very low density of rabbits that was obvious even in the most favorable habitats in 1941-42 favored extensive, rather than intensive, methods.

The rabbits' practice of holing up for periods of time and, also, of sneaking out far ahead of an observer made counts based on direct observation unreliable even as rough estimates. The number of individuals seen per walking hour was recorded regularly, however, and these counts were combined with estimates of abundance made by tracking in fresh snow. The combination of direct counts with counts of tracks, forms, and burrows was adopted as the simplest and most reliable method of estimating rabbit populations, even though it involved some personal error.

After fresh snows, rabbit cover was examined by two or more observers. Track concentrations, as well as burrows and frequently visited brush piles, old foundations, discarded bundles of barbed wire, and other havens of refuge were plotted on field maps. Jumped rabbits were recorded, care being taken not to duplicate counts. At the end of the day, the data were immediately interpreted in terms of the number counted, the number of ground and brush dens present, and the number indicated by track concentrations in feeding areas. The dispersion in reference to food and dens was of special significance and the key to accurate interpretation. One rabbit can make many tracks in a single night and enter numerous burrows and brush piles. Therefore, any estimate is subject to no little error. However, an attempt was made to estimate conservatively, and we believe that the population figures were probably low for the areas studied. Furthermore, the accuracy of the methods used varies inversely with increase in the size of the population; the smaller the population, the more accurate the estimate is likely to be because less confusion between individuals occurs. The rabbit population in 1941-42 was low. Estimates made in the winters of 1946-47 and 1948 indicated much higher populations that might have been difficult to evaluate without the experience in 1942.

Using the method described, the population of Area A in 1942 was estimated to be 14 rabbits and that of Area B to be 28 rabbits. On Areas A and B, no more than one rabbit was seen per walking hour in good rabbit habitat; whereas, in contrast, on protected areas outside of the study area, four to five rabbits several times were recorded per walking hour. Such protected areas were small and not representative of the country in general, but they emphasize the fact that the population in the township was low.

Another approach to the problem of estimating the rabbit population was tested. During the winter the rabbits were associated intimately with small, brushy kettle holes. Buttonbush kettles, particularly those in or near woodlots, were favored and used by one, or sometimes more, rabbits during the winter. Close observation of a few strategically located kettles indicated an average of one rabbit each.

The track and direct counts indicated a possible maximum of two rabbits per swamp or woodlot.

Area A contained six kettles, well located in relation to other cover, and 12 swamps and woodlots. Area B contained 42 kettles and nine swamps and woodlots.

On the basis of one rabbit per kettle and two per swamp or woodlot, Area A would have had an approximate rabbit population of 30 and Area B a population of 60. We believe that these numbers represent maximums and that the actual population for Areas A and B was somewhere between the numbers determined by the two methods used. Thus the rabbit population is expressed:

Area A—Winter population between 14 and 30,

Area B—Winter population between 28 and 60,

Areas A and B:A combined average population of between 42 and 90, or for the township between 190 and 405 rabbits. Let us say approximately 300 for the 36-square-mile area.

There was an average density of 5 to 11 rabbits per square mile. When this density is compared with a combined pheasant and Bob-white density of 41.5 birds per square mile and an average meadow mouse density of approximately 8,300, it becomes immediately evident that rabbits could not have been an important prey species in 1942. This assumption is borne out by pellet evidence (Table 33). Though the density of rabbits does reflect in general their availability as prey, their habit of concentrating in favorable habitats increased density in local areas, thus making them more vulnerable than their numbers on the study area would indicate (Fig. 6).

In 1948 similar observations were made, and, as a result of track counts on Areas A and B and estimates based on cover type use, indicated a winter population three to four times as great as that of 1942, or approximately 1,200 rabbits for the area. Predation on rabbits likewise was greater (Table 33 and Plate 46).

FOX SQUIRREL POPULATIONS

In the fall of 1941 the fox squirrel population was noticeably low in the vicinity of Ann Arbor, particularly on the study area. D. L. Allen (1943) stated concerning Michigan fox squirrels, "There evidently has been a progressive increase of fox squirrels in most areas from 1937 to 1942. The exception to this trend was 1941, when the animals received a setback in some habitats." Because of the apparent low population, intensive census methods were not used, but lack of such evidence as sight records, field signs, and track counts in fresh snow indicated the complete absence of squirrels in some suitable woodlots on Areas A and B, and there was no evidence that any woodlot supported more than a half-dozen squirrels. On the basis of estimates for woodlots, an approximate population of 300 fox squirrels in the township was arrived at.

In 1948 the situation was quite different. Squirrels were observed in nearly every woodlot, and tracks indicated high populations in the more favorable habitats (Plate 40). A time-area count (Goodrum, P. D., 1937) in a beech-maple woodlot in Area A also indicated clearly a high population. On March 25, from 2 to 3 p.m., five observers took prearranged positions throughout the woods and counted squirrels within a given area. Each noted the time and location of squirrels observed on the border of his area adjacent to the next observer. The entire woods of 32 acres could be observed by the five men. The weather was clear and cold, with new snow on the ground. Each observer remained quiet and motionless and counted squirrels for one hour.

The count clearly indicated 30 squirrels. After correction for duplicate and questionable observations, this count was considered to indicate a population of at least one fox squirrel per acre. The population estimate for the township cannot be based on these observations alone, as tracks (Plate 40) and sight records disclosed that this population was higher than average. A comparison of sign and counts from this woodlot with those in other woodlots led to the conclusion that one squirrel per two acres of woodlot or roadside habitat was close to the average population density. On that basis the nine per cent of the township that provided such habitat contained a population of approximately 1,000 fox squirrels. These figures, though admittedly rough, do indicate relative populations of the two years and are sufficiently accurate to place the fox squirrel populations in a density scale including other prey populations (Table 34), which is all that is necessary for an evaluation of predation on this species.

D. L. Allen (1943) stated, "It seems likely that two squirrels per wooded acre in the fall is near the maximum that any large area, say a township, supports." On this basis the study area, with 2,053 acres of woods consisting of trees from 8 to 24 inches d.b.h., would support approximately 4,100 fox squirrels and it is possible that the fall or pre-hunting season population approached this figure.

SMALL BIRD POPULATIONS

It was evident that the number of small birds far exceeded the number of individuals of the other major prey groups except mice.

Starlings and English Sparrows occurred in flocks of 25 to 300, and nearly every barn harbored Pigeons. Counts of these three species produced an average of 70 per farm. One hundred and sixty-eight average-sized farms supported a population of approximately 11,700 English Sparrows, Starlings, and Pigeons, with Sparrows predominating.

Woodlot counts indicated that on each acre there were approximately two birds, chiefly of such species as Tufted Titmice, Nuthatches, Flickers, Downy Woodpeckers, Blue Jays, Chickadees, Cardinals, and Juncos; 5,000 woodlot birds in all.

In open country, Tree Sparrows and Juncos were frequently observed in flocks of 50 or more. As many as 200 Tree Sparrows were counted in one flock, and 125 in another. Snow Buntings, Longspurs, and Horned Larks also occurred sporadically in large flocks. Meadowlarks and Mourning Doves were locally distributed, generally on sunny banks and slopes, in numbers varying from single individuals to groups of ten to fifteen. Song Sparrows and Cardinals were observed in bushes, fence rows, and kettles. From counts and observations an average for open country of about one winter bird per three acres, or 6,800 in all, was estimated.

The entire small bird population was therefore about 23,500, a figure which harmonized very well with an independent estimate made by J. L. George, 1952. Following intensive banding and counting of small bird populations on a 171-acre farm in Section 14 of Superior Township during the winters of 1948 and 1949, he made the following estimate for Superior Township:

Woodlot Birds	4,000
Larks, Buntings and Longspurs	2,600 (flocks)
Tree Sparrows	4,000
English Sparrows	10,800
Pigeons and Starlings	1,500
Total	22,900

The two estimates indicate a population of about 23,000, and we believe this is sufficiently accurate to be used as an approximate population of small birds available to raptors during the winters of 1941-42 and 1947-48.

CROWS AND POULTRY

Crows were absent during the winter but reappeared in numbers toward the end of February. A few individuals were observed occasionally in the forenoons, but were seldom seen after 1 p.m. It is believed that they came from a congregation along the Detroit River, some 30-odd miles away.

On a sample of nine farms, poultry flocks varied from 0 to 300, with an average of 93 birds per farm.

SUMMARY OF PREY POPULATIONS

We now have made an estimate of the prey populations during two winter periods. We know what kinds of prey species and approximately how many of each theoretically were available to known raptor populations.

Meadow mice were the most abundant prey, occurring in densities as high as 140 and 90 per acre in the winters of 1942 and 1948, respectively. They had an average late-fall density of approximately 8,400 per square mile in 1942, and a late-winter density of about 2,800 per square mile in 1948. The approximate population was 303,000 and 75,000, respectively, for these years.

In both years, white-footed mice, including deer mice, were the second most abundant prey, with a higher population level in 1942 than in 1948. Open land habitats supported higher populations of both species in the earlier year. The approximate population was 33,000 for 1942 and 27,000 for 1948.

The non-game or small birds, as a group, were third in abundance, with winter populations of about 23,000. The greater number of these were around farms, the woodlots were next in average density, and the open land areas, though thinly populated, supported large mobile flocks of winter migrants.

The game birds (Pheasants and Bob-whites) were fourth in abundance in 1942, and fifth in 1948. They showed an average winter density of 42 and 32 birds per section for 1942 and 1948, respectively, and approximate populations of 1,500 and 1,100 birds for these years.

The rabbit population was very low in 1942 and markedly higher in 1948. There were fewer rabbits than game birds in 1942, with a rabbit population of about 300, but slightly more rabbits than game birds in 1948, with a rabbit population of approximately 1,200.

In both years the fox squirrel population after the hunting season was lower than that of any other species. The population was estimated at about 300 in 1942, which was a winter of very low density. By 1948 the population had so increased that there were approximately 1,000 squirrels after a heavy fall hunting kill.

We have here population data which, when correlated with the food habits of the raptor population, should provide information as to why one prey population is utilized to a greater extent than another. We likewise should be able to perceive the role that density plays in determining the vulnerability of a prey species.

CHAPTER 6

Winter Food of Raptors

IN PREVIOUS chapters we have determined the raptor populations and calculated the prey populations which are theoretically available as winter food for the raptors. The next step is to determine what the raptor population actually fed upon and then relate this diet to the prey populations in order to interpret the dynamics of predation.

Owls swallow entire most of the small and medium-sized prey they catch. Larger animals, such as rabbits and pheasants, are torn apart and then consumed. Hawks pluck their prey—that is, they remove some of the feathers or fur before swallowing. They also tear their food to bits and swallow it in relatively small pieces. Varying quantities of bone, feathers and fur are eaten with the meat. These items are not digested but are formed into a pellet, which is usually regurgitated or "cast up" prior to the next meal. By comparing skulls, teeth, bones, hair, feathers and claws found in the pellets with the same items in an identified collection, identification of most items is easy. When pellets are collected from a restricted area the species of birds or mammals that can be present are limited and with practice the investigator becomes adept at identifying and counting pellet contents. This method of determining food habits yields more representative data than does stomach analysis, as the sample for any area or period is greater. It is possible even to obtain a reasonably exact record of the diet of an individual raptor, and a large enough sample of pellets from all raptor species present reveals the feeding trend of a population.

Since the food habits data determined by pellet analysis are indispensable to an interpretation of predation, and since the pellets of all species of hawks and owls cannot be collected with equal ease or identified and analyzed with equal accuracy, a brief discussion of the methods employed is necessary.

METHODS OF DETERMINING RAPTOR FOOD HABITS

The food habits of the raptors were determined by the analysis of pellets (Plates 42 and 43) and to some extent by observations of kills.

Raptor pellets collected during the fall and winter of 1941-42 numbered 3,992; those collected in 1947-48 were 784; a total of 4,776. In addition, there was some broken pellet material of the Horned and Long-eared Owls, which was collected and analyzed separately. Pellets were collected only at roosts or at perches where they could be dated and assigned definitely to a specific raptor. No pellets for analytical purposes were collected at random, a limitation which eliminated any chance of pellet misidentification.

Certain field signs are frequently the best characters for identifying pellets of specific raptors. For example, though some pellets of the Short-eared Owl and Marsh Hawk look alike, they are easily identified at roosts where both birds are associated. The fecal material associated with the pellets is diagnostic of the species and thus of the pellets.

The shape, structure, condition of the fur, and osseous remains of Red-tailed Hawk and American Rough-legged Hawk pellets are so similar that they must be collected from known roosts or favorite perches in order to be identified specifically (Plate 43).

Several methods of recording pellet content data are in general use. They may be expressed as the number of times remains of prey animals are represented or as percentages of individual prey items found in the pellets.

In this study, both methods have been used wherever possible, but major reliance is placed on percentage of prey species rather than on frequency of representation.

Analysis of the winter pellets showed that the raptors ate practically no insects. During summer, however, they formed a more important part of the diet. They were an insignificant portion of the year-round diet of all raptors except the Sparrow Hawk and Screech Owl. Since the occurrence of insects in the diet could be recorded only as frequncy of occurrence and not as number of individuals represented in a pellet, the data were not considered sufficiently quantitative to bear interpretation, but the occurrence of insects will be found recorded in the diet of raptor species (Tables 26 and 28).

Where small rodents are the major food of hawks and owls, there generally is sufficient osseous evidence in the pellets to reveal accurately the number of individuals represented. Because a greater amount of bone is found in owl than in hawk pellets (Plates 42, 44 and 45), the number of individual prey items per owl pellet more closely represents the actual number of individuals consumed to form the pellet than is the case with hawks. Hawks digest a larger percentage of the osseous evidence. We found that for each individual prey item, 772 Short-eared Owl pellets averaged 2.24 grams, while 450 Marsh Hawk pellets averaged .92 grams. This difference is largely indicative of the great difference in osseous remains found in the two types of pellets.

We also determined that approximately 69 per cent of the rodents fed to a captive Marsh Hawk was evident in the pellets, whereas practically 100 per cent was found in the Short-eared Owl pellets. With the Horned Owl, of whose diet large species, such as rabbit and pheasant, may at times form a high percentage, it is difficult and at times impossible to determine the number of these items that are represented in a pellet collection. There frequently is a carry-over from one pellet to another, and a single individual may serve for more than a single feeding. A more nearly accurate count of large prey animals can be obtained if pellets are gathered regularly at specific sites rather than at random. For example, if a group of ten pellets collected at one site during a fortnight shows skeletal remains of two rabbits, but rabbit remains are found in five of the pellets, it is safe to conclude that only two rabbits are represented. If the ten pellets had been gathered at random, no such conclusion could be made. Accordingly, Horned Owl pellets were gathered regularly at specific sites.

In consequence of the fact that the Horned Owls were feeding largely on small rodents (Table 25), it was possible to determine the number of small prey actually represented in their pellets (Plate 45). The approximate number of large prey individuals was determined from the amount of skeletal remains per containing-pellets. Thus the data presented for all raptors represent the number of individuals observed in the pellets, as well as the frequency of occurrence of prey items. In the case of the Short-eared and Long-eared Owls, the number of individuals represented in each pellet is very nearly the same as the number of prey individuals consumed to form the pellet. In the case of the Horned Owl, the number of individual small rodents tabulated corresponds very closely to the actual number of small prey individuals consumed (Plate 45), but because of greater bone destruction and digestion and the fact that the pellets of this bird are frequently broken, this relation is not as accurate as the similar one for Short-eared Owls, which digest little or no bone. The calculated number of large prey in the Great Horned Owl pellets is subject to some error.

As for the hawk pellets, where a large per cent of the bone is digested, the number of individuals computed per pellet represents only a portion of the number of prey individuals consumed to form the pellet. The number of individuals was determined by sorting and pairing incisors (Plates 42 and 44). Present data indicate that the percentage of the consumed prey that is evident in Buteo pellets is smaller than the corresponding percentage in Marsh Hawk pellets. This is perhaps a result of longer pellet retention by Buteos, rather than of more active digestion of bone.

Provided that the pellets are relatively fresh, the degree to which the number of individuals tabulated represents the number consumed to form the pellet varies with:

1. The degree of bone digestion.
2. The size of the prey consumed.
3. The heterogeneity of the prey species represented.
4. The durability of the pellet.

A combination of a high percentage of bone digestion and a wide range of prey, such as characterizes the large bird hawks, results in pellets that are difficult to evaluate quantitatively. A lower degree of bone digestion, combined with large and varied prey, such as generally characterizes the Horned Owl, produces pellets less difficult to evaluate quantitatively than pellets of bird hawks. On the other hand, a combination of little or no bone digestion and small prey of a few species, such as characterizes the Long-eared, Barn, and Short-eared Owls, produces pellets that can be analyzed very accurately in quantitative terms and expressed in number of individuals per pellet.

Pellets that are in good condition because of their compact structure, their small to medium size, and the short distance they fall to the ground lend themselves to determinations more accurate than those obtained from loosely constructed pellets of large size, or pellets that drop some distance through branches to the ground.

Raptor pellets which are usually found in good condition because of reasons mentioned above are those from the Marsh Hawk, Short-eared Owl, Barn Owl, Screech Owl, and Sparrow Hawk. Those from the Horned Owl, Long-eared Owl, and Buteos (Plate 43) are frequently broken in some degree.

The fact that for various reasons pellets of raptors are not equally satisfactory for analysis does not alter the fact that very useful qualitative data were obtained from all pellets and accurate quantitative data from the owl pellets. There is little doubt that analysis of a sufficient number of pellets will show the feeding trend of raptors in winter. Certainly no other known method will do this so well and yet leave the population undisturbed.

While it is possible in many cases to determine quite accurately the number of individuals consumed to form a pellet, lack of information on the number of pellets ejected per day by various raptors, as well as other conditions, has prevented accurate estimates, on the basis of pellet examination, of the daily ration of any given raptor. Even in the case of the Barn Owl and Short-eared Owl, where it is possible to recover in the pellet evidence of each prey animal consumed to form the pellet

(Chitty, 1938; Wallace, 1948), no accurate estimate of total food consumption has been possible.

PELLET SAMPLES

Because pellets are not found easily and those from different species are not equally available, it was impossible to collect from each species a sample sufficiently large to be proportionate to the numerical representation of each raptor species in the population. Therefore the procedure was to collect as large a sample of pellets for each species as possible. While this produced more food data for some than for others, the larger populations are in general represented by a larger proportion of pellets.

Table 30 shows the number of hawks and owls of each species and the corresponding number of pellets analyzed to determine their diets.

WINTER DIET OF RAPTOR SPECIES

The food data recorded for the fall and winter periods of 1941-42 and 1947-48 (Tables 25-28) will be referred to simply as winter food, since there was no significant variation in food habits of the raptor population from fall through winter. The following discussion will deal primarily with qualitative aspects of the raptor diets.

Red-tailed Hawk

The Red-tailed Hawks (Table 27 & Plate 17) consumed a high percentage of meadow mice (89 and 84 per cent) during both winter periods, and also comparable percentages of white-footed mice (7 and 6 per cent). Rats and shrews played a more prominent role in the 1942 diet. Rabbits, pheasants, and small birds (0.6 and 7.8 per cent), though higher in 1948 than in 1942, still represent a small portion of the diet. It is significant that though meadow mouse populations were lower in 1948, the percentages consumed were high during both years. This density relationship between Red-tailed Hawks and meadow mice cannot be explained in terms of a single predator preying on a single prey species since it is the result of the density relations of numerous prey species one to another, and also to varied densities of numerous predator species.

Red-shouldered Hawk

One Red-shouldered Hawk wintered in 1942, and three in 1948. The few pellet data accumulated indicate a diet of 94 and 100 per cent meadow mice during the two winter periods (Table 27).

American Rough-legged Hawk

The diet of the American Rough-legged Hawks in 1942 was 98 per cent mice, rats, and shrews. Meadow mice formed 84 per cent of the diet. None of these birds wintered on the area in 1948. An average of 30 Rough-legged Hawks hunted the area in the fall of 1942, and an average of 19 remained through the winter. The presence of these birds in numbers indicates high populations of small rodents. Thus their presence in relatively large numbers in 1942 and their absence during the winter of 1948 reflects the meadow mouse populations of the two periods. This illustrates one type of density relation—that of an increase or decrease of a single predator species along with the increase or decrease of its major prey. This is typical of restricted feeders (p. 182).

Buteos generally

The Buteos as a group consumed mice almost exclusively in both winters: 87.4 per cent meadow mice and 5.2 per cent white-footed mice in 1942, as compared with 85.3 per cent and 5.8 per cent, respectively, in 1948 (Table 27). Only the Red-tailed Hawk showed a tendency to shift

TABLE 25

COMPOSITION OF WINTER FOOD OF THE HORNED OWL AS DETERMINED BY PELLET ANALYSIS

1942 and 1948

1942—167 whole pellets + 75 broken pellets—11 owls
1948—297 pellets—14 owls

Species	Number of Prey Individuals		Per cent of Prey Individuals			Number of Whole Pellets Containing Prey Remains	
	1942	1948	1942		1948	1942	1948
White-footed mice ..	438	587	49.4 ⎫		58.2 ⎫	109	138
				91.0			
Meadow mouse	369	282	41.6 ⎭		28.0 ⎭ 86.2	110	123
Bog lemming	1	..	.1		..	1	..
Norway rat	24	22	2.7 ⎫		2.2 ⎫	18	29
Shrews	15	3	1.7 ⎬ 4.7		.3 ⎪	6	3
Moles	1	2	.1 ⎪		.2 ⎬ 2.8	1	2
Weasels	1	1	.1 ⎭		.1 ⎭	1	1
Rabbit	17	43	1.9 ⎫		4.3 ⎫	27	80
Muskrat	3	4	.3 ⎬ 2.4		.4 ⎪	3	5
Fox squirrel	2	3	.2 ⎭		.3 ⎬ 5.1	3	3
Opossum	..	1	..		.1 ⎭	..	1
Pheasant	4	15	.5 ⎫		1.5 ⎫	6	22
Bob-White	1	..	.1 ⎪		.. ⎪	1	1
Duck	1	..	.1 ⎬ 1.7		.. ⎬ 4.8	1	1
Small and medium sized birds	9	33	1.0 ⎭		3.3 ⎭	6	27
Screech Owl	..	4	..		.4 ⎫	..	4
Hawk	..	1	..		.1 ⎬ 1.2	..	1
Unidentified	..	7	..		.7 ⎭	..	7
Total	886	1008	99.8		100.1		

from a nearly pure mouse diet. With three times as many Buteos on the area in 1942 as in 1948, it might appear that game species would have suffered heavier losses in 1942, or, conversely, it might be reasoned that with the meadow mouse population at one-fourth of the 1942 level, game species would have suffered greater losses in 1948. Actually, the Buteo population came into equilibrium with its major food supply, so that the relative effect of the Buteos was similar in both winters, although the densities of prey and predator varied. We have here an illustration of biotic equilibrium: the number of meadow mice was to the number of Buteos in 1942 as the number of meadow mice to the number of Buteos in 1948.

We may conclude that since the Buteos have high mobility they can maintain equilibrium between themselves and their prey. We shall see later that studies carried on through two winters and observations during three additional winters leave little doubt that they do this on any given area in southern Michigan, winter after winter. Individual Buteo predators were observed to deviate from a high rodent diet by taking a higher percentage of game species, but in this region meadow mice are the staple food of Buteos and the population of these birds will adjust itself to the density of those animals. Although the percentage of such items as rabbit, pheasant, and small birds may fluctuate somewhat from year to year with changes in major prey populations, the fluctuation will not be significant. When meadow mice, the staple food, are low in numbers it is to be expected that the Buteo population will be low, even though other prey populations are high. Thus it is evident that small rodents seldom, if ever, play an important role as buffers [alternative food—Leopold (1937)] between Buteos and game species in southern Michigan. They are the staple diet rather than alternative foods.

Marsh Hawk

Concentrations of Marsh Hawks during winter indicate high densities of small rodents. In 1942 an average of 37 Marsh Hawks wintered. Not one spent the entire winter of 1947-48, although sufficient food data were gathered in early fall and late winter to show the feeding trend of that period. In 1942 meadow mice were 93 per cent of the food of the Marsh Hawks; in 1948, 98 per cent (Table 28). Mice and shrews formed 98.4 and 99.7 per cent of their food in 1942 and in 1948, respectively. Again the significant fact is that meadow mice formed the staple food and composed similar percentages of the diets during the two winters, irrespective of the fact that these animals were less abundant in 1948. Although 26 pellets contained rabbit remains, the individual rabbits represented are undoubtedly less than 26. Most pellets contained only fur evidence. It is also highly probable that some of those represented were shot or wounded during the hunting season and then consumed by the hawks.

TABLE 26

COMPOSITION OF WINTER FOOD OF THE SCREECH OWL AS DETERMINED BY PELLET ANALYSIS

1942— 51 pellets—7 owls
1948—122 pellets—9 owls   Winters of 1941-42 and 1947-48

Prey Species	Number of Prey Individuals		Per cent of Prey Individuals		Number of Whole Pellets Containing Prey Remains	
	1942	1948	1942	1948	1942	1948
Meadow mouse	39	61	45.3	40.9	34	57
White-footed mice ..	43	69	50.0	46.3	27	58
Shrews	3	..	3.5	...	3	..
Moles	..	1	...	.7	..	1
Small birds	1	17	1.2	11.4	1	17
Insects	..	1	...	.7	2	1
Total	86	149	100.0	100.0	..	..

TABLE 27

COMPOSITION OF WINTER FOOD OF THE RED-TAILED HAWK AS DETERMINED BY PELLET ANALYSIS

1942—126 pellets— 3 hawks   1948—103 pellets—10 hawks

Prey Species	Number of Prey Individuals in Pellets		Percentages of all Individuals in Pellets		Number of Pellets Containing Prey Remains	
	1942	1948	1942	1948	1942	1948
Meadow mouse	307	174	89.0 ⎤	84.1 ⎤	125	102
White-footed mice ..	23	13	6.7	6.3	15	12
Norway rat	2	..	.6 ⎬ 99.4	.. ⎬ 92.4	3	..
Short-tailed shrew ..	6	2	1.7	1.0	6	2
Other shrews	5	..	1.4 ⎦	.. ⎦	5	..
Moles	..	1	..	.5	..	1
Weasels	..	1	..	.5	..	1
Rabbit	1	4	.3 ⎤	2.0 ⎤	1	6
Pheasant	..	1	.. ⎬	.5 ⎬ 7.8	..	..
Small birds	1	11	.3 ⎦	5.3 ⎦	1	11
Total	345	207	100.0	100.2	..	..

1942—Average 2.7 individuals per pellet
1948—Average 2.0 individuals per pellet

COMPOSITION OF WINTER FOOD OF THE RED-SHOULDERED HAWK AS DETERMINED BY PELLET ANALYSIS

1942—24 pellets—1 hawk   1948—13 pellets—1 hawk

	1942	1948	1942	1948	1942	1948
Meadow mouse	32	17	94.0	100.0	24	13
White-footed mice ..	1	..	3.0	..	1	..
Small birds	1	..	3.0	..	1	..
Total	34	17	100.0	100.0	..	..

COMPOSITION OF WINTER FOOD OF BUTEOS AS DETERMINED BY PELLET ANALYSIS

1942—216 pellets—10 hawks   1948—116 pellets—11 hawks

Prey Species	Number of Prey Individuals in Pellets		Percentages of all Individuals in Pellets		Number of Pellets Containing Prey Remains	
	1942	1948	1942	1948	1942	1948
Meadow mouse	509	191	87.4	85.3	214	115
White-footed mice ..	30	13	5.2	5.8	29	12
Norway rat	4	..	.7 } 98.9	.. } 92.1	4	..
Short-tailed shrew ..	16	2	2.8	1.0	13	2
Other shrews	16	..	2.8	..	16	..
Moles	..	1	..	.4 } .8	..	1
Weasels	..	1	..	.4	..	1
Rabbit	1	4	.2	2.0 } 7.4	1	6
Pheasant	..	1	.. } 1.2	.4	..	1
Small birds	6	11	1.0	5.0	6	11
Total	582	224	100.1	100.3	..	..

COMPOSITION OF WINTER FOOD OF THE AMERICAN ROUGH-LEGGED HAWK AS DETERMINED BY PELLET ANALYSIS

1942—66 pellets—6 hawks   1948—No birds on area

	1942	1942	1942
Meadow mouse	170	83.7	65
White-footed mice ..	6	3.0	5
Norway rat	2	1.0	2
Short-tailed shrew ..	10	5.0	7
Other shrews	11	5.4	11
Passerine birds	4	1.9	4
Total	203	100.0	..

TABLE 28

COMPOSITION OF WINTER FOOD OF THE SHORT-EARED OWL AS DETERMINED BY PELLET ANALYSIS

1942—772 pellets—28 owls   1948—50 pellets—4 owls

Prey Species	Number of Prey Individuals in Pellets		Percentages of all Individuals in Pellets		Number of Pellets Containing Prey Remains	
	1942	1948	1942	1948	1942	1948
Meadow mouse	1307	82	87.7	97.6	748	50
White-footed mice	116	0	7.8	..	77	..
Other mice	18	..	1.2	..	15	..
Shrews	45	1	3.0	1.2	41	1
Small Birds	3	1	.2	1.2	3	1
Total	1489	84	99.9	100.0		

COMPOSITION OF WINTER FOOD OF THE LONG-EARED OWL AS DETERMINED BY
PELLET ANALYSIS

1942—approx. 443 pellets—12 owls   1948—no birds on area

Prey Species	Number of Individuals in Pellets	Percentages of all Individuals in Pellets
	1942	1942
Meadow mouse	843	88.6 ⎫
White-footed mice	86	9.0 ⎬ 97.6
Bog lemming	15	1.6 ⎭
House mouse	1	.1
Short-tailed shrew	3	.3
Other shrews	4	.4
Total	952	100.0

COMPOSITION OF WINTER FOOD OF THE SPARROW HAWK AS DETERMINED BY
PELLET ANALYSIS

1942—32 pellets—2 hawks   1948—52 pellets—2 hawks

Prey Species	Number of Prey Individuals in Pellets		Percentages of all Individuals in Pellets		Number of Pellets Containing Prey Remains	
	1942	1948	1942	1948	1942	1948
Meadow mouse	26	23	66.7 ⎫	52.3 ⎫	26	23
White-footed mice ...	7	18	18.0 ⎬ 87.2	41.0 ⎬ 93.3	7	18
Short-tailed shrew ...	1	..	2.5 ⎭	... ⎭	1	..
Small birds	5	3	12.8	7.0	5	3
Insects	..	..	..	..		
Total	39	44	100.0	100.3		

COMPOSITION OF WINTER FOOD OF THE MARSH HAWK AS DETERMINED BY
PELLET ANALYSIS

1942—2311 pellets—48 hawks   1948— 147 pellets— 5 hawks

Prey Species	Number of Prey Individuals in Pellets		Percentages of all Individuals in Pellets		Number of Pellets Containing Prey Remains	
	1942	1948	1942	1948	1942	1948
Meadow mouse	7011	424	93.2 ⎫	98.1 ⎫	2306	147
White-footed mice ...	256	7	3.4 ⎬ 98.4	1.6 ⎬ 99.7	230	7
Other mice	7	..	.1 ⎭		5	..
Shrews	128	..	1.7 ⎫		115	..
Rabbit	*26	..	.34 ⎬ 1.34		26	..
Small birds	76	..	1.0 ⎭		76	..
Garter snakes	21	1	.27	.3	21	1
Total	7525	432	100.1	100.0		

* Indicates number of representations in pellets

Sparrow Hawk

Five Sparrow Hawks wintered in 1942 and in 1948. The diets were 87.2 and 93.3 per cent mice and shrews (Table 28) during the two winters. The higher percentage of small birds shown in the 1942 diet may result from the fact that the 1948 pellet sample was obtained largely from one hawk. Kill observations made during both years confirmed a diet largely of meadow mice. (p. 148).

Long-eared Owl

In 1942 mice comprised 97.8 per cent of the diet of the Long-eared Owl, meadow mice being 88.6 per cent and white-footed mice 9 per cent (Table 28). No birds were found in the diet. An average of seven Long-eared Owls wintered in 1942, but none was present in 1948. Winter pellets we collected in 1939 from two widely separated Long-eared Owl roosts in the vicinity of Ann Arbor also revealed little or no bird predation by these owls. Fifty-three whole pellets, plus broken pellet material, contained 161 prey items—96 meadow mice, 52 white-footed mice, 2 house mice, 7 bog lemmings, 3 shrews, and only one small bird.

Short-eared Owl

Thirty-one Short-eared Owls wintered in 1942, and none remained through the winter of 1948. This difference in population density again reflected the difference in density of meadow mice between the two years and is a density-dependent relation between this species and its major prey. In 1942 and 1948 (Table 28), meadow mice were, respectively, 87.7 and 97.6 per cent of the diet. It should be pointed out that the diet of the Marsh Hawks and Short-eared Owls actually contained higher percentages of meadow mice in 1948 than in 1942, when these mice were at a population peak. This was largely because of the higher white-footed mouse and shrew populations in 1942, which were also utilized. It is an illustration of the manner in which the densities of one or more prey species affect predation on another, and again shows why it is difficult and misleading to interpret predation in terms of single predator or prey species.

Screech Owl

As Table 26 indicates, the major food items in the Screech Owl diet during the two years remained about the same. Meadow mice and white-footed mice combined were 95.3 per cent in 1942 and 87.2 per cent in 1948. The percentage of small birds (1.2 and 11.4 per cent) showed a substantial increase in the latter winter. This trend toward more small birds in the diet reflected the change in the relation of meadow mouse to small bird densities of the latter year.

Great Horned Owl

White-footed mice were the greatest proportion of individuals in the

Horned Owl diet. Except in a few instances, common white-footed mice could not be distinguished in the pellets from prairie deer mice. It is significant that in both years white-footed mice formed a higher proportion of the diet than the far more numerous meadow mice. This would seem to indicate that probably a large proportion of the white-footed mice in the pellets were of the woodland species and that wood-lots were hunted more than open country.

During both years, meadow mice and white-footed mice formed the staple food of these owls. They were 91 per cent of the diet in 1942 and 86 per cent in 1948 (Table 25). The proportion of meadow mice dropped in 1948 with the decrease in their population density, and there was a corresponding increase in the proportion of white-footed mice taken, although the lower prairie deer mouse population of this year made it probable that the combined population of common white-footed mouse and prairie deer mouse also was lower. Errington (1940) found in the diet of this owl a cold weather ratio of 27.7 meadow mice to 72.3 white-footed mice, indicating that the latter generally play the more important role.

Other small rodents, shrews, moles, and weasels showed a decrease from 4.7 per cent of the food in 1942 to 2.8 per cent in 1948. Rats ranked third in the diet in 1942, and fifth in 1948.

As Table 29 indicates, non-game food composed 96 per cent of the Horned Owl diet in 1942 and 89 per cent in 1948. This reduction in the proportion of non-game food taken was largely due to the smaller meadow mouse population of the latter year.

In 1948 (Table 25) there was an increase in the proportion of rabbits taken that appears to correspond with the greater density of their population. There was a marked increase in the proportion of pheasants and small birds in the diet (Table 25), which, in the case of the pheasant at least, was not correlated with an increase in their numbers on the study area but may have been correlated with the pheasant density within the respective owl ranges.

Errington (1940), in his studies of the Horned Owls' prey in the North Central states, found that rabbits and hares were represented in 68.5 per cent of 4,838 pellets and stomachs and constituted the staple fare of this raptor over most of its range. In our study area rabbits were represented in 20 per cent of 539 winter pellets. This is consistent with the relatively low rabbit population on the area during 1942 and 1948. Observations and pellet analysis in 1947 and 1949 indicate that rabbits were not staple foods during those years either. The relatively low rabbit density and the relatively high mouse density generally characteristic of this area probably account for a high consumption of small rodents and a low consumption of rabbits during four winters. The most distinctive feature of the Horned Owl diet, as compared with that of other raptors, is the wide range of prey species that it includes.

Cooper's Hawk

Unfortunately, few pellet data were obtained for the Cooper's Hawk, and thus the statement of its winter diet must be based solely on field observation. The dearth of pheasant, rabbit, and squirrel kills attributable to the Cooper's Hawk in 1942 showed that this hawk was not feeding principally on these species. On the eight sections where prey population studies were made, a tally of Cooper's Hawk kills showed that 3.4 per cent of the Bob-white population were taken by these predators in 1942 and 12.5 per cent in 1948. No pheasants were known to have been taken by Cooper's Hawks in 1942, but over 3 per cent of the pheasant population was taken in 1948.

In 1942 Cooper's Hawks were observed hunting meadow mouse habitats much more frequently than in 1948, and it is quite probable that they took more of these mice in the former year. Small birds composed the greater part of this Hawk's winter diet in both years, which is not surprising, as the density of this population was relatively high, and, in consequence of the Cooper's Hawk's flight adaptations, they were vulnerable prey. (p. 179).

WINTER FOOD HABITS OF THE RAPTOR POPULATIONS

In Table 29, prey species are placed in two categories, non-game food and game food. The latter is composed of species generally considered either beneficial or neutral from a human standpoint, though some are not usually classed as game.

In 1942 the raptors consumed 11,404 individuals of non-game species, amounting to 98.7 per cent of their diet, and 155 individuals of game or desirable species, forming 1.3 per cent. In 1948 they consumed 1,793 individuals of non-game species, forming 92.4 per cent, and 148 of game or desirable species, making up 7.6 per cent.

For the two years the non-game food in the raptor diet averaged 95.5 per cent; the game food, 4.5 per cent. It is obvious that this winter diet of the raptor population was not directly harmful to man's interests.

It is of special interest that no poultry was found in the pellets. The number of chickens on the farms was high. An allowance of 93 for each farm would give an estimated population of more than 15,000. It is evident that because of protection accorded them by man and numerous other conditions, their vulnerability was low.

THE RELATION OF THE DIETS OF
RAPTORS TO PREY DENSITIES

We have already described the dispersion and movement of hawks and owls that allowed members of the population to exert maximum hunting pressure on prey species. This raptor population has been shown to have hunted chiefly the open-field habitats. Only the Cooper's Hawk, Horned Owl, and Screech Owl hunted more in woodlots than in the fields. The tables recording the diet of each raptor show that each preyed most intensively on the most abundant prey in its hunting habitat. In other words, predation by a population of any hawk or owl on available prey was the result of a density relation. Thus, meadow mice rated first in the diet of hawks and owls hunting open-land habitats; white-footed mice first in the diet of the only nocturnal woodland raptors—the Horned Owl and the Screech Owl; and birds first in that of the only diurnal woodland species—the Cooper's Hawk.

It is obvious that no raptor can hunt all habitats equally well, but that efficient hunting of each species is limited by physical adaptations to rather specific habitats, either by day or by night. In other words, the raptors are to a certain degree specialists, evolved to occupy various ecologic niches. In the case of an individual hawk, these niches or habitats narrow down to specific fields, swamps, or woods within its established range. Thus the diet of an individual raptor will tend to reflect the relative local prey densities of the species within its range that it is adapted to catch. This may not, and frequently does not (Table 25), show a correlation with the relative densities of these same prey species over a larger inclusive land area. This fact has caused much mis-

TABLE 29

FALL AND WINTER FEEDING TREND OF RAPTORS ON AREA
1942-1948

Predator Species	Non-Game Food				Game Food			
	Mice, Shrews, Rats, Weasels, Snakes, Moles				Rabbit, Muskrat, Fox Squirrel, Pheasant, Bob-white, Small and Medium-sized Birds			
	Number of Prey Individuals in pellets		Per cent of diet		Number or Prey Individuals in pellets		Per cent of diet	
	1942	1948	1942	1948	1942	1948	1942	1948
Red-tailed Hawk .	343	191	99.4	92.4	2	16	.6	7.8
Red-shouldered Hawk	33	17	97.0	100.0	1	0	3.0	0
American Rough-legged Hawk ..	199	...	98.0	...	4	..	2.0	..
Marsh Hawk ...	7,423	432	98.6	100.0	102	0	1.74	0
Sparrow Hawk ..	34	41	87.2	93.3	5	3	12.8	7.0
Great Horned Owl	849	897	95.8	89.0	37	111	4.2	11.0
Long-eared Owl .	952	...	100.0	...	0	..	0	..
Short-eared Owl .	1,486	83	99.8	98.8	3	1	.2	1.2
Screech Owl	85	131	98.8	88.6	1	17	1.2	11.4
Totals	11,404	1,792	98.7	92.4	155	148	1.3	7.6

95.5% = average per cent of non-game food for two-year period
4.5% = average per cent of game food for two-year period
13,500 = total number of prey individuals

understanding in interpreting raptor predation phenomena on the basis
of raptor food habits.

TABLE 30

PELLET COLLECTION DATA

Hawk and Owl Species	Average Winter Population of Hawks and Owls on Township		Number of Hawks and Owls Represented by Pellets		Number of Locations where Pellets Were Collected		Number of Pellets	
	1942	1948	1942	1948	1942	1948	1942	1948
Red-tailed Hawk	24	13	3	10	3	6	126	103
Red-shouldered Hawk ...	1	3	1	1	2	1	24	13
American Rough-legged Hawk	19	0	6	0	5	0	66	0
Marsh Hawk	37	1	48	5	1	1	2,311	147
Sparrow Hawk	5	5	2	2	2	2	32	52
Great Horned Owl	11	13	11	14	21	28	167+	297
Long-eared Owl	7	0	12	0	3	0	433	0
Short-eared Owl	21	4	28	4	3	1	772	50
Screech Owl	14	19	7	9	5	9	51	122
Total							3,992	784
						Total =	4,776	

Note: Cooper's Hawk pellets and kills omitted.

CHAPTER 7

The Dynamics of Predation

W E NOW know, as accurately as it has been possible to determine, how many hawks and owls wintered and what kind of prey and roughly how many of each species were available. We also know much about the raptors' movements and their food habits. This information is essential to understanding raptor predation, but considered separately it tells us very little about it. Not until these data are combined and analyzed will they show what raptor predation is and how it works. What follows is not a philosophically presented theory of the dynamics of predation but rather findings resulting from analysis of the data. The outcome is not one that was previously evident to the authors, even though they had spent more than 15 years in observing, photographing and studying raptors. It is not predation as seen from day to day by the farmer, the sportsman, or even the research man. There is, then, little wonder that there have been almost as many ideas on the subject of predation as there have been interested observers. Once we see and understand the function and mechanics of raptor predation, many isolated and seemingly conflicting observations will fit into the pattern.

PREDATION BY A COLLECTIVE RAPTOR POPULATION
ON A POPULATION OF A SINGLE PREY SPECIES

A clarification of certain terms and concepts is necessary before analyzing previously presented data and interpreting the winter food habits of the raptor population to show their biological and ecological significance. McAtee (1932) concluded that "With availability as the controlling factor it follows that in the long run, and on the average, losses to predators will be closely in proportion to the relative abundance of the group concerned." Errington (1936) suggested that predation in some cases may be proportional to overpopulation, and he concluded that the basic principle underlying predation is availability of prey. Tinbergen (1946) believed that the risk run by a prey species determines the intensity of predation upon it by the old world Sparrow Hawk. Risk is partly determined by habitat preference, exposure, conspicuousness, and escape reactions.

The term "vulnerability" frequently is found in predation literature, and there seems to be confusion of usage with respect to this term and the term "availability." Much of this confusion has arisen through laxity in defining the terms. Accordingly, no attempt will be made to define them as used by other authors, but their use and meaning in this study will be made clear.

Availability is used to mean that prey organisms are present.

Vulnerability is used as a term embracing a multiplicity of things that influence predation. It expresses the condition of a prey species (including prey density) or a prey population in regard to the extent of predation it suffers or the risk it runs of being preyed upon by raptors. Vulnerability of a prey species is the result of all the physical and biological conditions that cause one species to be preyed on more heavily than another. In this respect it is analogous to "availability," as used by McAtee (1932).

In the following section, we shall attempt to isolate and evaluate the principal conditions that are most influential in causing predation on the meadow mouse. With this as a working basis, we shall interpret the food habits of all the raptors, correspondingly expanding our concept of prey vulnerability until we can evaluate each causative condition and discuss predation as a phenomenon of populations.

RELATION OF RAPTOR DENSITY TO MEADOW MOUSE DENSITY

In determining the conditions that rendered the 1942 meadow mouse population vulnerable to predation and in analyzing the relation between density of raptors and that of a single prey species, nine sites or areas, along the hawk census route were selected. Both raptor and prey densities on these sites were determined (Map 11).

Evaluations of raptor density are based on the number of hawks recorded hunting the various study sites during the censuses (Chap. 3) and are thus comparable. The raptor densities (Table 31) represent the sum of the raptors seen on any site during all censuses. A concentration may represent numerous individuals recorded infrequently in consequence of movement on and off the strip or one or two individuals recorded frequently because they hunted one site almost exclusively. The number of times a raptor was recorded on a given site, rather than the number of individuals, expresses the raptor density and the hunting pressure. Taking the census by car gave a common time denominator and allowed observation of all nine sites within a period of three hours, thus almost eliminating variations due to time differentials, weather conditions, and raptor movement.

The meadow mouse densities were ascertained by trapping and by population studies (Chap. 5). The relationship of raptor densities to prey densities and the role of vulnerability will be considered only for the principal prey species, the meadow mouse. Each of the nine study sites on Map 11 represents an area of extensive meadow mouse habitat and the raptor pressure associated with it. All nine sites supported late fall densities of approximately 140 meadow mice per acre. The distribution and relation of meadow mouse and raptor densities within the census strip are shown on Map 11. Six of the nine sites supported raptor concentrations throughout all or part of the fall, winter, and early spring. Three sites with high densities of meadow mice did not support high densities of raptors (Map 11).

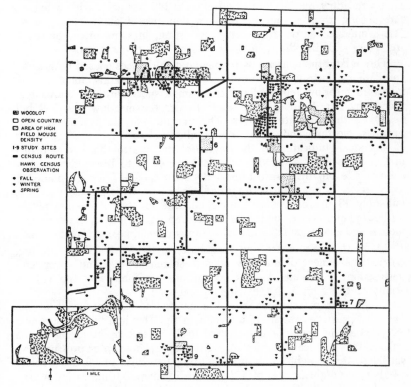

MAP 11. Relation of hawk density to meadow mouse density

. Forty-eight per cent of all hawk observations made on the census strip during fall, 57 per cent of those in winter, and 53 per cent of those in spring were concentrated on the nine sites, which were a small proportion of the 18-square-mile census area (Map 11). The number and composition of raptors on six of these sites are recorded in Table 32. Fifty per cent of all the hawks observed on the census trips were recorded in six raptor concentrations where meadow mouse populations were high (Map 11). Thus, raptor concentrations were definitely associated with high meadow mouse densities but high meadow mouse densities were not always associated with high raptor densities. Nevertheless, density of the prey species was most important in determining vulnerability.

Conditions Influencing Meadow Mouse Vulnerability

A discussion of the relation of the raptors to meadow mice on the several study sites illustrates the complexity of some of the causes of vulnerability of a prey population. In late October, Site 1 supported a high meadow mouse population of over 140 per acre, with extensive surrounding areas supporting medium densities of about 50 per acre. The ground cover was protective, consisting, in some areas, of sedge and grass and, in others, of heavy redtop and couch-grass.

Throughout the winter, environmental disturbances caused by grazing and flooding destroyed or altered meadow mouse habitats. Such changes increased meadow mouse densities in small spots within the area embraced by Site 1. In some of these locales the meadow mice all but destroyed the vegetative cover. The ground was honeycombed with burrows, and the woody plants were girdled (Plates 5 and 6). The extent of damage is indicated by the degree of damage to the woody plants in a transect, 13 feet by 660 feet, laid out so as to bisect the area. A count made on this transect showed that out of 487 woody plants of five species, 78 per cent were girdled, 9 per cent were girlded partially, and only 13 per cent were untouched. The greatest diameters of the girdled plants were: sumac, 2 inches; apple, 5 inches; hawthorn, 3 inches; and ash, 1 inch. Some of the shrubs were almost stripped of bark.

Many things, animate and inanimate, including flooding, freezing, grazing, movement of meadow mice in search of food and cover, and destruction of habitat by mice, contributed to increasing the vulnerability of these mice as the winter progressed.

The raptor pressure on Site 1 amounted to 13.6 per cent of the observed pressure exerted on the census strip. This hunting pressure was 8.7 per cent in winter and increased to 27.3 per cent in spring as the vulnerability of meadow mice rose. After the spring raptor pressure had subsided, three nights of trapping (using two lines of 66 traps each) yielded trap-night indices of .015 and .025 (Chap. 5), indicating a very low population for this area, which was originally estimated to support

TABLE 31

RAPTOR DENSITY AS INDICATED BY OBSERVATIONS OF RAPTORS ON NINE MEADOW MOUSE SITES OF HIGH DENSITY

Per Cent of Total Census

Site	1		2		3		4		5		6		7		8		9	
	No. Obs.	%	No. Obs.	%	No. Obs.	%	No. Obs.	%	No. Obs.	%	No. Obs.	%	No. Obs.	%	No. Obs.	%	No. Obs.	%
FALL 48 Obs. ..	..	..	11	10.9	6	5.9	0	0	1	.99	0	0	12	11.8	15	14.8	3	2.97
WINTER 192 Obs.	29	8.7	57	17.1	19	5.7	5	1.5	12	3.6	3	.9	17	5.1	14	4.2	36	10.8
SPRING 62 Obs.	32	27.3	8	6.8	1	.85	0	0	0	0	3	2.5	14	12	1	.84	3	2.56
TOTAL 302 Obs.	61	13.6	76	13.8	26	4.7	5	.9	13	2.35	6	1.1	43	7.8	30	5.4	42	7.62

Fall Census total for area .... 101 (Nov. 10 census omitted)
Winter Census total for area 333
Spring Census total for area 117
Census Grand Total 551

TABLE 32

COMPOSITION OF HAWK CONCENTRATIONS ON SIX SITES WITH HIGH DENSITY OF MEADOW MICE

Site	1		2		3		7		8		9		
	No.	%	No.	%	No.	%	No.	%	No.	%	No.	%	% Total
FALL													
Buteos	..	..	10	91	6	100	3	25	15	100	2	67	
Other Hawks	..	..	1	9	..	..	9	75	0	0	1	33	
WINTER													
Buteos	25	86	50	88	11	58	6	35	10	71	26	72	
Other Hawks	4	14	7	12	8	42	11	65	4	29	10	28	
SPRING													
Buteos	30	94	5	63	0	..	0	..	1	100	3	100	
Other Hawks	2	6	3	37	1	100	14	100	0	..	0	0	
TOTAL EACH SITE													
Buteos	55	90	65	86	17	65	9	21	26	87	31	74	
Other Hawks	6	10	11	14	9	35	34	79	4	13	11	26	
TOTAL	61		76		26		43		30		42		278

a meadow mouse population of approximately 140 per acre.

While meadow mouse density decreased from winter to spring, raptor hunting pressure increased. This was because grazing, flooding, freezing, destruction of cover, and meadow mouse movements in response to these changes increased vulnerability. The obvious conclusion is that vulnerability, though greatly influenced by meadow mouse density, was the result of numerous conditions. The destruction of cover by these mice was one of the more important factors in increasing vulnerability.

Site II (Map 11) supported a raptor pressure equal to that of Site I (Table 31, Map 11), and a comparable fall population of meadow mice, but this site lacked heavy ground cover and was subject to no drastic environmental changes. Two trap lines, operated for four nights, October 21 to 24, yielded trap-night indices of .433 and .340 for meadow mice and .492 and .410 for all small rodents and insectivores.

On March 7, after four and one-half months of constant predatory pressure, 79 traps, run for three consecutive nights, yielded only one meadow mouse and one white-footed (prairie deer) mouse, or trap-night indices of .004 each. Meadow mouse vulnerability on this site was attributed chiefly to scarcity of heavy ground cover and gradual reduction of this cover throughout the winter. Raptor predation pressure declined from a maximum of 17 per cent in winter to 7 per cent in spring (Table 31). Thus, on this site, conditions were sufficiently favorable for small rodents to permit a high fall population but were not adequate to give protection from predation when cover was reduced.

Site III supported its highest raptor density during the fall and winter and a very low density in spring (Table 31). The recorded hawk concentration was about one-third less than for Sites I and II, but the fall density of meadow mice was comparable.

The role played by meteorological conditions in determining vulnerability of prey was illustrated on this site. A late winter thaw covered the ground with slush and water, and this then was blanketed by snow. A sudden drop in temperature formed ice over entire fields. Runways and tunnels of meadow mice were filled with ice, making it necessary for the animals to travel on the snow, greatly increasing their vulnerability. A few days later the disappearance of the snow blanket on southern exposures left the mice without snow cover and with their tunnels still filled with ice. Under these conditions, there was little habitat protection and vulnerability of the mice increased still more. On one such exposure, six American Rough-legged Hawks were flushed from the ground and 10 kills of meadow mice were located. Three American Rough-legged Hawks were observed to kill three meadow mice in 10 minutes. In two hours, 49 fresh kills of meadow mice were found on about 100 acres of bare or snow-covered ground. None was believed to be more than six days old.

About two weeks later, a trap line was run for three consecutive nights. Three meadow mice were caught, giving a trap-night index of .037. Low though this index is, it is higher than the average for Site III because the trapping was done where observations indicated the highest residual population of meadow mice. On some other areas of Site III, meadow mice almost may have been extirpated.

Doubtless, conditions other than predation contributed heavily to the decrease of meadow mice on Site III, but meteorological conditions initiated changes that increased vulnerability of these mice to a point where they could be taken at will by raptors. Only on this site were partially eaten meadow mice observed, good evidence that the hawks actually were catching more than they could consume.

It is interesting to note that only one out of 117 hawks observed on the spring censuses (Table 31) was on Site III. This was a Cooper's Hawk. From winter to spring there was a reduction of hawks on Site III from 8 per cent of the total included in the census to less than 1 per cent. Thus the reduction in raptor density corresponded to the reduction in prey density.

Environmental destruction by meadow mice on Site IV was extensive, though not as widespread as on Sites I and III. Sixty-three sumac stems in a clump of 77 were girdled. This was fairly representative of the condition of the girdled woody plants on Site IV, but the damage was not general over the entire area.

On January 22, 33 meadow mouse kills (none much more than a week old) were counted in an area of 0.25 acres. During the following month, 56 additional kills were observed on the same area. This was not a complete record; nevertheless, it represented an average kill of two animals per day. Observations indicated that one Sparrow Hawk and one Red-tailed Hawk were responsible.

A trap line for March 6, 7, and 8 took 12 meadow mice and gave a trap-night index of .125, indicating that the meadow mouse population on Site IV, where raptor pressure was low (Table 31), had not been reduced to the same extent as on Sites I, II, and III. On this site hawks did not concentrate to take full advantage of meadow mouse vulnerability because the site was located close to a main highway. A male Sparrow Hawk hunted the site constantly from a favored perch and alone accounted for most of the predation.

Only six hawks, or 1.1 per cent of the known population (Table 31), were recorded on Site VI during fall, winter, and spring. The site included a young apple orchard with an extremely heavy mat of meadow-grass, redtop, and couch-grass.

Meadow mouse populations were high, but no trapping was done in the fall. By early March, 300, or 30 per cent, of 1,000 young trees were girdled completely or partially. A reexamination in February, 1947, indicated that approximately 10 per cent of the trees had died and 20 per cent had suffered permanent injury.

The Raptors

PLATE VII. Adult Bald Eagle at nest. Our national bird is a raptor, a majestic, keen-eyed symbol of freedom.

PLATE VIII. Adult Cooper's Hawk and young. The nest of this bird is most readily found before the trees leaf. Small birds are its staple food.

151

PLATE IX. Young Osprey. Juvenile hawks remain within limited areas during the breeding season.

PLATE X. An immature female Sharp-shinned Hawk forty-one days old and completely feathered. The Sharp-shinned can fly very fast for short distances.

PLATE XI. Immature male Sparrow Hawk. This little falcon was observed to have small daily ranges but quite large seasonal ranges. Falcons have long-pointed wings, dark eyes, and a notched beak.

PLATE XII. Immature Red-tailed Hawk. The Red-tailed Hawk population of the winter 1941-42 was composed largely of immatures.

PLATE XIII. Cooper's Hawk nests are usually flat-topped with a shallow cup. They may be distinguished from most other hawk nests by the bark lining. Sharp-shinned Hawk nests are constructed more delicately than the nests of Cooper's Hawks and they are lined with small twigs.

PLATE XIV. The Marsh Hawk nests on the ground. Nests are frequently destroyed, but if incubation is not advanced, the birds will renest. The productivity of the raptor population was determined by making egg and fledgling counts.

154

PLATE xv. Raccoon in the act of destroying the eggs of a Red-tailed Hawk in Superior Township in 1949. Destruction of eggs and young was frequent and usually occurred at night.

PLATE xvi. Nest and eggs of the Red-shouldered Hawk. The nest is generally lined with fresh green leaves or evergreen needles.

155

PLATE XVII. Mature Red-tailed Hawk. High perching and the silhouetting effect of a level landscape enabled this bird and other Buteos to be accurately counted.

PLATE XVIII. Red-shouldered Hawk feeding a frog to her young. Frogs are so completely digested and there are so few remnants left at the nests that they undoubtedly form a larger part of the diet than is indicated by the data recorded.

PLATE XIX. Young Ferruginous Rough-legged Hawks. This species nested near, but not on, the Wyoming study area. It is a frequent winter resident in nearby desert areas.

PLATE XX. Young Red-shouldered Hawks. Sex can be determined at this early age.

157

PLATE XXI. Young Prairie Falcons. Young hawks and owls generally remain within the range of their parents until fall migration. No juvenile banded hawks of any species were recorded as returning to the vicinity of their parents' nests following their first fall migration.

PLATE XXII. One young Red-tail has flown and the other is about to leave the nest. If undisturbed, young birds remain near the nest and go through a period of parental care and training.

158

PLATE XXIII. Red-tailed Hawk and young. This Hawk is more powerful and more aggressive than the Red-shouldered Hawk, and takes larger prey. Two is the usual number of young, but three and even four were recorded.

PLATE XXIV. Marsh Hawk feeding young. Because small birds as well as mice usually are abundant in the vicinity of Marsh Hawk nests, they are well represented in the spring and summer diet of this species.

PLATE XXVI. Osprey returning to its nest. These birds may have isolated nests and ranges or they may nest in colonies containing 30 or 40 pairs where each pair defends only a small area in the immediate vicinity of the nest and ranges over a large communal hunting locality.

PLATE XXV. Duck Hawks usually defend a cliff or nest site and hunt along a river or over a large valley. Their hunting range is extensive. It is quite common for this bird to attack and strike human intruders at its nest or eyrie.

PLATE XXVII. Adult Prairie Falcon returning to feed young. Prairie Falcons, Ravens, and Horned Owls often nest on the same cliff, and live in harmony, hunting overlapping ranges, but usually hunting different prey, or the same prey at different times.

PLATE XXVIII. Western Goshawks kill many rodents, but they also take toll of game birds. When food is abundant, they frequently cache some of it near the nest. This bird repeatedly attacked the authors.

161

PLATE xxix. Swainson's Hawk defending its nest. This bird struck the authors on numerous occasions. Individual birds vary greatly in the degree of defense exhibited at the nest; this variation makes it possible to recognize individual birds.

162

PLATE xxx. Short-eared Owl flying above winter communal roost after being flushed during a roost count.

PLATE xxxi. Marsh Hawk flying above winter communal roost. Counts were made as the hawks arrived and departed from the roost.

PLATE XXXII. Nest of a Short-eared Owl. This species wintered in Superior Township, but did not nest in the area. It wintered and nested in Jackson Hole, Wyoming. This species, like the Marsh Hawk, is a ground nester.

PLATE XXXIII. Banded young Marsh Hawks at the nest. At about this age the young birds hide in the surrounding vegetation. To gather food data, the hawks must be tethered near the nest.

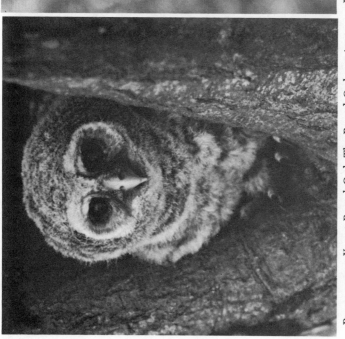

PLATE XXXIV. Young Barred Owl. The Barred Owl requires extensive woodlots containing hollow trees for nest sites. Where such sites are lacking, as in Superior Township, this owl is not able to compete with the more powerful and more aggressive Horned Owl. At times it uses stick nests.

PLATE XXXV. Young Barn Owl at nesting hollow. Barn Owls are efficient mousers. During periods of meadow mouse abundance they kill more of these animals than they can eat. Generally migratory in the Ann Arbor region; they will winter if sufficient food is available.

PLATE XXXVII. It is not difficult to make a census of Long-eared Owls because they generally roost in coniferous stands or thick tangled vegetation. Also they seldom flush until closely approached, usually return to the same roost, and, when present in numbers, roost communally.

PLATE XXXVI. Young Short-eared Owl. The migratory habits and the gregariousness of this species enable it to exert very effective pressure on high rodent populations.

166

PLATE XXXVIII. Young Horned Owls. Pheasants and cottontail rabbits are vulnerable to the Horned Owl, particularly in early spring. Prey species found in the nests of these birds represent only a part of the year-round diet. Conclusions as to the economic status of these owls should be drawn from data obtained throughout the year.

PLATE XXXIX. Horned Owls use trees, ledges, and crevices in cliffs for nest sites. In desert regions they may nest on slight ground elevations. In many ways this powerful owl dominates a community of prey and other raptor species.

167

This area of high meadow mouse population was not hunted by either nocturnal or diurnal raptors. The hazards offered by the young trees to the normal hunting of Rough-legged Hawks, Marsh Hawks, and Short-eared Owls partially explained the situation, but the dense ground cover played the greater role in making the prey secure. They were so well protected that raptors were not even attracted to the site. Access to ample food probably reduced meadow mouse movement. The bark of the young apple trees was consumed in preference to the grass, thus aiding in preserving cover and, consequently, security.

On March 14, trap lines of 64 traps took 26 meadow mice, giving a trapnight index for one night of .406, convincing evidence that a lack of predatory action, resulting from an especially protective environment, had allowed the meadow mouse density to remain high.

Similar but less intensive studies, carried out on raptor concentrations during the winter of 1947-48, substantiated the conclusions drawn from the data gathered in 1942.

SUMMARY OF CONDITIONS DETERMINING THE VULNERABILITY OF MEADOW MICE

The diverse conditions of these sites illustrate the complexity of predator-prey relations, but certain facts appear to be fundamental.

High raptor densities almost always were associated with high meadow mouse densities, but high local meadow mouse densities were not necessarily associated with high raptor densities. Raptor numbers were directly related to prey vulnerability. Meadow mouse density and the nature of the habitat were basic conditions influencing vulnerability. Snow, ice, flooding, grazing by stock, and plowing all affected the habitat in such a way as to increase vulnerability, either by exposing the animals or by initiating movements that increased prey density through local crowding. When consumption of food supplies destroyed the cover, it necessitated still greater movement and caused exposure. The rate and extent to which these reactions progressed influenced vulnerability. Habitat changes, resulting from flooding and freezing, heavy grazing and plowing, or from combinations of any of these phenomena, increased meadow mouse vulnerability suddenly. The slow consumption of cover by the mice, movements in search of food, and lack of a sufficiently protective environment in the first place gradually increased vulnerability as the winter progressed. Frequently several forces operated simultaneously on a given area, as on Site I, where the melting of a snow blanket and the consumption of the environment by the mice exposed them to predation as effectively as did more sudden changes. In other words, increased vulnerability caused by a combination of several slowly working forces was comparable with the vulnerability changes caused by a sudden destruction of habitat.

Thus, it would seem that the vulnerability of meadow mice is determined in part by density, in part by the intensity and extent of movement, and in part by the degree of exposure resulting from the destruction of cover.

Vulnerability is, however, more than a condition of a prey population: it is a relation between predator and prey; and the action of the predator or predator population upon prey is predation. The raptor population on the study sites reacted in direct proportion to the degree of prey vulnerability. Thus, on areas of high vulnerability they destroyed the prey, reducing their numbers as winter progressed. Since density was an important factor in determining vulnerability, it would be natural to assume that, as density of meadow mice decreased through predatory action, their vulnerability also would decrease. This was not entirely the case, since the gradual deterioration of protective cover tended to offset the decrease in vulnerability caused by a decrease in prey density.

In addition to the study of sites where density of meadow mice was high, observations were made on areas where raptor numbers were relatively high, but populations of these mice relatively low. Such conditions invariably were related to extensive open fields of medium to poor cover. These conditions appeared to increase the hunting effectiveness of Marsh Hawks and American Rough-legged Hawks in particular. Poor meadow mouse protection compensated for low numbers and thus the vulnerability of a local prey population was higher than the density alone would indicate.

These observations in areas of both low and high meadow mouse numbers indicated some of the basic components of vulnerability.

These are: density, lack of adequate cover, and the sum of all things, animate and inanimate, that intermittently cause an increase in meadow mouse numbers or movement, or a decrease in cover security. Numerous combinations of these basic components continually change the vulnerability of local meadow mouse populations.

We seem justified in stating that, in general, local density is a major item in determining the vulnerability of any single prey species to specific raptors, but that other conditions are also important. These other conditions may at times make a species vulnerable in spite of its low population.

It should be evident now that a complete analysis of vulnerability would disclose the nature of predation itself. Since predation is essentially a phenomenon of populations, vulnerability can be analyzed in terms of predator and prey populations.

The difficulty of interpreting the dynamics of predation in terms of one or two prey species or a single predator species (as has been the procedure in the past) appears obvious when we consider that different prey species have different degrees of vulnerability to the same predators and that some prey species are vulnerable to a wide range of predators, while others are vulnerable to only one or a few, and still others are practically invulnerable to all predators under favorable cover conditions.

PREDATION BY A COLLECTIVE RAPTOR POPULATION ON A COLLECTIVE PREY POPULATION

By definition, the vulnerability of a prey species is a combination of all the conditions which make it susceptible to predation. We already have disclosed a number of these conditions. With this concept of vulnerability, we see that, in a representative sample of predators' food, the percentages of prey taken will naturally indicate their relative vulnerability ratings. The sample from which percentages of diet (Tables 33 and 34) have been derived has not been weighted to correspond to the hawk populations and consequently the percentages may not be entirely representative of the food of the entire raptor population as a unit. If weighted, the relative percentage of each prey species in the raptor diet, however, would still remain the same. When the Marsh Hawk sample (large in proportion to those of other species) was reduced to 900 pellets, making a total sample from the raptor population of 2,581 pellets, the percentages indicating inclusion of prey species did not change

TABLE 33

SUMMARY OF FALL AND WINTER FOOD OF RAPTOR POPULATION AS DETERMINED BY PELLET ANALYSIS

1942–1948

Prey Species	Number of Individuals in Pellets		Percentage of all Individuals in Pellets			
	1942	1948	1942		1948	
Meadow mouse	10104	1063	87.4 ⎱		54.8 ⎱	
White-footed mice ..	976	694	8.4 ⎰ 98.4		35.8 ⎰ 92.2	
Other mice, rats, shrews, and moles	302	32	2.6 ⎰		1.6 ⎰	
Rabbit	44*	47	.3 ⎱		2.4 ⎱	
Fox squirrel	2	3	.02 ⎰		.16 ⎰	
				1.21		6.66
Pheasant	4	16	.03 ⎰		.8 ⎰	
Small and medium-sized birds	100	65	.86 ⎰		3.3 ⎰	
Weasels	1	2	⎱		⎱	
Muskrat	3	4				
Bob-white	1	0		.39		1.14
Snakes	21	1				
Other items	1	14	⎰		⎰	
Total	11559	1941	100.00		100.00	

* 26 of the 44 are representations in the Marsh Hawk pellets and thus indicate the maximum number of rabbits that could possibly have occurred in these pellets.

TABLE 34

THE RELATION OF THE PREDATION OF A COLLECTIVE WINTER RAPTOR POPULATION TO
RELATIVE PREY DENSITIES

Superior Township, Michigan

Prey Species	Approximate Prey Populations for Superior Township		Vulnerability Rating	Percentage Representation of Prey in the Diet of the Collective Raptor Population		Raptor Hunting Pressure No. of Raptors to which each Prey Species was Vulnerable		Prey-raptor Density Ratio	
	1942	1948		1942	1948	1942	1948	1942	1948
Meadow mouse	(1) 303,000	(1) 75,000	1	87.4	54.8	159	59	1905:1	1271:1
White-footed mice	(2) 33,000	(2) 27,000	2	8.4	35.8	149	50	221:1	540:1
Small birds	(3) 23,000	(3) 23,000	3	0.9+	3.3+	152	55	151:1	418:1
Game birds	(4) 1,500	(5) 1,100	4	0.03+	0.8+	21	32	71:1	34:1
Rabbits	(5) 300	(4) 1,200	5	0.3	2.4	45	32	7:1	38:1
Fox squirrels	(6) 300	(6) 1,000	6	0.02	0.16	21	19	14:1	53:1

Note: The percentages of small birds and game birds in the raptor diet are low because of omission of the food of the Cooper's Hawk.

materially, and the relative utilization of each prey species remained the same. This relation, corresponding to actual numbers of prey species taken, or to percentages, is the result of the combined forces that make one prey more vulnerable than another. Hence a vulnerability rating (Table 34) can be assigned in the order of relative vulnerability of the prey species, a rating of 1 being high. Only those prey species for which population estimates were obtained have been rated. During both years meadow mice rated 1, being most vulnerable, white-footed mice 2, small birds 3, rabbits 4, game birds 5, and fox squirrels 6. Rats and shrews are omitted because of lack of population data.

In Table 34, we bring to a focus the population work already discussed. The approximate prey populations are numbered in the order of their densities. The percentage of each prey species in the raptor diet likewise is shown, and these percentages indicate the vulnerability ratings. The number of raptor individuals that to some degree fed on each prey species is listed. This is based on the occurrence of the prey in the pellets of the raptors. The entire local population of each raptor is counted as hunting a prey if it was found in the raptor's diet. In the case of the Cooper's Hawk, such inclusion is based on observations and kills. For reasons already discussed, Marsh Hawks were not included as rabbit predators. These figures indicate the raptor hunting pressure applicable to any single prey species. The intensity of this hunting pressure is apparent from the extent of inclusion of each prey species in the diet of each raptor species (Tables 25-28). Thus all the Red-tailed Hawks are shown as exerting hunting pressure on rabbits in 1942, although actually the intensity was low. In the last column of Table 34 we have expressed the approximate ratio of prey to raptor, based on the raptor population known to hunt the various species.

If we now compare the prey population figures of Table 34 with the vulnerability ratings, we see that, in general, they are in the same order. Game birds in 1942 appear to be an exception, being less vulnerable than rabbits, although the population was somewhat higher. If the influence of the Cooper's Hawk were shown, the percentages of Bobwhites and pheasants attributable to this species would reveal that game birds also are vulnerable to a collective raptor population in the order of their relative densities. The vulnerability rating of 3 for small birds likewise would be established by a higher diet representation were the Cooper's Hawk's diet of small birds included. This is conclusive evidence that predation upon prey species by a collective raptor population is, in general, roughly proportional to the relative prey densities.

By determining the vulnerability of the prey by study of the diet of each raptor, we eliminate the possibility of overemphasizing the diet of any one species, as would be likely when combining varied diet samples for the entire raptor group. In Table 35 we arrive at another vulnerability rating by use of a different method and find that the ratings of Tables 34

TABLE 35

VULNERABILITY RATINGS DETERMINED BY RELATING PREY INDICES TO THE NUMBER OF RAPTOR SPECIES TO WHICH THE PREY WERE VULNERABLE

Vulnerability Index of Prey Species in Diets of Raptor Species (6 high)	Number of Raptor Species Applicable to Index											
	Meadow Mouse (6)		White-footed Mice (5)		Small Birds (4)		Rabbits (3)		Game Birds (2)		Fox Squirrel (1)	
	1942	1948	1942	1948	1942	1948	1942	1948	1942	1948	1942	1948
6	7	5	2	2								
5	2	2	7	3		1						
4					7	3	1	1				
3					1	1	1	1		1		
2									1	1		
1											1	1
Numerical Ratings Sum of Products of Indices X	52	40	47	27	31	20	7	7	2	5	1	1
Number of Raptor Species	6		5		4		3		2		1	
Vulnerability Rating	1		2		3		4		5		6	

Note: Multiply each figure in every column by the vulnerability indices opposite each at left and add the resulting products, column by column, to obtain the numerical ratings. These are then converted to vulnerability ratings, with 1 rating highest.

and 35 agree. In this instance, each prey species has been given an index corresponding to its vulnerability to each raptor. For convenience, an index of 1 has been used to designate a low vulnerability to a given raptor, and 6 indicates the highest vulnerability. The sum of the products of the vulnerability indices (of each prey species) and the number of raptors for which the index applied gives a figure indicative of the vulnerability of each prey species to all the raptors that preyed on it. Thus, the meadow mouse in 1942 was most vulnerable to seven raptors (having an index of 6 for seven raptor species) and was next most vulnerable to two other raptors, having a vulnerability index of 5 for two species. This gives us $7 \times 6 + 2 \times 5 = 52$, which is indicative of the vulnerability of the meadow mouse to the raptor population. This figure, when compared with those for other prey species, can be converted to a vulnerability rating. The meadow mouse has a vulnerability rating of 1 for both years and the ratings of other prey species obtained in this way agree with those in Table 34, which were determined on a basis of the per cent of representation in the collective raptor diet.

If we look at the hunting pressure (Table 34) exerted on each prey species by the raptor population and expressed by the number of hawks that took a particular prey (indicated by diet), we see that in both years the meadow mouse population was hunted by all the raptors present. The degree of this hunting pressure varied with raptor species and even with individuals. White-footed mice were preyed on by 149 raptors in 1942, and, as shown in Table 34, the intensity of hunting pressure was less than that on meadow mice. Only the Horned Owl and the Screech Owl exerted sufficiently intense hunting pressure on white-footed mice to make them the most vulnerable prey in their diets (Tables 25 and 26).

The raptor hunting pressure on each prey species (Table 34) is roughly proportional to the vulnerability ratings and would correspond even more closely were the raptors treated individually. From the prey-raptor ratios in the same table, we find that the most vulnerable prey populations had more individuals available per raptor; hence chance alone would give expectation of a more intense predation on these denser populations. Although various risks tended to alter more noticeably the influence of density among the small birds and the larger and less dense prey species, it remains true, even with these species, that the vulnerability of any one prey species to a raptor population is largely influenced by the relative population densities of all the other prey species.

McAtee (1932) showed that, among the larger animal classifications, the groups with the greater number of species will suffer greater losses to predators.

We have shown that a similar density relationship applies also to the populations of prey species existing simultaneously on any given area of land. This does not mean simply, however, that raptors subsist on the most numerous prey species—that they eat whatever is abundant. The significant fact is that in the case of each major prey species or group the ratio of the total population of that species or group to the total of the collective prey population approximates the ratio of the kill from the species or group by a collective population of raptors to the total kill from the collective prey population by the same raptor population (Table 36). If, for example, meadow mice are 90 per cent of the total available prey population, they will be approximately 90 per cent of the kill by a collective raptor population.

ANALYSIS OF PREY VULNERABILITY

A thorough analysis of prey vulnerability should disclose the nature of predation. Much of the confusion and many of the seemingly contradictory conclusions concerning the nature of predation have come from working with only a few of the conditions that determine prey vulnerability. As a result, the conditions have not been properly evaluated or inter-related.

Prey vulnerability is a result of two important conditions: prey density and prey risk. We can define prey risk as all the forces operating with prey density to make a prey vulnerable. Prey density is the basic condition and its effect is continually modified by prey risk (Fig. 6). The role of prey density in determining vulnerability of meadow mice has been shown.

Prey Risk

Protective cover: Cover protection for prey species or populations is one of the most important things determining the degree of risk to predation. As brought out in the discussion of meadow mouse vulnerability, a sufficiently protective cover will modify the effect of a high prey density to such an extent that the prey enjoy a high degree of security. On the other hand, destruction of cover will render vulnerable even low population densities of certain prey species. This point is so obvious and the literature cites so many examples that it requires no

TABLE 36

RELATION OF THE PERCENTAGE OF EACH MAJOR PREY SPECIES OR GROUP IN THE TOTAL PREY POPULATION TO THE PERCENTAGE OF EACH SUCH PREY SPECIES OR GROUP IN THE TOTAL DIET SAMPLE.

Prey Species or Groups	1941-42 % of Total Population	1941-42 % of Total Diet Sample	1947-48 % of Total Population	1947-48 % of Total Diet Sample
Meadow Mouse	83.9	87.4	58.4	54.8
White-footed Mice	9.1	8.4	21.0	35.8
Small Birds	6.4	0.9	17.9	3.3
Game Birds	0.4	0.03	0.9	0.8
Rabbit	0.08	0.3	0.9	2.4
Fox Squirrel	0.08	0.02	0.9	0.16

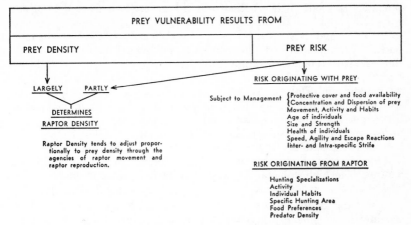

FIGURE 6. Prey vulnerability

further discussion. In passing, we merely draw attention to the fact that cover manipulation offers the best approach to managing predation.

Concentration and Dispersion: Concentration or dispersion of prey species may operate to increase or decrease their vulnerability. The concentration of pheasants into flocks for winter roosting tends to reduce their vulnerability to Horned Owls, whereas their dispersion in spring greatly increases it. Predation on pheasants in early spring greatly exceeded winter predation, in spite of the fact that their population densities were lower at that time (Chap. 12). Various conditions such as increased movement in spring and reduction of cover are effective simultaneously, so that the degree of influence of each condition is difficult to measure. Concentration of rabbits in and near woodlots made them more vulnerable to Horned Owls than their numbers would indicate.

In general, concentrations increase vulnerability by increasing local density. This is particularly true when such concentrations occur within the range of a raptor specifically adapted to catch the prey species involved. If, however, concentrations are small and sufficiently dispersed so that the chance of contact between predator and prey is reduced, then vulnerability also is reduced.

Movement: Movement on the part of any prey increases the risk it runs from predators by making the prey more conspicuous and exposing it more frequently. The same applies to populations. Since increased movement on the part of a population frequently is associated with a high population density, it is difficult to determine the relative degree that each increases vulnerability. In poor habitat, where movement in search of food is increased and prolonged, the risk of capture by a predator rises, but in such habitat this risk generally is more than offset by a low population density. Many prey species "freeze" when pursued by raptors, and this escape reaction is especially effective for game birds. The immobility of Pheasants and Bob-whites in roosts reduces the risk from nocturnal raptor predation in spite of increased local density.

Activity and Habits: Activities of prey species continually modify the degree of vulnerability that their population densities impose upon them. For example, white-footed mice, because they are largely nocturnal, are not highly vulnerable to diurnal raptors (Tables 27 and 28), but their vulnerability to the Horned Owl and Screech Owl (Tables 25 and 26) is increased because they are active when these birds are hunting. The high vulnerability of the cottontail rabbit to the Horned Owl is accounted for partly by its nocturnal activity (Table 25).

The habit of white-footed mice of running on top of snow, while meadow mice generally follow tunnels and runways beneath it, undoubtedly affects the relative vulnerability of these species to certain raptors.

Tinbergen (1940) cites numerous examples of how the activity and habits of a bird increase the risk it runs of being preyed upon by the old world Sparrow Hawk. He points out that the noisiness and social habits tend to increase the vulnerability of the English and European Tree Sparrows, while the Great Tit is more vulnerable than the Blue Tit because of its greater size, color, and noisiness. The Redstart is vulnerable because of its habit of choosing exposed places for singing and feeding, while the Wren is far less vulnerable because it rarely exposes itself.

Observation of food at more than 20 Duck Hawk eyries indicates that the flash patterns of Meadowlarks, Red-wings, and Blue Jays and the conspicuous flight of Flickers may increase their vulnerability to this large falcon.

The role of seasonal activity is obvious but, nevertheless, should be mentioned. In Michigan, frogs and snakes are highly vulnerable to the Red-shouldered Hawk in spring (Chap. 12), but because of hibernation are invulnerable in winter, in spite of a high population density. Ground squirrels are highly vulnerable during the summer months but, hiberating in burrows, they are invulnerable to raptors in winter. The examples that could be cited here are innumerable and many are too familiar to require mentioning.

Type of Habitat: Raptors are adapted to hunt nearly all terrestrial habitats; therefore, it is doubtful if the habitat of most prey species insures them any special security. Even mammals living largely in underground habitats, such as moles and pocket gophers, are heavily preyed on by certain raptor species. Purely aquatic prey, such as fish, are taken not only by ospreys which are peculiarly adapted to their capture, but also by Bald Eagles, Horned Owls, Red-shouldered Hawks, and other raptors.

Size and Strength: The very small birds and mammals are less conspicuous than larger forms and therefore are less vulnerable to predation. On the other hand, large prey animals, because of their size and strength, are better able to escape predators. The fox squirrel is an excellent example of a prey that is available to a large number of raptors, yet is relatively invulnerable, partially because of its size and toughness and its ability to bite when captured.

Speed, Agility, and Escape Reactions: There is no question but that speed and agility reduce the risk of predation. The small-bird population did not show the density relationship as clearly as the meadow mice and white-footed mice (Table 36) largely because speed and agility reduced the vulnerability imposed by density. The Bob-white is less vulnerable to most Buteos than to bird hawks. Its flight speed is not great enough to render it safe from the Cooper's or the Sharp-shinned Hawk. Fast, shifty birds, such as plover (in fact, most of the shore birds), are not nearly as vulnerable to the Peregrine Falcon as are speedy birds, such as ducks, that cannot shift readily in flight to avoid a falcon's stoop. Swallows are invulnerable to most North American raptors except the Pigeon Hawk (J. and F. Craighead, 1940). Fast-climbing birds, such as Ibis, Crows, Short-eared Owls, and Herons, can escape the much faster Peregrine Falcon by outclimbing it (Plate 25).

Protectively colored birds may escape pursuit from various predators by remaining motionless and blending with their environment.

Such prey animals as skunks, porcupines and tortoises possess specialized protective devices that render them invulnerable to most raptors, though not to all predators. We have records of predation on tortoises by Golden Eagles.

Hunting Specializations: The vulnerability of a collective prey population is greater to some predators than to others. For example, a larger range of prey is vulnerable to the Horned Owl than to the Long-eared Owl. Likewise, any prey species is more vulnerable to some predators than to others. This is due to adaptation and hunting specializations. Specializations of various raptor species in relation to abilities of prey groups determine, for example, that small birds run a greater risk of capture from Cooper's Hawks and Sharp-shinned Hawks than from Buteos and that mice run greater risks from Buteos than from bird hawks. The adaptations of raptors are such that small rodents are vulnerable to a far greater number of species than are small birds, game birds, or game mammals (Plates 9, 10, 11, 12).

Specific Hunting Area: It is obvious that prey species or populations situated largely outside of the hunting ranges of a predator will not be highly vulnerable to that predator. For example, in this study the most abundant prey (meadow mice) was not the most frequently represented prey in the Horned Owl diet (Table 25). The Horned Owl has a relatively small hunting range and in this region hunts primarily in the swamps, marshes, and woodlots. During the winter, the hunting range of an individual Horned Owl did not exceed a square mile, which would indicate that less than one-third of the land area was hunted by the Horned Owl population, and since that area did not include a high percentage of meadow mouse habitat it follows that, in spite of relatively high meadow mouse populations, this prey did not run the same risk of capture as did the woodland white-footed mouse. Rabbits, however, though scarce, were confined largely to Horned Owl habitat and thus their vulnerability was high.

Similar examples could be cited for other raptor species and for individuals.

Individual Habits and Abilities: Individual hunting habits may influence prey vulnerability locally. The hunting habits of raptors follow definite patterns. A Cooper's Hawk will return again and again to harry a covey of Bob-whites. A Red-tailed Hawk will habitually hunt favorite spots within its range. Some individual hawks consistently will take prey larger than that taken by other members of the same species. There is much difference in individual abilities. The diet of a stronger, faster, more aggressive individual will reflect its superiority. Innumerable examples of superior abilities of individual hawks could be cited from falconry experiences and literature (Plate 65).

Food Preferences: Uttendorfer (1939) thought that raptors show individual preference for certain species of prey, while L. Tinbergen (1932) has denied this possibility. With so many inter-related conditions determining the vulnerability of a prey, it is certain that selective hunting could be only a minor factor in determining prey vulnerability, probably evident only in the diet of some individual birds.

Highly specialized and efficient raptors, such as the Peregrine Falcon and Cooper's Hawk, preying on many birds, would be most likely to practice selective hunting. In captivity, the Peregrine Falcon (Plate 65) and the bird hawks (Plate 10) exhibit definite food preferences. The former cannot be induced readily to feed on fish-eating ducks or Crows: Pigeons, Doves, and small passerines are preferred to adult Starlings. In training a falcon to hunt Rooks or Crows, special care is taken to prevent the falcon from feeding on the Crow; instead it is rewarded with a freshly killed Pigeon. A choice meal on a preferred food will maintain the falcon's interest in the quarry for future flights, whereas a feeding of Crow will dampen its ardor. A falcon can be taught to fly at a particular type of quarry (J. and F. Craighead, 1942), indicating that habit formation and food preference could very well lead to selective hunting by wild falcons. The degree to which this factor normally would determine prey vulnerability would, however, be extremely small.

RAPTOR DENSITY

Within limits, prey vulnerability increases directly with raptor density, but since raptor density itself is determined chiefly by prey density, it naturally follows that prey vulnerability is largely a relation of predator and prey densities (Fig. 6, Table 34).

All or any number of the conditions of prey risk just discussed, operating simultaneously and collectively, tend to modify continuously the combined effect of predator and prey densities and determine vulnerability. No single condition is effective independently of all others, and thus an exact measure of the effect of each condition cannot be made. At a given time or place, however, some of these conditions markedly affect the vulnerability of a prey species or a population (at times locally overcoming the effect of the relation between predator and prey densities); but, broadly, they tend only to modify the density relation, which is the major condition determining prey vulnerability.

As has been pointed out, predation tends to be divided among the prey species in relation to their relative densities. All components of a mixed prey population are taken simultaneously and each species or prey group bears a reduction that is roughly proportional to the ratio of its numbers to those of all prey species present at the same time and in the same area. For example, in 1942 meadow mice represented approximately 84 per cent of the prey population and formed 87.4 per cent of the raptor diet. In the same year, white-footed mice formed approximately 9 per cent of the population and 8.4 per cent of the diet. Small birds represented about 6 per cent of the prey population and 1 per cent of the diet. The same general trend was observable in 1947-48 (See Table 36).

Each component does not provide the precise fraction of the take that it forms of the entire prey population present, but there is a tendency toward such a proportion. There are variations from this major effect of density relations, and they are the result of the many delicately interacting and interrelated prey risks just discussed. This concept of simultaneous and proportionate reduction of prey populations is the key to evaluating predation as a regulatory force.

THE MOVEMENT OF RAPTOR POPULATIONS IN RESPONSE TO THE DENSITIES OF PREY SPECIES

Since certain species run a greater risk of capture from some raptors than from others, we shall now attempt to classify raptors according to the variety of species upon which they prey.

In general, the raptors studied fall into two groups—those that by their physical structure are adapted for catching only a few species, and those that because of one or more physical endowments of size, foot strength, speed, and courage are capable of taking a greater variety of prey. The first group will be called *restricted feeders* and the latter group *general feeders*. The restricted feeders include the Marsh Hawk, the Short-eared and Long-eared Owls, and all the Buteos with the exception of adult Red-tailed Hawks. Immature Red-tailed Hawks were included in the restricted group because they subsisted largely on one or two prey species. The general feeders include the Horned Owl, the Cooper's Hawk, the Screech Owl, the adult Red-tailed Hawk, and the Sparrow Hawk. These raptors can take a large variety of prey species but may confine their predation to several prey species forming the largest prey populations. For example, the Screech Owl can take the meadow mouse, white-footed mouse and small birds—members of the three most dense prey populations. The two general feeders with the widest prey range are the Horned Owl and the Cooper's Hawk, in the order stated. The adult Red-tailed Hawk, because of its size and strength, can be and commonly is a general feeder. Under certain conditions, however, it reacts to prey populations in the manner of a restricted feeder. This grouping, though generally applicable, is based on the ability of the various species to take food on the area studied. We have seen that the vulnerability of any kind of prey varies for each raptor. Likewise, it is evident that as the most numerous prey species (in this case meadow mouse) declines in numbers, it will become less vulnerable to those raptors most poorly adapted to catch it and will continue to be relatively more vulnerable to those species best fitted to catch it. As either the density or the vulnerability of the most numerous prey species declines, other species will be utilized to a greater extent by the raptor capable of catching them. Thus their vulnerability increases with a decline in vulnerability of the major prey species. Because of their ability to take varied prey, those raptors that are general feeders will be better fitted to remain as permanent residents than those restricted feeders dependent upon the vulnerability of one or two prey species. Thus, restricted feeders have to move and do move when the vulnerability of the staple prey species drops to a point where that part of the raptor population cannot readily feed itself.

This is well illustrated by the raptor and prey population figures of the two study years (Table 34). In 1942, when the meadow mice were about four times as abundant as in 1948, there were 96 wintering hawks as compared with 27 in 1948, and 44 Buteos as compared with 15. There were 159 raptors in 1942, compared with 59 in 1948. The higher 1942 population largely was the consequence of a congregating of restricted feeders that moved in in response to meadow mouse vulnerability. For example, there were approximately 19 American Rough-legged Hawks in the area in 1942, none in 1948. There were 37 Marsh Hawks in 1942, one in 1948. There were 31 Short-eared Owls and 7 Long-eared Owls as compared with none of either species in 1948. Thus 94 raptors moved into the area and remained in response to meadow mouse vulnerability in 1942, as compared to only one Marsh Hawk in 1948, when the population of these mice was lower. Of the residents, or species that consistently winter and nest, there were 11 Horned Owls in 1942 as compared with 13 in 1948; 14 Screech Owls as compared with 19; 17 adult Red-tailed Hawks as compared with 12; 10 Cooper's Hawks compared with 6; and 5 Sparrow Hawks both years—or more or less similar totals of 57 and 55 residents. Evidence indicates that the Screech Owl population actually was about the same in both years, but even if we assume this, the totals will still be reasonably comparable.

If the 94 restricted feeders that moved in in response to meadow mouse vulnerability are added to the 56 raptors for the winter of 1948 that did not respond, but are generally year-round residents, the total of 150 closely approaches that of 159 in 1942. This indicates that the higher raptor population in the winter of 1942 was due largely to those species that readily move in response to meadow mouse vulnerability. These species—the American Rough-legged Hawk, Marsh Hawk, Long-eared Owl and Short-eared Owl—are not adapted for catching either large or agile prey. Thus the total vulnerability of prey on any area for this group of birds in winter is dependent on the vulnerability or density of only one or a few species, in this case the meadow mouse, the capture of which is within their powers, rather than on the presence of a wide range of prey species.

The raptors which were present during both winters and whose numbers remained most constant were those capable of a wider range of prey selection, and thus the vulnerability of prey for them was not linked chiefly to the vulnerability of meadow mice.

It was evident that the population density of restricted feeders dropped when meadow mouse vulnerability was low. The population density of the general feeders and year-round residents remained constant. They did not move when meadow mouse vulnerability was lower, but their diets changed, indicating the changing relative vulnerability of a large range of prey. The change was not reflected in a relatively great decrease in the proportion of the major prey species eaten, but the corresponding increase in the proportion of minor prey species was relatively great. For example, meadow mice were 89 per cent of the food of Red-tailed Hawks in the winter of 1941-42 and 84 per cent in the winter of 1947-48. Meadow mice were the greater part of the winter food in each of these winters, but the percentage of rabbits, pheasants and small birds increased from 0.6 in the first winter to 7.8 in the second. This is a relatively large change in these minor items. A similar change is recorded for the Horned Owls and Screech Owls (Tables 25 and 26).

It is highly probable that had the densities of all prey species been low, the year-round residents (general feeders) would have responded to the lower prey vulnerability just as the restricted feeders did—by movement to another region. This appears to be the condition that initiates large-scale winter movements or invasions of general feeders like the Goshawk, Horned Owl, and Snowy Owl. No figures indicating raptor and prey densities in areas of such winter invasions were gathered. Some information was compiled, however, on areas vacated in winter by general feeders. In Jackson Hole, Wyoming, where long winters are characterized by deep snow and severe cold, as in more northern latitudes, the general feeders, such as the Red-tailed Hawk, the Goshawk, and the Horned Owl, emigrate. In the winter of 1947, no Red-tailed Hawks wintered, and a large portion of the Horned Owl and Goshawk population left. Winter food was not abundant. Ruffed Grouse populations were relatively high in localized areas, but snowshoe hares were exceedingly scarce; ground squirrels, pocket gophers, chipmunks, small birds, and even mice either were not available or very difficult to catch.

If these general feeders make adjustments during their migratory movements as the restricted feeders did in 1942 and 1948 in Superior Township, they will concentrate (in the areas they migrate to and through) in proportion to the vulnerable prey. Under such conditions, the density of the winter population of raptor migrants should not only reflect low prey densities in the areas vacated, but should give an indication of prey density in the areas invaded. Thus, a high invading population of Snowy Owls, Horned Owls, or Goshawks could be used as a clue to the density of prey—in some cases, game species, such as rabbits, grouse, pheasants, squirrels, or even Bob-whites. The movement of raptors in response to prey fluctuations is one key to their ability to balance prey populations.

PREDATION BALANCES PREY POPULATIONS

There are misconceptions over both the use and the meaning of the terms, biotic balance, biotic equilibrium and "balance of nature." The existence of such a balance is seriously questioned by many biologists. The use of the term balance is helpful and appropriate in describing a natural phenomenon, but it must be defined carefully. Data indicating the existence and operation of balancing forces have been presented. Biotic balance or biotic equilibrium means a dynamic state in which interrelated animal population densities fluctuate about a mean in such a manner that extremes either do not occur or if they do are quickly limited by natural regulating forces of the environment. In other words, we refer to a relation between collective animal populations in which they adjust to an approximate harmonious inter-density relation, though not necessarily to a constant density level. (See also pages 245 and 309.)

Because a predator population takes its prey roughly in proportion to prey densities, the threshold of security from predation (that point at which security for a prey species or population is attained by virtue of low density and equilibrium of prey with habitat, so that vulnerability is zero or nearly so) theoretically would be attained by all prey species at about the same time (Fig. 12). The tendency of predation then, is to balance populations while reducing them to their threshold of security, or to reduce those populations most capable of enduring, and in many cases most needing reduction. The fact that one prey population is more vulnerable than another is evidence that its numbers are higher than can be carried satisfactorily by the environment. If the population over and above the threshold of security is considered surplus (Errington, 1936), then it is evident that predators in almost all cases prey on surpluses.

The threshold of security for any species and the carrying capacity of the area for the same species, which depends on those properties of the area that determine the average number of animals it can support over a long period of time and that limit population densities in critical periods, may, however, occur at the same population level (Fig. 12). The relatively small numbers of rabbits and fox squirrels killed during both years indicated that their populations were at the threshold of security existing under the particular environmental conditions. It is doubtful, however, if the populations of these mammals approached the carrying capacity, as both populations increased markedly in 1948 and yet still, as indicated by the raptor diet, were not much above the threshold of security. The population of raptors capable of taking such prey had, on the other hand, not fluctuated. In the case of the pheasants, the similar populations of 1942 and 1948, as well as a corresponding one in 1947, would indicate that this species may have been at a winter population level approaching the carrying capacity of the area and yet was not much above the threshold of security. For this reason it might be well to limit the use of the term "surplus population" to those above the carrying capacity of an area and assign the term "vulnerable population" to those above the threshold of security.

The effect of predation on vulnerable populations can have no important influence on the survival of prey species. It therefore appears that predation on vulnerable or surplus populations affects man largely in proportion to the economic value he attaches to these surpluses. In the case of mice, predation is usually considered of value, while in the case of predation on pheasants, it is usually considered injurious. The actual good or bad that results should not be interpreted solely in terms of man's self-interest; predation must be viewed as a biological function tending to keep prey populations balanced within the limits of their vulnerability.

It is obvious that any land area is limited in the organisms it can support. Prey populations, however, do increase when local environments (continually changing) temporarily change to the benefit of one or more species. The force of reproduction then outweighs the resisting forces in the environment. Adverse conditions in the environment, however, will eventually balance the increasing productivity. When one such adverse condition is more influential than others in effectively limiting a population it becomes a "limiting factor" (Leopold, 1933). Predation is one adverse condition in the environment and, like many of these conditions, it can for short times be a limiting factor, determining maximum population levels. Errington (1946), in summing up his work on predation and vertebrate populations, concluded that, "Predation looks ineffective as a limiting factor to the extent that intra-specific self-limiting mechanisms basically determine population levels maintained by the prey." The data so far presented in this book indicate that where raptor populations have not been persecuted and drastically reduced in composition and number by man, the effect of predation can be controlling. Certainly the published accounts of increases of prey populations following predator destruction tend to support this thought. In the winter of 1942, predation by raptors was a limiting factor in checking the meadow mouse populations. The effect of predation in controlling a population in this instance hardly can be doubted when we consider that mammalian predation, probably equal in effect and observed to be directed primarily on meadow mice, was simultaneously operative. Had bird and mammal predation not been the limiting factor, some other force naturally would have been or have become limiting, but probably would have been less generally and proportionately regulatory. The point is that predation was controlling. Raptor predation operated to decrease the prey populations most in need of reduction and exercised a continuous and proportional pressure on all populations, helping to limit fluctuations.

Because prey populations with the highest ability to increase rapidly in numbers are those usually exhibiting marked fluctuations (Dymond, 1947), predator pressure is not able to keep pace with these population increases through increased predator productivity alone (Chap. 10). Thus we see that the great mobility of raptors, which enables them to concentrate from distant regions and thereby rapidly increase hunting pressure on dense prey populations, is significant in controlling, during winter, prey populations that have increased during spring and summer.

Dymond (1947), in dicussing periodic fluctuations in animal populations of Canada, stated, "Briefly, the theory here offered to account for fluctuations of the populations of lemmings, hares, voles, grouse, etc., is that the biotic potential of these animals is definitely greater than the normal resistance of the environment in which they live. In other words, they are endowed with too high a reproductive capacity, and it is the monotony or relative uniformity of the Arctic and far northern environment that accounts for the regularity with which they increase to the unstable level of numbers which brings about the periodic decline." He also stated that, "So far as the predator part of the lemming's environment is concerned, it must exert a fairly uniform pressure, the

Arctic foxes and Snowy Owls increasing in fairly direct ratio to the lemmings . . ." Dymond then evaluated the full effects of predation on periodic prey fluctuations by accepting Errington's conclusions (1946), and thus did not attempt to show possible relations existing between the forces of reproduction and predation (a part of environmental resistance) that might be a contributing cause of periodic prey fluctuations. We shall later see that a dearth of winter predator pressure in the far north may have an important effect on prey fluctuations.

SUMMARY OF WINTER RAPTOR PREDATION

The abundance of individuals of prey species on a large land area is controlled by the carrying capacity of the land, but fluctuates from year to year with biological, physical, and land use changes. The density and composition of a raptor population is largely dependent upon the density of the prey species. Conversely, the densities of prey species are very definitely influenced by raptor predation. One part of the winter raptor population is resident and, because of equilibrium with the environment, tends to remain constant numerically. Because the resident species are capable of taking a wide variety of prey forms, their diet may change with variations in the density of the various prey species. The other part of the population is dependent upon the densities of a few prey species, largely meadow mice in the case studied, and therefore these raptors must move to new locations when their staple prey become scarce. Raptors that are year-round residents in one region may be migrants in another, depending on the species of prey available and their densities. Thus migration enables raptors to adjust their numbers to those of the prey, thereby exerting the greatest pressure on the densest prey populations. During winter an entire raptor population adjusts to the prey densities and each raptor hunts only a restricted area, or range. Ranges that embrace habitats that each particular hawk or owl hunts in accordance with its adaptations are established. Thus a stable winter raptor population whose composition is such that hunting pressure is exerted on all major prey forms, evolves. The local movement of any one raptor influences the movement of others, the spatial adjustments of one to others tending to distribute hunting pressure in proportion to the immediate prey vulnerability. This compensatory movement alleviates direct competition for food among the raptors, thus permitting maximum hunting pressure to be exerted on an area of land.

A collective population of raptors comprising numerous species, each adapted to hunt varied habitats at different times, exerts hunting pressure upon the available prey of any large area in such a way that each prey species tends to be preyed upon in proportion to its relative density.

Since any species of raptor is adapted to hunt effectively over only a portion of the varied habitats found in any large land area, and many can take only a limited range of prey species, a population of any single raptor cannot exert hunting pressure upon all members of the prey population but only upon certain species within its hunting habitat. Thus, the diet of one raptor species may not reflect the relative prey densities found over a large area, but will reflect the prey densities of those sepcies that are available within its habitat and vulnerable to it.

Although predation is roughly proportional to prey densities, a complex of risks continually alters this relation, thus modifying the vulnerability of prey. *Predation, then, is a relation between predator and prey and between predator populations and prey populations, operating so that the prey tend to be taken by the predators generally in proportion to their relative densities, this density relation being modified by variable risks that help to determine prey vulnerability.*

PART II
Ecology of Raptor Predation in Spring and Summer

CHAPTER 8

The Study Areas and Research
Technics

WINTER has gone and spring is here. We have pieced together the story of raptor predation during the fall and winter, but this is only half the story—a story of the steady reduction of animal life. Now the almost explosive force of reproduction appears. Seemingly as if to compensate for its seasonal failure to keep pace with the decimating force of predation, it bursts into action, with all its power and vigor, to create more mice, more pheasants, more rabbits and, as though in afterthought, also more hawks and owls. What will be the outcome of this frenzied multiplying of animal life? Will the force of predation temporarily be overpowered? Will it continue to operate basically as it did during the cold months of the year? How will this new activity fit with what we have already considered? Will the two combine to form a satisfactory whole or shall we get poorly co-ordinated or even entirely distinct concepts?

These are natural and proper questions, but they must be put aside while we concentrate on gathering the data needed. The census of hawks and owls and the major prey species must be continued, and their

191

reproductive increase must be measured, for the problem is still basically a relation of numbers. With the change in seasons, study technics change. Instead of counting hawks from a car, nests are located and eggs and young counted. To gather the food data will require climbing hundreds of trees, for no longer can pellets be secured under favorite roosts or perches. To get a sufficiently broad understanding of what is taking place, operations will be expanded to other townships in Michigan and to an area in Wyoming. The objective will be to gather basic facts and information that will make it possible to correlate raptor nesting populations and their productivity with prey densities and their reproductive increments, and in turn to correlate the population characteristics of the raptors and their prey with the diet of the collective raptor population. This approach might throw light on such questions as: How is raptor hunting pressure applied during spring and summer; is it altered significantly from what it is in winter and if so, why; is this pressure a condition limiting various prey populations and, if so, when; what part does raptor reproduction play in maintaining hunting pressure on prey populations; what are the natural biotic controls that tend to check expansion of raptor populations?

In the nesting season, investigations were conducted largely in Superior Township, Michigan, and in a 12-square-mile area near Moose, Wyoming. For comparative purposes the densities of nesting raptors were also determined in a 36-square-mile check area west of Ann Arbor, Michigan. The investigations were carried on during three seasons in Superior Township and during two seasons in the western area. Throughout this book the study areas will be referred to as the Western Area, Superior Township, and the Check Area. A description of Superior Township has been given in Chapter 1 with details in Appendix A.

GENERAL CHARACTERISTICS OF THE WESTERN AREA

The Western Area was selected for study to ascertain whether some of the findings recorded in Michigan were applicable to predatory birds in an entirely different region. Data from the two areas allowed the comparison of predation phenomena in a semi-wilderness only slightly affected by agriculture and lumbering with those in a heavily populated, intensively farmed locality.

The Western Area, situated near Moose, 13 miles north of Jackson, in Teton County, Wyoming, lies in the intermontane valley of Jackson Hole. The narrow valley is 48 miles long and has a maximum width of 12 miles; its area is about 400 square miles. The valley floor is from 6,000 to 7,000 feet above sea level and the surrounding mountains attain elevations of 9,000 to almost 14,000 feet (Plate 3).

It is drained chiefly by the Snake River, which flows southward into Jackson Hole and empties into Jackson Lake. After emerging from Jackson Lake, the river traverses the length of the valley and escapes

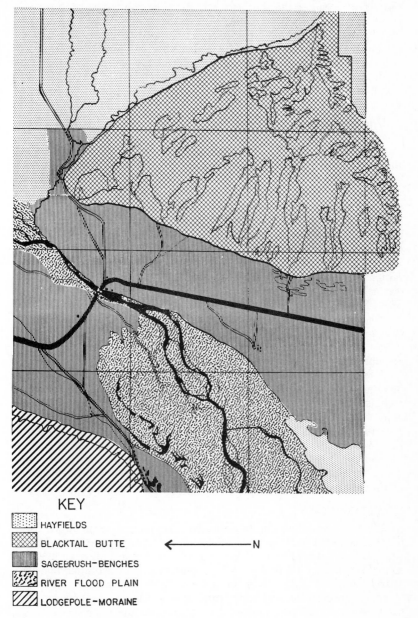

KEY

▦ HAYFIELDS

▧ BLACKTAIL BUTTE   ←————————— N

▥ SAGEBRUSH—BENCHES

▨ RIVER FLOOD PLAIN

▨ LODGEPOLE—MORAINE

MAP 12. Ecological land divisions, Moose, Wyoming

through a gorge at the southern extremity. Numerous tributaries enter the basin from the surrounding mountains.

The drainage and relief features of Jackson Hole basin are explained largely by the glacial and post-glacial history of the region. For a detailed and interesting account of the glacial history of the valley, the reader is referred to Fryxell's work (1930).

PHYSIOGRAPHIC FEATURES

The physiographic features affect the type and distribution of the vegetation, and all have a direct bearing on certain aspects of the predation study. The chief physiographic features of the study area are Blacktail Butte, the Snake River terraces, the flat river flood plain and a glacial moraine (Map 12 and Plates 3 and 4). The butte, the moraine, and the river bottom are timbered areas and furnish the principal raptor nesting sites. The river benches paralleling the flood plain support sagebrush and are utilized as hunting areas by raptorial birds. Rather extensive hay fields, including alfalfa (originally sagebrush) lie east of Blacktail Butte and are likewise a very productive hunting area. These divisions, as indicated on Map 12, are closely connected with the raptor ecology and for purposes of simplicity will be referred to in the text as the lodgepole moraine, the sagebrush benches, the river flood plain, Blacktail Butte, and the hayfields

Blacktail Butte, with an elevation of 7,676 feet, comprises more than one-fourth of the area, rising island-like from the valley floor to a height of 1,300 feet above the Snake River. It is a remnant of the ancient pre-glacial topography.

The northwest corner of the area (Map 12) embraces an extensive moraine, some 150 to 300 feet above the Snake River. Its steep eastern face is the result of river erosion, the moraine itself being a river bench. The sagebrush terraces between this lodgepole moraine and the Snake River are an expanse of gravel flats that rise 20 to 150 feet above the river.

The bench between Blacktail Butte and the river is an outwash plain.

The present river flood plain is composed of recent alluvial deposits in the form of low terraces and bars of sand and gravel.

METEOROLOGY

Weather conditions on the Western Area vary considerably from those of Superior Township. The mean annual temperature is 36.54° F. and the mean annual precipitation is 26.21 inches, including an average annual snowfall of 180.96 inches. Evaporation is high and humidity low. Yearly maximum and minimum temperatures vary greatly, with the maximum usually occurring in July, the minimum in January. The maximum for a 13-year period was 94° F. and the minimum -41° F. The mean monthly temperatures of the spring and early summer months are almost identical with those occurring in Superior Township a month earlier. This difference influences significantly the nesting activities and predation phenomena in the two regions. The heavy snowfall, a maximum of 265 inches in 1936 and a minimum for the same 13-year period of 114 in 1947, is likewise important in determining the effects of avian predation through the year in this western area.

HISTORY AND GENERAL LAND USE

The use to which man puts any area of land influences its vegetation and the composition and numbers of its animal populations. These in turn affect predation. It is interesting to note that Jackson Hole was discovered by John Coulter in 1807, at almost the same time that Superior Township was being settled by French traders. Prior to its discovery, it was utilized by Indians in summer as a travel route and a hunting and fishing ground. Following its discovery, the valley was explored and some trapping took place. Intensive fur trapping began about 1824 and continued until its decline around 1842. Trapping and hunting operations undoubtedly affected the animal populations, but no drastic land-use changes occurred during this time. The first settlers arrived in 1884 and, by 1907, ranches were scattered throughout the valley (Watson, 1935). Cultivation and grazing altered the wilderness character of much of the valley floor. The forest areas came under the supervision of the U. S. Forest Service in 1897, when Teton Forest Reservation was created. The Teton National Park was established in 1929. Both private and government land is utilized at present largely for cattle ranching, dude ranching, and outdoor recreation.

Changes in the study area itself since its original settlement have been minor. Some logging was conducted on Blacktail Butte and the entire area has been grazed by both sheep and cattle. A limited number of the latter still graze the land today. Though no longer a wilderness, the study area can be considered a semi-wilderness, in marked contrast to the highly populated and cultivated Superior Township.

GENERAL INVESTIGATIVE PROCEDURE AND METHODS

The investigations fundamentally were studies of nesting raptor populations in definite land areas, but they can be divided into the following phases:

1. The determination of raptor populations and their productivity.
2. Studies of raptor movements, activities, nesting ranges, and interrelationships.
3. Studies of prey populations.
4. Raptor food habits studies.

All of the raptor nesting data were gathered by systematized field observations that involved more than 2,000 miles of walking and approximately 1,200 ascents of trees, including repeated ascents of some individual trees and visits to Crow nests in Superior Township. One hundred forty raptor nests were studied intensively in the two localities during three breeding seasons, and 291 additional raptor nests were found and observed in these same areas and the surrounding vicinities during four breeding seasons. A previous study of several hundred raptor nests throughout the United States, which had been carried on during an eight-year period, served as a working background.

Direct observations were used to obtain as complete data about the raptor populations as possible. From the standpoint of the biologist who must gather all essential data about a mixed bird population, the nesting season appears as an inverted pyramid that starts with the first early nesters and expands as more and more species begin to breed. As the number of nesting birds under observation increases with the advancing season, the activity of the biologist increases proportionately. It is necessary in this type of study to set no limitations on the time and energy to be expended in gathering information. The task is a constant battle against time, since collecting the data that can and should be gathered far exceeds the physical limitations of one or several individuals. The results obtained are in direct proportion to the amount of time and energy expended in systematic field work. Any study in synecology is limited by the time available, and work must be planned with this in mind. Time-consuming technics were used only where necessary. Banding was confined to nestlings. Nesting ranges and movements were determined without the use of bands or marking devices, since these technics were neither feasible nor necessary.

What contributed most to the accuracy and completeness of this study was the care taken to locate all the raptor nests and secure as complete knowledge of the nesting population as possible. It should be emphasized that there is a great difference between locating *most* of the raptor nests and obtaining a *complete* count. Not only does the latter involve much more careful and prolonged work, but the deductions and conclusions that can be drawn from full nesting information exceed anything that can be drawn from partial population data. When every nest and bird is located in respect to every other one on the area, observations on ranges, movement, inter- and intra-specific strife, food habits, raptor nesting mortality, and numerous interrelations can be interpreted on a foundation of fact, with a minimum of supposition. All the major items of information necessary for reaching valid conclusions are present.

NEST FINDING TECHNICS

Since the analysis of predation during the spring and summer periods will emphasize relations between members of a collective raptor population and not a single species, the methods and procedures employed in locating nests should be discussed.

Methodical systematic searches for nests and paired birds were begun in late winter. A census was first taken on each area, with 100 per cent coverage for the earliest nesters, the Horned Owls, and a similar census was then made for the Buteos. The second search revealed either nests or sites of still later nesting species and was a check on any Horned Owls that might have been overlooked. Systematic searching for nests, including specific searches for individual nests, continued through the breeding season until nest locations correlated completely with sight observations recorded on the field maps.

Approximately 85 per cent of all raptor nests were found with relative ease, but finding the remaining 15 per cent, to make the count complete, involved numerous searches and the recording and mapping of all field observations. It was essential to find nests at the time of building or early laying in order to obtain data on territorial selection, first layings, clutch sizes, first hatchings, etc. (Tables 37 and 38). Nests at times are deserted or destroyed early in the nesting cycle and unless they are found prior to such mishaps it is difficult to determine if a nest actually existed. Averages of random nest productivity usually will not include these early mortalities and thus the productivity figures tend to be high and the population figure low. Errors of this type will bias population and mortality data. Taking nesting data on an areal basis reduced such error to a minimum in this study. If only the determination of a nesting population is desired, nest hunting can begin later and the majority of nests can be found prior to general leafing out. On any area nests can be found a second year with much greater ease and in less time. Discovery of each nest frequently was a problem in itself, but there were certain clues that proved extremely helpful.

Old Nests

Large, well-constructed stick nests observed in winter were plotted on maps and checked in the spring. Such nests were generally re-used, by either the same or a different raptor species, or a new nest was constructed near by. We have never observed the Horned Owl to build a nest. It either uses the old nest of some other bird or utilizes a cliff ledge, a hollow stub or other suitable site. In Michigan, Red-tailed Hawk nests seem to be preferred, probably because of their large size (Plate 52). Such a nest within the winter range of a Horned Owl was very likely to be utilized in spring.

Nest Site Activities as Indicators of Nest or Nesting Site

The most productive time to search for hawk and owl nests was found to be prior to general leafing out, during the early period of nest site selection and territorial defense. In general hawks exhibit especially conspicuous territorial defense behavior prior to egg laying and react to man's encroachment by circling, screaming and diving. Defense displays by either hawks or owls are excellent indicators that a nest exists or will be constructed near by. At this time estimates of raptor populations can be obtained with a high degree of accuracy by counting hooting owls (Baumgartner, 1939) or displaying hawks (Craighead, F. C., and J. J. Craighead, 1947).

During incubation, hawks become more secretive, making this the most difficult time to discover nests. After the eggs hatch the adults become less wary and their increased activity makes it easier to observe them and find their nests. This is not always true when hatching occurs after general leafing out.

Nest Spacing

Making a nesting census of raptors is simplified by the fact that almost without exception no two pairs of the same species will nest close to each other. Thus if a pair of Red-shouldered Hawks is found in a small to medium-sized woodlot, the possibility that another pair will be in the same woodlot can be ruled out. These spatial limitations will be discussed further in another chapter. The fact that the distance between nests of the same species is relatively great enables individual pairs to be distinguished and observed with a minimum of confusion.

Binoculars

Binoculars, always an aid in observing hawks and finding their nests, are especially useful in open or hilly country. Careful observation of birds such as Marsh Hawks and Sparrow Hawks often proved to be the best or only way of determining the exact location of their nests.

Field Signs

Field signs, such as pellets, kills, and "butcher blocks," are valuable clues. Food scrap or "whitewash" beneath nests is a sign of occupancy, (Plate 52), but the lack of such a sign is typical of raptor nests prior to hatching and thus is not a reliable indicator of non-occupancy during the early nesting period. Numerous winter pellets of the Screech Owl or the Horned Owl frequently are indicators of birds nesting nearby. A butcher block of bird hawks is usually quite close to a nest.

Nest and Sight Plottings

Systematic searches utilizing field clues revealed a large percentage of the raptor populations. To obtain a complete count of nesting pairs, to locate renesters, and to distinguish non-nesting pairs and single birds it was essential to record all nests and all raptor observations on an accurate cover map. A few examples of the use of the cover map will illustrate its value and the technic employed.

For example, paired Red-shouldered Hawks were observed several times over a swamp woodlot on the Michigan area, but a careful search failed to disclose a nest. In the course of time, observations accumulated and mapped made it evident that these birds were residents of the woods and not a visiting pair. It seemed possible that they were non-nesters, but before they were recorded as such, every alternative had to be eliminated. In a final combing of the woodlot before leaves appeared the nest was found. The female had used the broad crotch of a large oak and a few sticks for a nest not visible from the ground. Hours spent in watching the birds solved the problem, but the systematic recording of observations on the map and the relation of these observations to those of neighboring birds provided the clues that led to the intensive search.

A pair of Sharp-shinned Hawks on the Western Study Area caused similar difficulties. Observations recorded on the map indicated the general area of a nesting pair, but intensive search failed to find a nest. The Sharp-shinned Hawk generally reveals the position of its nest by very active territorial defense. No such behavior had been detected and yet the observations recorded on the map indicated that a pair frequented a certain area. Tall Engelmann spruce, the heavy cover and the extensive area in which the pair might nest hindered direct observation and therefore a search was made for butcher blocks. Several were found, thus narrowing the search to their immediate vicinity. The nest finally was discovered by carefully examining the upper trunks of the spruce trees with field glasses until the protruding tail of the female was seen.

After four years of experience in applying these methods to locating raptor nests, it can be stated confidently that they were almost completely effective with the exception of the Screech Owl and possibly the Barn Owl.

CHAPTER 9

Activity and Movement

PERHAPS one of the most fascinating and complex fields of natural history is the study of animal behavior. The scientist must ask himself continually what, how and why. Why does a Horned Owl nest two months earlier than a Cooper's Hawk? What effect does the presence of Horned Owls have on the productivity of Crow and hawk populations? How do a pair of hawks establish and maintain a nesting territory? Fortunately, we know a great deal about the behavior of the various species of hawks and owls, as they have long been a subject of study. There is still much to be learned, but enough information is available for use in a study of the more complex behavior of a mixed population—specifically the major activities and movements of hawks and owls with relation to one another and to reproduction and getting food.

Practically nothing is known about the behavior of a mixed population of raptors. Data, observations and conclusions are presented here in an attempt to portray the behavior of such a population and its significance in relation to the phenomenon of predation.

Spring migration greatly alters the raptor population by changing both its numbers and its composition. It also alters the numerical relation of small bird populations to prey groups. These changes affect predation pressure on a collective prey population.

CHRONOLOGICAL EVENTS ALTERING RAPTOR POPULATIONS AND RAPTOR ACTIVITY IN SUPERIOR TOWNSHIP, MICHIGAN

In Superior Township changes in the stable wintering population of raptorial birds occurred during the last week in February. This marked the beginning of a transition period during which, though many winter residents remained to hunt within their winter ranges, others left and spring arrivals appeared. American Rough-legged Hawks were among the earliest arrivals and hunted definite areas until their departure in late April. They remained until April 30, 25, and 30 in 1942, 1948, and 1949, respectively (Map 18). Marsh Hawks also were conspicuous in the early influx. Some remained to nest and others, like the Rough-legged Hawks, hunted a while before moving on. The return migration of the Red-shouldered Hawk marked the beginning of the replacement of the winter population by a breeding one. This Red-shouldered Hawk influx started on February 28 in 1942 and on March 8, 10, and 5 in 1947, 1948, and 1949, respectively. Some Red-tailed Hawks also moved into the area in late February and early March, though others were resident there. A number of Sparrow Hawks and Cooper's Hawks both wintered and nested, while others arrived in early March and began to defend nesting sites.

By the middle of March, the raptor population was exhibiting the activities associated with the nesting season. Winter residents that would not nest remained, but they no longer dominated either the population or its activity. The township was nearly devoid of Crows during the winter, but these birds returned in numbers during February, making their first appearances on February 14 and 17 in 1942 and 1948, respectively.

Prior to the period of transition from a wintering to a nesting population of hawks, the Horned Owl initiated the nesting season by laying its eggs. The earliest dates of laying of this bird were February 21 and 12 in 1942 and 1948, respectively, and February 4 in 1949. The Red-tailed and Red-shouldered Hawks were the first hawks to lay but they did not begin until the last of March (Table 37).

The breeding season of the raptor population was a period of great activity, characterized by courtship, territorial selection and defense, nest building, egg laying, incubating, brooding, hunting, and feeding of young. Following the fledging of the offspring, the center of activities shifted from the nest to hunting areas, where family groups frequently were observed. Young birds discovered and perfected their flight abilities. Adults fed their young but, by gradually diminished feeding, forced them to hunt for themselves. The young learned where and how to catch prey.

The young of all raptors studied were observed to confine their activities generally within the hunting range of the parents until toward

TABLE 37

OBSERVED DATES OF REPRODUCTIVE ACTIVITY OF RAPTORS ON MICHIGAN STUDY AREA
Superior Township, Washtenaw County, Mich.

Species	1st Selection of Nesting Territory		Earliest Laying Date		Latest Laying Date		Average Laying Date		Earliest Hatching Date		Latest Hatching Date	
	1942	1948	1942	1948	1942	1948	1942	1948	1942	1948	1942	1948
Red-shouldered Hawk	2/27	3/10	3/26	3/31	4/17	4/8	4/5	4/5	4/26	5/8	6/10	6/3
Red-tailed Hawk	2/28	2/18	3/28	3/26	....	4/2	....	3/28	5/2	4/25	....	4/26
Cooper's Hawk ...	3/24	3/31	4/26	4/28	5/3	5/2	4/29	4/30	5/29	6/5	6/1	6/6
Marsh Hawk	3/26	3/28	5/5	4/26	5/20	5/9	5/13	5/1	6/5	5/31	6/26	5/31
Sparrow Hawk ...	3/23	4/5	4/12	5/1	....	....	....	....	*5/11	*5/18	....	....
Great Horned Owl	1/25	1/21	2/21	2/12	3/19	2/25	3/6	2/19	3/23	3/15	4/17	4/3
Long-eared Owl	....	....	*3/28	....	....	....	....	....	*4/21	....	....	....
Screech Owl	2/31	2/27	....	4/18	....	....	....	....	5/2	....	....	....
Barn Owl	....	....	4/20	....	....	....	....	....	....	....	....	....

Breeding Seasons for the collective raptor population $\left\{\begin{array}{l}\text{1/25/42–8/1} = 190 \text{ days} \\ \text{1/21/48–7/10} = 172 \text{ days}\end{array}\right.$
* Computed dates

TABLE 37, Continued

	Average Hatching Date		Earliest Brood Departure Date		Latest Brood Departure Date		Average Brood Departure Date		Span of Nesting Period (Days)		Av. No. Days in Nest	Breeding Season Span (Days)	
	1942	1948	1942	1948	1942	1948	1942	1948	1942	1948		1942	1948
Red-shouldered Hawk	5/10	5/14	6/6	6/10	7/16	7/10	6/23	6/17	82	64	39	141	123
Red-tailed Hawk	....	4/26	6/12	6/5	6/12	6/7	....	6/6	42	44	41	106	111
Cooper's Hawk .	5/30	6/6	6/27	6/26	8/1	7/7	7/8	7/3	65	33	27	131	99
Marsh Hawk ...	6/16	5/31	7/21	....	7/27	....	7/24	....	53	..	31	124	..
Sparrow Hawk .	....	....	6/10	6/17	....	....	....	....	31	..	31	80	74
Great Horned Owl	4/5	3/27	4/17	4/24	5/24	4/29	4/31	4/27	63	46	44	120	100
Long-eared Owl	....	....	5/19	....	....	....	....	....	29	..	29	..	..
Screech Owl ...	....	....	5/13	....	....	6/19	....	6/1	48	..	30	..	..
Barn Owl	....	....	....	....	....	....	....	....	..	..	.	..	..

the close of summer. Young Red-shouldered Hawks were observed within the parental range for two months and Horned Owls for three months. In no cases did fledglings move suddenly out of the nesting vicinity, nor were there any records of distant occurrences before late summer.

During the summer, following fledging and the early phase of learning by the young raptors, there was a period of relative inactivity on the part of all bird predators. Both adults and young were seen less frequently than during the nesting period but were nevertheless observed to be present and within their nesting ranges (Tables 50 and 51). This apparent inactivity was due in part to the difficulty of making observations under full foliage conditions and at a time when the raptors were no longer confined to the nest site. The decrease in activity, however, was real. Hunting by the hawks took place in the cooler hours of early morning and late evening. Food requirements were easily and rather quickly supplied. The increase in vulnerability of prey species, as a result of reproduction, shortened the periods of hunting. Moreover, higher temperatures decreased the amount of food necessary to maintain body weight. (Feeding experiments that will be discussed later show that hawks and owls require less food in summer than they do at colder seasons.) The reduced food requirements and the ease of satisfying them in summer permitted the hawks to spend many hours of each day perched quietly with full crops in cool, secluded retreats. Not until fall brought lower temperatures did activity again increase.

By the first week in September, Red-tailed, Cooper's and Sharp-shinned Hawks were migrating into and out of the township, thus altering the summer populations. Crows were also gathering in fall flocks.

CHRONOLOGICAL EVENTS ALTERING RAPTOR POPULATIONS AND RAPTOR ACTIVITY IN THE WESTERN STUDY AREA

Mean monthly temperatures in the Western Study Area in late winter and early spring were comparable to those occurring a month earlier in Superior Township. This difference in temperature affected migration dates, nesting dates, and the duration of the breeding season (Table 38, Fig. 7). Very few raptors wintered in the western Study Area, though some species, notably non-resident American Rough-legged Hawks, wintered in the lower end of the Jackson Hole valley. Spring migrants entered the lower valley as the snow on southern exposures melted and gradually worked northward as the snow receded. Red-tailed Hawks appeared in the valley on March 13 and by March 20 the first arrival was observed on the Study Area. The first Marsh Hawk was observed on April 3, the first Goshawk on April 10, and the first returning Prairie Falcon and the first Sparrow Hawk on April 19. The collective breeding season population was well established by the middle of April. Table

TABLE 38

OBSERVED DATES OF REPRODUCTIVE ACTIVITY OF RAPTORS ON WESTERN STUDY AREA
Moose, Wyo., 1947

Species	1st Selection of Nesting Territory	Earliest Laying Date	Earliest Hatching Date	Latest Hatching Date	Earliest Brood Departure Date	Latest Brood Departure Date	Average Brood Departure Date	Span of Nestling Period (Days)	Av. No. Days in Nest	Breeding Season Span (Days)
Red-tailed Hawk	3/28	4/20	5/14	5/27	6/20	7/10	6/29	58	42	105
Swainson's Hawk	4/19	5/15	6/16	7/3*	7/27	7/30	7/29	47	42	103
Cooper's Hawk	...	6/1*	7/4	...	7/28	...	...	25	24	89
Sharp-shinned Hawk	5/5	6/16*	7/12	....	8/5	8/8	8/6	28	21	96
Goshawk	4/1	5/6*	6/3*	....	7/10	7/25	7/18	28	28	116
Sparrow Hawk	4/24	5/20	6/20	6/22	7/15	7/20	7/18	31	29	88
Prairie Falcon	4/24	5/1*	6/5	...	7/16	...	...	42	42	84
Horned Owl	2/27	3/12*	4/14	...	5/29	6/4	...	52	46	99
Long-eared Owl	...	5/10*	6/3*	...	7/1	7/3	7/2	30	29	...
Screech Owl	3/25	5/4*	6/1*	...	7/1	...	...	31	31	121
Gt. Gray Owl	...	4/1*	4/30*	...	6/8	...	...	40	40	...
Saw-whet Owl	4/28	...	...	...	7/17	...	...	...	...	...
Raven	3/5	...	5/3	5/8	6/5	6/7	6/6	36	36	94

Breeding Season for the collective raptor population 2/27–8/8=163 days.
Breeding Season for each species is considered as terminating with departure of fledglings from nests.
*=Computed dates.

38 shows the dates of various activities observed in the course of the breeding season.

Concentrations of hawks in fields of high meadow mouse densities heralded the fall migration, which was first noticeable during the last of August and continued intermittently through October. Migrants, such as Duck Hawks, Pigeon Hawks, Sharp-shinned Hawks and Cooper's Hawks, were seen in the first few weeks of September.

DISCUSSION OF ACTIVITY TABLES

The dates at which different raptor species initiate and terminate various reproductive activities affect predation pressure on prey species. Tables 37 and 38 show most of the significant dates as determined by observation in both the Michigan and the Wyoming area. These dates are derived from observations on the entire raptor populations. Earliest dates of selection of nesting territory or laying, for example, are the earliest actual observed dates for the species. In some cases, earlier dates could have been computed. Because we are dealing not with individual hawks or owls but with populations of each species, the earliest laying may have been observed for one bird and the earliest hatching for a different individual of the same species. For this reason the incubation period cannot be determined from these two dates. The average dates of laying, hatching, and departure are obtained from at least two and generally many more observations. The number used to determine the average has not been indicated in the tables, but it is roughly proportional to the number of nesting pairs of each species.

The span of the nesting period indicates the number of days that nestlings of a species could be found and observed. It includes the period from the first observed hatching to the last date of nest departure (Plates 22 and 52), and is consequently longer than the average number of days that the young are in the nest. The greater the number of nesting pairs of a species, the more chance for variation in these dates, and thus the nestling period generally appears prolonged for the predominant species. The beginning, ending, and duration of this period vary for different climatic regions.

The breeding season of a species has been designated arbitrarily as the period between the first observed selection of territory and the latest brood departure. It is the time during which activity is centered around the nest.

The number of days during which young of a species were found to remain in the nest varied with circumstances. Birds whose nests were visited frequently or visited shortly before the nestlings were about to fly usually left a few days earlier than unmolested birds.

In some cases in which only one or two nesting pairs represent the population of a species, missing dates (that were not obtained from birds on the area) were filled in by dates obtained from birds of the same species nesting adjacent to the area. The starred dates for the earliest egg laying and earliest hatching were computed by adding or subtracting the period of incubation. When data on incubation periods were lacking, the information was obtained from Bent (1937, 1938).

SIGNIFICANCE OF ACTIVITY DATES

Later reference to Tables 37 and 38 will indicate the usefulness of these observed dates in interpreting predation. At present the discussion will be limited to comparisons of the nesting season activities of the collective raptor populations in Michigan and Wyoming.

Since the average temperatures in the Wyoming area were almost exactly the same as those occurring a month earlier in Michigan, it is of interest to note that the difference between the earliest layings in the two areas, which were those of Horned Owls, was exactly one month (Fig. 7). The time from the deposition of the first raptor egg to the last brood departure is longer in Michigan than in Wyoming (Fig. 7). Although egg laying starts a month earlier in Michigan, the last brood departure dates are very nearly the same for both localities. This telescoping of the vital breeding activities of a collective raptor population in a region characterized by shorter seasons would be expected. It implies that for each individual, activity dates, such as those of laying and hatching, must more nearly conform to the norm for the species; that there can be less pronounced variation and little or no renesting. This is actually the case, as would be demonstrable for each species were a comparable number of nesting pairs treated. It is perhaps best illustrated in Fig. 7 by a comparison of the breeding seasons of the Horned Owls in the two regions and by a comparison of the Red-shouldered Hawks in Michigan with the Western Red-tailed Hawks. These Buteos have comparable periods for incubation and fledging and each was the predominant species in its respective region, yet there is a marked telescoping of the breeding season in the colder region.

It is also significant that in both areas the Horned Owl started laying more than a month before the next raptor species began. This has a marked influence on the dominance of this owl over other raptors.

EFFECTS OF THE TIMING OF NESTING ACTIVITIES ON RELATIONS BETWEEN RAPTORS

Before nesting populations are discussed, some of the most important interactions among nesting species will be mentioned, with particular attention to the manner in which the physical attributes and activity of the Horned Owl enable it to dominate a nesting population of raptors. Because, in the populations studied, the Horned Owl is the most powerful bird, as well as the earliest nester, it has preference as to its nest location and cannot be evicted by other raptors. It also can hunt or defend itself equally well by day or by night. This is not true of the large hawks that might possibly dispute the owl's dominance. Adult raptors are particularly vulnerable to predation by this powerful owl from the onset of incubation to the departure of the young, because they are then closely attached to a nest. Tables 37 and 38 show that this period of vulnerability of the raptor population to Horned Owl predation begins just after the hatching of the young owls. In other words, the remaining raptor population is most vulnerable, particularly at night, during a period when the Horned Owl is extremely active in securing food for its young (Plates 38 and 39). Raptors and Crows nesting within the range of this species are affected very noticeably.

In three nesting seasons in Superior Township, the Horned Owl is known to have been responsible for the destruction of four hawk nests and may have destroyed others. Two of the four were Red-shouldered Hawk nests, one belonged to Red-tailed Hawks and one to Cooper's Hawks. In three cases the young were eaten. One adult Red-shouldered Hawk was also killed and eaten. Five pairs of hawks constructed nests, but, because of the disturbing influence of neighboring Horned Owls,

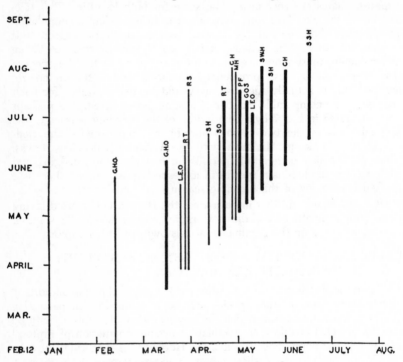

FIGURE 7. Raptor nesting season chronology
Note: Thin line = Michigan area
Heavy line = Wyoming area

they did not lay. Three of these were Red-shouldered and two were Red-tailed Hawks. One Cooper's Hawk nest within the home range of a pair of Horned Owls was deserted because of their presence. These Cooper's Hawks nested in a woods that was used as a winter roost by a pair of owls, but that was vacated early by them for a nesting territory in another woodlot. After the female Horned Owl was shot on her nest, the male eventually moved back to the winter roost. In the meantime a pair of Cooper's Hawks and a pair of Red-tailed Hawks had nested, the latter birds having hatched young. Following the owl's return, one of these nests was destroyed and the other deserted. Nests recorded as destroyed by Horned Owls had owl feathers at the scene of destruction and in some cases remains of the young hawks were also found at the owls' nest.

Maps 13 and 14 show that in most cases neither the Red-shouldered nor the Red-tailed Hawks nested close to or in the same woods with the Horned Owl. When they did so, their nests usually were destroyed or their nesting routine was so disrupted that there was no reproduction. The few successful Buteo nestings close to Horned Owls were cases in which these birds started nesting after a Horned Owl nest was destroyed and the owls were not restricted to a particular woodlot nor actively defending a territory. Hawks nesting in the same woods with a single non-nesting owl fared somewhat better, but even under this condition some nests were destroyed.

The Cooper's Hawk, though influenced in its nesting activities by the Horned Owl, is less affected than the Buteos, largely because of its later nesting and its utilization of nest sites, such as dense brushy woodlots, that are not suitable for the owls.

Screech Owls, whose nests and ranges are commonly within the nesting range of a Horned Owl, suffer their heaviest mortality after the young leave the protection of the nest hollow.

In the Western Study Area only one Red-tailed Hawk nest definitely was known to have been destroyed by Horned Owls. Hawk nests which were situated close to Horned Owl nests (Map 15) were often isolated from them by high ridges or differences in elevation.

So great was the effect of the Horned Owls' presence on nesting Crows that they ranked as a major factor in limiting Crow productivity. Adult Crows, as well as young, were killed and eaten on the nest at night. Some pairs built nests, but appeared to be so affected by the owls' presence that they did not lay. There were 119 pairs of nesting Crows in the Michigan Study Area in 1942 and 107 in 1948. This was determined by nest counts. These birds initiated nesting activities and laying simultaneously with the Red-shouldered and Red-tailed Hawks. A population of only 13 and 12 Horned Owls during the same seasons checked the Crow productivity to such an extent that the owls' ecological value in this regulatory capacity cannot be overlooked when weighing their desirability in any community. Conversely, next to man himself, the activities of Crows caused the greatest direct and indirect mortality of the nesting Horned Owls.

With respect to the Crow and other raptors, the Horned Owl is clearly a regulating force. Its influence is so evident that its dominance can be detected in many ways; it seems to radiate, web-like, through all raptor activities during the breeding season.

The drama of the countless and continuous effects of raptor species on one another does not fall within the scope of this work. It will be sufficient to point out that the cumulative effect on the various nesting species is measurable and significant; is influenced by the timing of nesting activities; and alters predation pressure through the tendency to limit raptor productivity. We should not, however, lose sight of the fact that the evolutionary adjustment of a collective raptor population's nesting activities, which reduces intra-specific strife, though perhaps less evident to the observer than is conflict, is nevertheless very real and significant in permitting numerous species of predatory birds to live and raise their young in relative harmony and security.

CHAPTER 10

Raptor Nesting Populations

THE WORK of gathering detailed information about all raptors nesting over a large area is an exhausting task. It is, paradoxically, also most exciting—a pleasure that stimulates at the time and lingers long in memory. Let us, therefore, digress a while from population details to visualize the land with its nesting hawks and owls—Superior Township as the investigators saw it during the breeding season. There was, for instance, the day in February with the temperature at zero when the first Horned Owl nest was located in an oak-hickory woodlot. The wind was swaying the brittle tree tops, the snow crunched under foot and Crows cawed from the far margin of a woodlot. On the near side, protruding above the snow-rimmed nest, were the ear tufts of a Horned Owl, which, upon discovery, became breeding raptor number one. We climbed to the nest and observed and recorded one white egg. A day later there were two eggs, the full clutch. A week passed before part of a rabbit was picked up beneath the nest. This initiated the recording of food for raptor number one. In the meantime, another Horned Owl nest was marked on the map. A bold female looked down from this large nest of sticks in a massive beech tree and did not fly until the climbing spurs bit into the trunk just beneath her. This was nest number two, its history short. On the second visit we found the frozen body of the mother owl still protecting two solidly frozen eggs. We dropped

TABLE 39

SUMMARY OF NESTING OBSERVATIONS
1942
SUPERIOR TOWNSHIP + AA36
37-square-mile area

Species	No. of pairs	No. of nesting pairs	No. sq. miles per pair	Pairs failing to lay eggs		Pairs renesting	Nesting pairs incomplete data	No. complete clutches under observation
				No nesting attempt	Nest built Eggs not laid			
Great Horned Owl	6	4	6.2	2	..	..	..	4
Long-eared Owl	1	1	37	..	..	..	..	1
Barn Owl	1	1	37	..	..	..	..	1
Screech Owl*	13	13	2.9	..	..	..	10	3
Red-shouldered Hawk	22	22	1.7	..	..	1	..	23
Red-tailed Hawk	2	2	18.5	..	1	..	..	1
Cooper's Hawk	9	9	4.1	..	3	..	..	6
Marsh Hawk	7	7	5.3	..	..	..	2	5
Sparrow Hawk	2	2	18.5	..	..	..	..	2
Total	63	61	.59	2	4	1	12	46

Single birds 1 Great Horned Owl
      2 Immature Red-shouldered
       Hawks   1.7 pairs of hawks and owls per square
      1 Immature Red-tailed   mile
       Hawk   Survival = 2 young per pair of adults.
      3 Sparrow Hawks
Total Population—133

the body to the ground, where it came to rest beside a conspicuous red shotgun shell. Through the stillness of the woods came the soft melancholy hoot of the male. It seemed to voice a resignation to fate, an acceptance of the failure to produce, this year, more of his kind.

At the north end of the township, raptor nest number six rested high in a huge hollow stub—a Horned Owl nest seemingly so placed as to keep investigators away. The old stub shook with each jab of the spurs and the soft rotten wood too frequently gave way, leaving the climber hanging by a flip rope and one insecure spur. The stub also housed bees that invariably made the ascent more hazardous and each time raised a question as to whether the data on this particular nest would ever prove to be worth the price.

As the cold days gave way to warmer ones and the varied green draperies of spring were pulled across the gray, open-windowed woods of winter, more and more nests were found and accompanying notes grew from recordings on a few sheets to filed stacks of cards. By this time the young Horned Owls were fledging. Twenty-two Red-shouldered

TABLE 39, *Continued*

	Eggs			Nestlings		Potential	Productivity	
	Av. clutch	Max. clutch observed	Number produced	Number hatched	Number to survive	Based on av. clutch	Based on max. clutch	Number of young produced* per pair of adults
Great Horned Owl	2	2	8	8	4	12	12	.7
Long-eared Owl	4	4	4	4	4	4	4	4
Barn Owl	6	6	6	0	0	6	6	0
Screech Owl* .	4	5	52	34	34	52	65	2.6
Red-shouldered Hawk	3.5	5	81	66	41	77	110	1.9
Red-tailed Hawk	2	2	2	2	1	4	4	.5
Cooper's Hawk .	4.3	5	26	18	18	39	45	2
Marsh Hawk ..	4.6	6	23	18	16	32	42	2.3
Sparrow Hawk .	4.5	5	9	8	8	9	10	4
Total	..	..	211	158	126	235	298	2.0

* Screech Owl Number of nesting pairs is based on the location of 13 occupied nesting sites
Method of computing Screech Owl productivity
66% of 12 eggs known to have been produced & hatched
100% of 8 that hatched & survived to leave nest
52 eggs produced based on av. clutch of 4
66% of 52 = av. no. to hatch = 34
100% of 34 = 34—No. to survive
* Survival to the flying stage

Hawk nests were plotted and their histories were evolving. The data files filled with nesting and hatching dates, nests destroyed, number of young to fledge, causes of mortality, animal forms eaten, behavior notes and new nests located. Before long, the various nests and their occupants not only had their own individual histories but showed relationships with nests nearby. This of course gave promise of significant glimpses into the community and its activities—information that when tabulated and interpreted might throw light on such questions as:

1. What is the abundance and composition of a population of nesting raptors on a given land area?
2. How much predation pressure is applied by a collective population of breeding raptors on prey species throughout a large land area?
3. What is the productivity of a collective raptor population?
4. What degree of nesting stability is exhibited by a collective raptor population?
5. What are the natural biotic controls operating to keep raptor populations in check?

These raptor population studies were conducted for three nesting seasons in Superior Township, for two nesting seasons in the Check Area

TABLE 40

SUMMARY OF NESTING OBSERVATIONS
1948
SUPERIOR TOWNSHIP + AA36

Species	No. of pairs	No. of nesting pairs	No. sq. miles per pair	No nesting attempt	Pairs failing to lay eggs		Pairs renesting	Nesting pairs incomplete data	No. complete clutches under observation
					Nest built Eggs not laid				
Great Horned Owl	7	6	5.3	1	..	..	..	6
Screech Owl*	15	15	2.5	..	..	..	14	1
Red-shouldered Hawk	18	17	2.1	1	..	1	1	17
Red-tailed Hawk	5	5	7.4	..	..	..	..	5
Cooper's Hawk	8	8	4.6	..	1	..	..	7
Marsh Hawk	9	9	4.1	..	1	1	..	8
Sparrow Hawk	4	4	9.3	..	..	..	..	4
Total	66	64	.56	2	2	2	15	48

Single birds 1 Barn Owl
    1 Long-eared Owl
    3 Great Horned Owls   1.8 pairs of hawks and owls per square
    1 Barred Owl   mile.
    2 Immature Red-shouldered   Survival = 1.6 young per pair of adults.
      Hawks
    1 Sparrow Hawk
    5 Immature Red-tailed
      Hawks
    2 Immature Cooper's Hawks
Total Population—148
* Screech Owl—Number nesting pairs is based on location of 15 nest sites.
For computation of Screech Owl productivity, see Table 53.

	Eggs			Nestlings		Potential Productivity		Number of young produced per pair of adults
	Av. clutch	Max. clutch observed	Number produced	Number hatched	Number to survive	Based on av. clutch	Based on max. clutch	
Great Horned Owl	1.8	2	11	5	3	13	14	.4
Screech Owl*	4	5	60	40	40	60	75	2.7
Red-shouldered Hawk	3.3	4	56	36	30	59	72	1.7
Red-tailed Hawk	2	3	10	6	4	10	15	.8
Cooper's Hawk .	4	5	28	22	18	32	40	2.3
Marsh Hawk ..	4.1	6	33	0	0	37	54	0
Sparrow Hawk .	4.3	5	17	11	11	17	20	2.8
Total	..	..	215	120	106	228	290	1.6

TABLE 41

SUMMARY OF NESTING OBSERVATIONS
1947
JACKSON HOLE, WYOMING
12-square-mile area

Species	No. of pairs	No. of nesting pairs	No. sq. miles per pair	Pairs failing to lay eggs			Nesting pairs in-complete data	No. com-plete clutches under observation
				No nesting attempt	Nest built Eggs not laid	Pairs renest-ing		
Great Horned Owl	4	4	3	..	..	..	..	4
Long-eared Owl	3	3	4	..	..	..	..	3
Screech Owl	3	3	4	..	..	..	2	1
Red-tailed Hawk	12	12	1	..	2	..	..	10
Swainson's Hawk	5	5	2.4	..	1	..	..	4
Cooper's Hawk	1	1	12	..	..	..	..	1
Sharp-shinned Hawk	2	2	6	..	..	..	..	2
Sparrow Hawk	11	11	1.1	..	..	..	1	10
Prairie Falcon	1	1	12	..	..	..	..	1
Raven	3	3	4	..	..	..	..	3
Total	45	45	.27	0	3	0	3	39

Single birds 1 Immature Swainson's Hawk 3.8 Hawk and Owl nests per square mile.
Total Population—91   Survival = 2.6 young per pair of adults.
Osprey 1 pair—not included in table because of specialized feeding habits.

	Eggs			Nestlings		Potential	Productivity	
	Av. clutch	Max. clutch observed	Number pro-duced	Num-ber hatched	Num-ber to survive	Based on av. clutch	Based on max. clutch	Number of young produced per pair of adults
Great Horned Owl	2.2	3	9	8	8	9	12	2
Long-eared Owl	4.7	5	14	13	13	14	15	4.3
Screech Owl ..	4	4	12	9	9	12	12	3.0
Red-tailed Hawk	2.3	4	23	21	17	28	48	1.4
Swainson's Hawk	1.8	2	7	5	3	9	10	.6
Cooper's Hawk .	5	5	5	5	5	5	5	5
Sharp-shinned Hawk	3.5	4	7	7	7	7	8	3.5
Sparrow Hawk .	4.4	5	44	43	42	48	55	3.8
Prairie Falcon .	5	5	5	5	4	5	5	4
Raven	5.7	7	17	15	10	17	21	3.3
Total	..	..	143	131	118	154	191	2.6

west of Ann Arbor, Michigan, and for two nesting seasons in the Wyoming study area.

COMPOSITION AND POPULATION DENSITY IN SUPERIOR TOWNSHIP AND SECTION 36, ANN ARBOR TOWNSHIP

In 1942, four species of owls—the Great Horned Owl, the Long-eared Owl, the Barn Owl, the Screech Owl—and five species of hawks —the Red-shouldered Hawk, the Red-tailed Hawk, the Cooper's Hawk, the Marsh Hawk, and the Sparrow Hawk—nested in the area (Table 39, Map 13).

In 1948 (Map 14), the composition and abundance of the collective raptor population was again determined. Of the Barn Owl and the Long-eared Owl, which in 1942 were represented by one nest each, only one bird each was present. In 1949 the composition of the nesting raptors remained unchanged as to species.

In 1942, the 37-square-mile area supported 63 pairs (61 nesting) and seven single bird predators, a total of 133 individuals. In 1948 there were 66 pairs, of which 64 pairs were nesting, and 16 single birds, a total of 148 individuals. In 1949, the same area supported 65 pairs and six single birds, making a population of 136 individuals (Table 42).

In all three years the Red-shouldered Hawk (Plates 18 and 20) was the predominant species, being 35, 27, and 26 per cent, respectively, of the annual nesting populations. The Screech Owl was second in numbers, being 21, 23, and 22 per cent. The Horned Owl was the dominant bird, although it was only 10 per cent of the population each year. The Buteos were 38, 35, and 35 per cent of the populations. Detailed species composition of the populations is recorded in Tables 39, 40, and 41 and on Maps 13 and 14. Thus after a seven-year interval the raptor populations in Superior Township and Section 36 of Ann Arbor Township were changed by only 15 birds, some of which were unmated (Tables 39 and 40). In 1942, 1948, and 1949 there were, respectively, 1.7, 1.8, and 1.8 pairs of hawks and owls per square mile. Raptor populations for Superior Township alone are presented in Table 16.

The glaciated topography, with numerous kettles, swamps, and marsh areas (Plate 2) is ideal habitat for the Red-shouldered Hawk and has a major influence in determining its numerical predominance. That the combination of small, isolated woodlots, marshes, and grasslands (Map 1) enabled the Red-shouldered Hawks to meet their biological requirements within limited areas is attested by their uniform nesting distribution and their relatively small hunting ranges. (Maps 13 and 17).

The small proportion of woodland (11 per cent) limited the winter population of Horned Owls and their nesting density.

It appeared that lack of suitable nesting sites may have limited the Sparrow Hawk population. The distribution and population density

TABLE 42

TOTAL NESTING RAPTOR POPULATION IN
SUPERIOR TOWNSHIP & SECTION 36, ANN ARBOR TOWNSHIP
1949

Species	Pairs
Red-shouldered Hawk	17
Red-tailed Hawk	6
Cooper's Hawk	8
Marsh Hawk	8
Sparrow Hawk	5
Great Horned Owl	6
Screech Owl	14
Barred Owl	1
	—
Total Pairs	65
Single Birds	
Immature Red-tailed Hawks	5
Immature Red-shouldered Hawk	1
	—
Total Population	136

of a single species, however, were dependent on more than the physical environment. This will be discussed later in detail.

The outstanding change in the composition of the raptor population was the increase in the number of nesting Red-tailed Hawks (Plate 23, Table 45 and Maps 13 and 14) from two pairs in 1942 to six pairs in 1949. It is possible that the increase (Tables 39 and 42) is only a temporary fluctuation; it seems significant, however, that it occurred simultaneously with drainage of swamps, cutting of woodlots, and generally more intensive farm practices, all of which reduced the habitat of the Red-shouldered Hawk, which prefers deep woods and swamps. This population change appears to be a gradual response whereby the component species of a collective raptor population adjust themselves to changing prey densities resulting from changes in land-use practices.

Buteo populations of 53, 53, and 52 during a three-year period may indicate that these figures represent the carrying capacity for Buteos and that the Red-shouldered Hawk perhaps is being replaced gradually by the Red-tailed Hawk, which is better adapted to small woodlots and cultivated land. Because the Red-tailed Hawk is a more aggressive bird and a permanent resident, establishing a year-round home range, it has a decided advantage over the Red-shouldered Hawk. The latter bird seems milder in temperament and is more migratory in this region; consequently it must obtain nesting territories anew each season. Each spring the number of Red-shouldered Hawks that can nest is partially dependent on the number of Red-tailed Hawk pairs already in established territories.

The replacement of the Barred Owl (Plate 34), inhabitant of deep

HAWK AND OWL NEST MAP
SUPERIOR TOWNSHIP

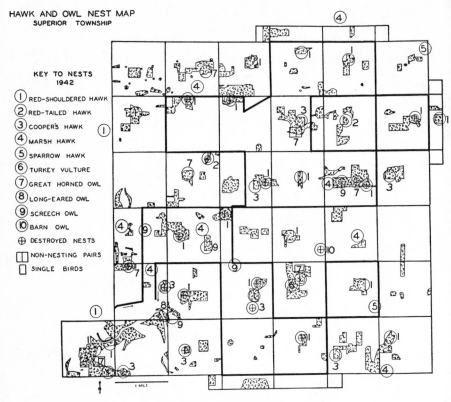

MAP 13. Raptor nests, Superior Tp., 1942

woods, by the more adaptable Horned Owl appears to have been completed.

The Michigan Check Area

In order to determine whether the populations in Superior Township were representative of larger areas in southern Michigan, comparison was made with an area of the same size and similar characteristics west of Ann Arbor. A census of this check area was made in 1947 and 1948, with the help of R. L. Patterson and W. H. Lawrence. Table 43 shows that it supported a slightly higher population of the raptors counted. This is a result of the generally larger size of the woodlots. It is significant, however, that the population composition and density of the two areas agree rather closely.

In the Check Area, as in Superior Township, the Red-shouldered Hawk was the predominant species, its population densities on the two areas being alike. The primary difference in composition was the pressence of Barred Owls in the Check Area and their absence in Superior Township.

With minor exceptions the population density and composition of the two areas were similar and the data for both areas seem reasonably representative of more extensive regions in broken hardwoods of southern Michigan.

COMPOSITION AND POPULATION DENSITY ON THE WYOMING STUDY AREA

How did the raptor composition and population density found in Michigan compare with those in other regions? One answer was obtained in the semi-wilderness in Wyoming, which supported 10 raptor species. There were three owls, the Great Horned Owl, the Screech Owl, and the

HAWK AND OWL NEST MAP
SUPERIOR TOWNSHIP

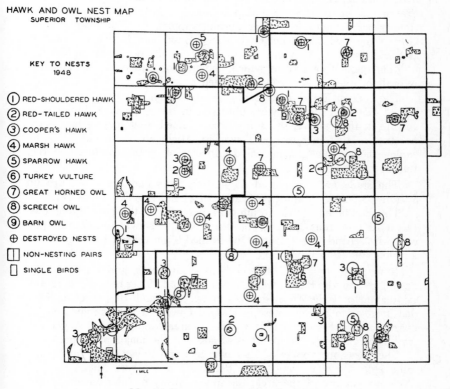

KEY TO NESTS
1948

① RED-SHOULDERED HAWK
② RED-TAILED HAWK
③ COOPER'S HAWK
④ MARSH HAWK
⑤ SPARROW HAWK
⑥ TURKEY VULTURE
⑦ GREAT HORNED OWL
⑧ SCREECH OWL
⑨ BARN OWL
⊕ DESTROYED NESTS
▯ NON-NESTING PAIRS
▯ SINGLE BIRDS

MAP 14. Raptor nests, Superior Tp., 1948

TABLE 43

COMPARISON OF NESTING RAPTOR POPULATIONS
ON TWO 36-SQUARE-MILE AREAS IN SOUTHERN MICHIGAN

Species	Superior Twp. 36 sq. miles			Check Area 36 sq. miles	
	1942	1948	1949	1947	1948
Red-shouldered Hawk	19	16	14	17	18
Red-tailed Hawk	2	5	6	9	12
Cooper's Hawk	8	7	7	6	6
Great Horned Owl	5	7	6	8	9
Long-eared Owl	1	0	0	0	0
Barn Owl	1	0	0	0	0
Barred Owl	0	0	0	3	2
Marsh Hawk	7	9	8	5	6
Total pairs	43	44	41	48	53

Long-eared Owl; and seven hawks, the Western Red-tailed Hawk,
Swainson's Hawk, Cooper's Hawk, Sharp-shinned Hawk, Sparrow Hawk,
Prairie Falcon, and Osprey (Map 15 and Table 41). In the immediate
vicinity were found five other nesting raptors; the Short-eared Owl,
Great Gray Owl, Saw-whet Owl, Marsh Hawk and Goshawk. Thus an
area of less than 15 square miles supported 15 species of raptors. In
addition there were Ravens which, though not taxonomically classified
as birds of prey, ecologically functioned as such.

The Golden Eagle, Bald Eagle, (Plate 7), and Ferruginous Rough-
legged Hawk (Plate 19) were observed in the study area, but nested
outside it. If, however, the study area selected in Wyoming had been
as large as that in Superior Township, Michigan, we could have included them as nesters, making 18 species of nesting raptors and the
Raven.

The study area supported 45 nesting pairs and one immature of
Swainson's Hawk, a raptor population of 91 individuals (Table 41).

The Red-tailed Hawk, forming 27 per cent of the population, was
the most numerous species, and the Sparrow Hawk, forming 24 per
cent, was second. The Horned Owl formed 9 per cent of the population
and the Buteos 38 per cent.

It may be significant that in both study areas, and during every
nesting season, the Buteos were between 35 and 38 per cent of the raptor
population and the Horned Owls between 9 and 10 per cent. This
density relation between nesting Horned Owls and Buteos may well be
characteristic of much larger woodland areas and does, in a general
way, reflect the relative densities of some of the major prey species.

There was an average of 3.8 hawk, owl and Raven nests per square
mile, a density more than twice that found in southern Michigan. The
wilderness nature of the area, the low intensity of interference, direct
or indirect, by man, the numerous nest sites provided by a timbered

butte and a wooded river bottom, as well as a plentiful food supply, were factors contributing to the high population density in the Wyoming study area.

POPULATION STABILITY

All raptors studied exhibited strong tendencies to reoccupy nesting territories in consecutive years. In the course of a sufficient length

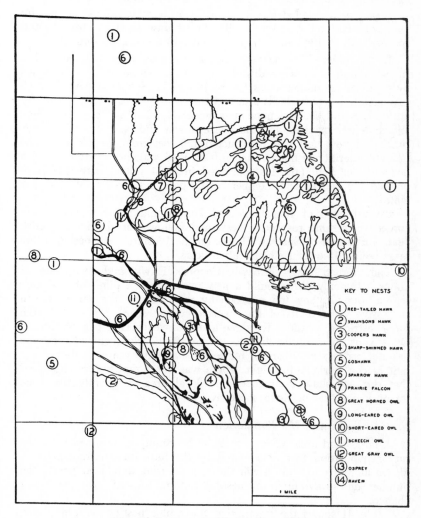

Map 15. Raptor nests, Western Area, 1947

of time this continued occupancy of a territory involves birds other than the original nesting pair, but accumulated evidence leaves little doubt that a pair, or at least one member of it, returns annually to the same nesting area, frequently to the same nest, until death or some drastic change in the environment breaks the pattern. Many individuals could be identified by characteristic markings or unusual behavior patterns (Plate 29).

Reoccupancy of Nests and Territories by the Same Species

A pair of Red-tailed Hawks in Michigan occupied the same nest for three consecutive years. Another Red-tailed Hawk nest was used by this species both in 1942 and in 1948. (Maps 13 and 14, Section 11).

Red-shouldered Hawks used the same nest in 1940, 1941, and 1942. In 1947 and in 1948, when this nest was again examined, it was occupied by a pair of the same species (Maps 13 and 14—Section 36AA). In each year mentioned these birds attacked the investigators, while few of the other pairs did so, and none so aggressively. In 1942 five eggs were laid and four hatched; in 1948 one egg was laid, and it was infertile. In 1949 a new nest was constructed 150 feet from the old one. The probable sterility of one of the birds in 1948 and the construction of a new nest in 1949 (the first observed in eight years) may well have marked the end of one member of the pair and the arrival of a new mate.

Several pairs of Red-shouldered Hawks and Sparrow Hawks occupied their same respective nests and hollows for three consecutive years and a pair of Sparrow Hawks on the Western Area nested in the same hollow tree for six consecutive years.

The Horned Owl, Goshawk, Cooper's Hawk, and Sharp-shinned Hawk commonly either build a new nest or use a different one each season, though they too occasionally may occupy the same nest in successive years. Pairs of these species were recorded nesting in the same restricted localities for three and four years. One pair of Horned Owls, kept under observation to 1953, continuously occupied the same nesting territory in Wyoming for seven successive years, while another pair held a territory for eight. There was not a raptor species studied that did not show occupancy of a local area, frequently the same territory, for two or three successive years.

Nesting Pattern of the Population

A large land area reflects the same continuity of occupancy by the raptor population that is exhibited by the individual nesting pairs. A glance at Maps 13 and 14 will show that during a six-year period the number and distribution of nesting raptors showed little important change.

Table 44 shows the stability of the raptor nesting pattern during

TABLE 44

STABILITY OF THE RAPTOR POPULATION AS INDICATED FROM THE RELATION OF THE 1947 NESTING POPULATION TO THE 1948 POPULATION (36-SQUARE-MILE CHECK AREA, ANN ARBOR, MICHIGAN)

Species	Nests on Area 1947	Nests on Area 1948	No. 1947 nests reoccupied by same species in 1948	No. woodlots reoccupied by same species	No. sections reoccupied by same species	Total No. of 1947 pairs believed to have reoccupied nesting vicinity in 1948	1947 Pairs* failing to return to same locality	Pairs in new 1948 locations	Population increase
Red-shouldered Hawk	17	18	2	11	13	14	3	4	1
Red-tailed Hawk	9	12	2	4	6	7	2	5	3
Cooper's Hawk	6	6	0	4	5	5	1	1	0
Great Horned Owl	8	9	1	2	3	5	3	4	0
Barred Owl	3	2	0	1	1	1	2	1	0
Totals	43	47	5	22	28	32	11	15	4
Per Cent			12	51	65	74	26		

10 nests occupied in 1947 were utilized by the same or different species in 1948.

* Since no banding or marking of adult birds was undertaken, there could be no certainty that the pairs in 1948 were those present in 1947, but there was much indirect evidence to support the belief that they were.

TABLE 45

STABILITY OF THE RAPTOR POPULATION AS INDICATED FROM
THE RELATION OF THE 1947 NESTING POPULATION TO THE 1948 POPULATION,
MOOSE, WYOMING

Species	No. of Nests on Area 1947	No. of Nests on Area 1948	1947 nests re-occupied by same species in 1948	No. pairs utilizing same immediate area	Total No. of 1947 pairs believed to have re-occupied nesting vicinity in 1948	1947 Pairs failing to return to same locality	Pairs not rechecked
Red-tailed Hawk	12	11	2	11	11	1	
Swainson's Hawk	5	1	0	1	1	4	
Cooper's Hawk	1	1	0	1	1	0	
Sharp-shinned Hawk	2	..	..	..	..	..	2 not rechecked thoroughly
Goshawk	1	1	0	0	1	0	
Sparrow Hawk	11	10	7	10	10	1	
Prairie Falcon	1	1	0	0	0	1*	
Great Horned Owl	4	4	0	4	4	0	
Long-eared Owl	3	2	.	2	2	..	
Short-eared Owl	1	.	..	..	..	..	1
Screech Owl	3	2	..	2	2	..	1
Great Gray Owl	1	1	0	1	1	0	1
Osprey	1	1	1	1	1	0	
Raven	3	3	2	3	3	0	
Totals	49	38+	12	36	37	7	5
Per cent	..	..	25	73	75	14	10

13 nests occupied in 1947 were utilized by the same or different species in 1948.

* Prairie Falcon moved to new site several miles along bluff.

+ In rechecking the nesting population in 1948, 5 nests were not rechecked and no attempt was made to locate new pairs. It is very probable that there was little change in the population density, certainly not as great a change as is indicated by the total of nests recorded.

Four species that nested on the periphery of the study area are included in this table for comparative purposes.

two consecutive years in the Ann Arbor region of southern Michigan. In the 36-square-mile Check Area there were four more nesting pairs in the raptor population in 1948 than in 1947. The most noticeable change was the increase in the Red-tailed Hawks from nine to twelve pairs. There was little change in Superior Township in 1948 and 1949.

Ten nests in the Check Area that were occupied in 1947 were re-occupied in 1948; five by the species that formerly used them and five by different species. The 1947 nesting pattern remained 74 per cent intact in 1948 (Table 44).

The same stability recorded in Michigan was characteristic of the raptor population in the Western Area (Table 45). The nesting pattern remained at least 75 per cent intact during a two-year period and had every nest been rechecked (five were not) the percentage might have been higher.

It is probable that the disturbance created by the observers in making the study caused some shifting of nest sites the following year. Under normal conditions the stability of nesting patterns from year to year is possibly even greater than has been recorded in this study.

It is apparent that there is a marked tendency for a raptor nesting pattern to remain unchanged year after year.

Many of the changes that do occur from year to year in the composition of the raptor population and in the spatial relationship of the nesting species can be explained. Year-round residents, such as the Horned Owls, exerted a strong influence towards both maintaining and altering the nesting pattern. As the earliest nester in Superior Township, the Horned Owl (Fig. 7) laid the foundation of the nesting pattern long before the other raptors started to nest. Yet the past nesting pattern of other raptors and that of Crows determined to some extent the annual nest pattern of the Horned Owls, since these owls (in Michigan) relied largely upon the old stick nests of hawks and crows. Thus each year the owls' nest sites were determined by the location of suitable nests within their home ranges. Any change in Horned Owl nest sites tended to require changes in the nest locations of raptors nesting later. Since the Horned Owl frequently took over a Red-tailed Hawk nest of the previous year, those hawks whose nests had been usurped had to alter their nesting ranges and each pair generally did so by moving to an adjacent woodlot. At times a pair apparently failed to find a suitable site and remained as non-nesters.

The reshuffling of these two species made slight changes in the pattern. By the time the migrant Red-shouldered Hawks had returned, a territorial pattern of Horned Owls and Red-tailed Hawks had been established and the Red-shouldered Hawks fitted into this pattern. Those Red-shouldered Hawks that had their previous nesting sites or territories altered by the Horned Owl and Red-tailed Hawk made adjustment by selecting new sites, generally near the old ones. The Cooper's Hawk,

Sparrow Hawk, Marsh Hawk, and other raptors were not so greatly affected by the presence or absence of the Horned Owl and Red-tailed Hawk, but they too made some adjustment.

The constancy of raptor composition and density (Tables 39, 40, 42, and 43) further supports the conclusion that though minor changes occurred in the nesting pattern, these changes were spatial adjustments by previous nesting pairs to altered distribution of early nesters. In brief, the raptor population as a whole tended strongly to maintain a uniform nesting pattern and the comparatively small changes that occurred from year to year were the extent of failure to do so.

Such activities of man on the area as heavy cutting of woodlots, shooting of adult nesting birds, sugar tapping, Crow hunting, and farming tended to alter the nesting pattern. Stability was, however, the dominant feature of the nesting pattern from year to year.

THE RELATION OF RAPTOR NESTS AND RAPTOR POPULATION DENSITY TO AVAILABILITY OF FOOD

Whether raptors select nesting sites in relation to an abundant food supply and whether an area will support a heavier nesting population when populations of a major prey species are high are questions that have often been asked.

In regard to the first question, we have seen that in the areas studied the nesting pattern of entire populations was largely determined by the pattern that existed the year before, each pair tending to reoccupy a territory previously utilized. The pairs apparently did not select a territory annually because of food or other desirable features, but automatically settled into their previous niche or a vacated one. Stability was thus a prominent characteristic of the nesting pattern.

No doubt many forces determine where a hawk or owl will nest. Available food is one of them, but there was no indication on the areas studied that a raptor selected a territory solely because mice, rabbits, or other prey were abundant there. In other words, all data indicate that a pair of Red-shouldered Hawks, for example, would not return to their territory of the previous year and then, because food was more abundant a mile or so away, shift to that area and establish a territory. To do so would involve encroaching on the territories of other raptors and, were this the general procedure, conflict would be extreme. There are many examples of hawks and eagles returning to previous nesting territories and attempting to raise broods where there was not sufficient food to support them.

The shifting of raptors within their territories to nest sites in the immediate vicinity of abundant food was observed. It may be that food was the major stimulus initiating such movements. Also, the more dominant members of a population usually settled and held the most productive areas, indicating that food availability may be a powerful force in initial settlement. In areas where prey forms are restricted in numbers, a scarcity of the major prey will apparently cause a marked reduction in the number of nesting raptors, since there are not enough other prey forms to support them. This appears to be true especially of the Snowy Owl, American Rough-legged Hawk and Short-eared Owl. It seems that in such cases food is the major factor determining where and in what numbers these raptors will nest. It appears, however, that in some other regions prey forms are so numerous that a decline in one species does not reduce drastically the total prey population. Such regions consistently will support raptor populations year after year, so that a stable nesting pattern is established. Territory is reoccupied and a nest constructed with little or no relation to the comparative availability of food in adjacent areas and territories.

In regard to the second question, it has been demonstrated that the nesting raptor populations studied, being in adjustment with the carrying capacity of the area for prey, did not change greatly from year to year. On an area of land such as Superior Township at a given time the population of certain prey species will normally be high while others will be low. The total prey during the breeding season is nearly always, however, in excess of the requirements of the raptor population. Furthermore, it appears that generally there is no great increase or decrease in the total quantity of prey during any one breeding season, an upward trend in one prey species being compensated for by a downward trend in another. If this is the case, we should not expect normally to find any great annual increase or decrease in the breeding raptor population. Likewise the opportunity is slight for new arrivals to break in and alter the established nesting pattern substantially.

When a prey species or a collective prey population is abundant, the fall migrating hawks and owls will, as we have seen, respond by settling and wintering in the area in proportion to the available food supply. It is significant, however, that in Superior Township the raptor nesting population was no greater following a winter of very high meadow mouse density than it was following a winter of much lower density (Maps 13 and 14). Similarly, there was no relation between raptor nesting densities and winter rabbit density. These conclusions hold true for the Western Study Area, where the raptor nesting densities were similar in 1947 and 1948, in spite of a high over-wintering meadow mouse density in 1947.

The predominant Red-shouldered Hawks arrived in Superior Township in the spring and established a territorial pattern long before there was any evidence of what the density of their major food supply would be. Frogs, snakes, crawfish, and small birds, which are their chief food in summer, were not available generally until long after their nesting territories had been established.

It seems safe to conclude that in north temperate latitudes the increase in any single staple prey species has no immediate effect on the raptor nesting density.

The carrying capacity of an area of land for nesting raptors evolves over a long period of time and is established as the result of many inter-acting biological forces. In other words, the area has proved its capacity to support a given population of raptors year after year, a condition which enables a community of raptors to evolve and to resist change in nesting density.

COMPARISON OF POPULATION DENSITIES

Since the nesting composition and pattern of a collective raptor population remain relatively stable for a period of years, it would be expected that the population densities should show little fluctuation. This we have found to be the case.

In 1942, 1948, and 1949 the Michigan study area supported 63, 66, and 65 pairs of hawks and owls, respectively, a fluctuation in nesting raptor density of three pairs in three seasons included in a seven-year period. Apart from the Sparrow Hawk and Screech Owl (Table 43), Superior Township supported 43, 44, and 41 pairs of raptors during 1942, 1948, and 1949 and the Michigan Check Area supported 48 pairs in 1947 and 53 pairs in 1948. The Wyoming area supported 45 nesting pairs in 1947, and the situation of 35 of these was redetermined in 1948. No attempt was made to rediscover five nests nor to get a count of all nesters. There was little variation, however, in the population density of three species of which a count was made. In the two consecutive years, the Red-tailed Hawk population was 12 and 11 pairs, the Sparrow Hawk population 11 and 10 pairs, and the Horned Owl population 4 and 4. There was strong indication that the population densities of the two years were similar (Table 45).

From the population data obtained in the Michigan and Wyoming areas, it appears that on a land area the size of a township, interspecific competition in raptor populations is kept to a minimum by the estab-lishment of territories and that this is significant in determining the carrying capacity of the area for raptors. The data show that raptors can maintain an average population density as high as one nesting pair per 0.27 square mile in ideal habitat and that even in areas of intensive land use they can attain an average density of one nesting pair per 0.56 square mile. In some restricted localities they can and do attain much heavier population densities. A maximum density of four nesting pairs of raptors per section was observed in Superior Township in both 1942 and 1948 (Maps 13 and 14), with maximum local concentrations of six pairs within close proximity of one another. In the Wyoming area a maximum local density of eight nesting pairs per section was ob-served, with a local concentration of ten in close association (Map 15 and Plate 55).

These local nesting concentrations were composed in Michigan of as many as five species—Red-tailed Hawk, Red-shouldered Hawk, Marsh Hawk, Cooper's Hawk and Screech Owl—and in Wyoming of eight species—Red-tailed Hawk, Swainson's Hawk, Sparrow Hawk, Cooper's Hawk, Sharp-shinned Hawk, Long-eared Owl, Great Horned Owl, and Screech Owl (Maps 13, 14 and 15). From this it would appear that direct competition between different species during the nesting season is not intensive.

Because of the stability of the composition, density and nesting pattern of raptor populations, we can conclude that on the areas studied at least one member of a pair, and in many cases both members, tend either to occupy a home range throughout the year (e.g., Red-tailed Hawks and Great Horned Owls) or to return to a definite area (frequently the same territory) to nest. These phenomena weld a raptor nesting population into an extremely stable community that changes little from year to year or even in the course of longer periods of time. Such phenomena also very probably reflect the ecological stability of the land area.

PAIRS FAILING TO LAY EGGS

Certain pairs of raptors did not lay eggs, although they defended territories and built and lined nests. Other pairs did not even make nesting attempts, though they confined themselves to a territory and weakly defended it.

During both years there were paired Horned Owls that made no nesting attempts (Tables 39 and 40). These paired birds might have been only a year old. In 1948 one such pair maintained a territory in Section 7 (Map 14). They remained through the winter of 1948-49, but nested for the first time the following spring. Had this pair been juveniles in 1948, as suspected, then the observations might suggest that the Horned Owl does not nest until the second year. The fact that three out of the four pairs of raptors that made no nesting attempt (1942-48) were Horned Owls and that non-nesting pairs of Horned Owls also were observed in the Check Area may be construed in harmony with this suggestion.

Certain Red-tailed Hawks, Cooper's Hawks, and Marsh Hawks were observed to build nests and defend territories, yet failed to lay eggs. Reasons for this behavior appeared to vary and were thought to include inter-specific intolerances, senility, and environmental disturbances. The Cooper's Hawk (Plate 8) was most prone to exhibit this behavior. Its habit of frequently building two or even three nests may reflect the extremely high-strung temperament which seems to be characteristic of the Cooper's Hawk in southern Michigan.

In 1942, 9.5 per cent of the raptor population (paired birds) either made no nesting attempts or failed to lay eggs (Table 39). In 1948,

6 per cent of the population was in this category, and, in the Western Area, 6.5 per cent (Tables 40 and 41).

RENESTING

Renesting attempts were few and, when made, followed destruction of eggs before incubation was well under way. A pair of Red-shouldered Hawks renested in 1942, and a pair of Red-shouldered Hawks and a pair of Marsh Hawks (Plate 14) renested in 1948. There were no renestings in the Western Area (Tables 39, 40, and 41).

It is well known (Bent 1937, 1938) that most birds of prey will lay two and even three sets of eggs if their eggs are taken before incubation is advanced. When, however, the nesting cycle is disrupted by destruction or loss of eggs in an advanced stage of development or by the destruction of young birds, renesting is seldom attempted.

Paired hawks of all species will continue the nesting-cycle behavior after the nest has been destroyed. This very strong tendency to maintain the nesting behavior of brooding, food exchange, food getting, and territorial defense after the destruction of eggs or young was observed in every species studied (J. and F. Craighead, 1939). The physiological conditions responsible for the behavior evidently must run their course, and, unless nest destruction occurs at the initiation of the physiological change associated with egg laying and early incubation, the cycle cannot be broken and restarted. Thus, renesting attempts are few and there were no records of double broods.

UNPAIRED BIRDS

Unpaired birds represented 5, 11, and 8 per cent of the population in Superior Township in 1942, 1948 and 1949, respectively, and 1 per cent in the Western Area. The species involved are recorded in Tables 39, 40 and 41. In the case of species in which immatures can be distinguished from adults in the field, all the unpaired birds were immature. It seems quite probable that a very high percentage of unpaired birds of all species were immature.

INCOMPLETE DATA

The heading "Nesting Pairs Incomplete Data" in Tables 39, 40 and 41 refers to nests that were found without obtaining data on the number of eggs and young produced. The majority of such nests were those of the Screech Owl. The population records of 13 pairs in 1942 and of 15 in 1948 are based on nests or broods actually found, and are believed to be rather accurate. Only those nests where the nesting hollow definitely was discovered were recorded on Maps 13 and 14, and consequently the maps do not indicate the number of nests of these owls.

In order to complete the study of the raptor population in the nesting season, the data on Screech Owl productivity and survival that were obtained were used as a sample for computing the probable productivity and survival of the population (Tables 39 and 40).

PRODUCTIVITY OF THE RAPTOR POPULATIONS

All raptor nests were visited three or more times to determine the number of eggs in a clutch, the number of young hatched and the number of young fledged (Tables 39, 40 and 41). In addition, many visits were made to collect food items and to determine egg-laying dates.

With few exceptions nests were found before the young were hatched in order to ascertain clutch size (Plate 38). The number of eggs produced (Tables 39, 40 and 41) was determined by egg counts at the nest. For those nests destroyed before egg counts could be made, average clutches were not computed. There was undoubtedly egg loss before some nests were found and the eggs counted (Plates 13, 14, 16); and therefore the number of eggs recorded as produced by each population (Tables 39, 40 and 41) was slightly lower than the actual number. The hatch recorded was possibly smaller than the actual hatch since some of the young may have died and been removed by the parents before a count of young was made. In a few cases nearly a week elapsed after hatching before the young were counted. Nevertheless, frequent visits reduced such error to a minimum and it is believed that the figures presented in Tables 39, 40 and 41 very closely approach the true facts. Special care to make counts at the fledgling stage was taken so all mortality would be recorded for each nest and a true productivity figure determined.

No nesting information was secured from the ground (Plate 54), since much error is inevitable if observations are made in that manner. This applies with special force to fledgling counts. Nesting trees were climbed with the aid of climbing spurs and flip rope, and in the Western Area cliffs and escarpments were scaled or descended with ropes (Plate 53.) The nesting hollows of some Sparrow Hawks were opened to allow observation, but in general, nests were disturbed as little as possible. In hiking to nests, varied routes were taken if feasible and when approaching ground-nesting raptors special care was exercised to conceal a trail. It was felt that an extensive study of all nests in a population would yield data on the natural history of the various species that would be proportionately less affected by the observers' activities than would intensive studies of a few nests. In spite of keeping the disruptive effects of observation to a minimum, it was apparent that the process of gathering data did affect the nesting mortality somewhat.

Egg Clutches

During the study, 133 complete egg clutches belonging to various species were recorded. Tables 39, 40 and 41 show the average clutch size determined for each species during each study year and the maximum clutch size observed.

In 1948 the average clutch of each species but the Screech Owl and Red-tailed Hawk was smaller than in 1942. The average clutch sizes of these two raptors remained the same.

In the Western Area both the maximum and the average clutch size of the Great Horned Owl and Red-tailed Hawk exceeded the average obtained for these species in Michigan, while the clutches of the Sparrow Hawk averaged about the same.

POPULATION PRODUCTIVITY AND MORTALITY

The number of eggs produced, the number of young hatched, and the number of young surviving were recorded for each species in the collective population and for each population unit (Tables 39, 40 and 41). The potential productivity was calculated for each species and for the collective population. For example, in 1942 the raptor population of Superior Township produced 211 eggs, of which 158 hatched, and from which 126 young survived to leave the nests (Table 39). On a basis of the average clutch observed for each species, the population was capable of producing 235 eggs, or, on a basis of the maximum clutch observed per species, 298 eggs. Among the 1942 population a deficit of 29 per cent resulted from incomplete clutches and failure to lay eggs. This figure is based on a potential of 298 eggs. Of the 211 eggs actually produced by the raptor population in Superior Township, 25 per cent failed to hatch. Since 80 per cent of the young birds that hatched survived to leave the nest, there was a juvenile mortality of 20 per cent. From the 211 eggs laid, 126 young fledged, a nesting success of 60 per cent. The survival to fledgling stage from a potential 298 birds was only 42 per cent.

Details of the mortality and nesting success of the populations of Superior Township and the Western Area can be obtained by consulting Tables 39, 40, 41, 46 and 47. Table 46 shows that the mortality of eggs in Superior Township was greater in 1948 than in 1942, but juvenile mortality was less. In 1948 the nesting success from the total of 215 eggs produced was 49 per cent, as compared with 60 per cent in 1942, and there was a 37 per cent survival to fledgling stage from a potential of 290 birds, as compared with 42 per cent in 1942.

The Western Area (Table 47) had a proportionately higher productivity. Ninety-two per cent of the 143 eggs produced hatched, and 90 per cent of those that hatched survived to leave the nests as fledglings. The nest success from the eggs laid was 83 per cent, which may be compared with 60 and 49 per cent in Superior Township.

Of 569 eggs laid in the two areas during three breeding seasons, 71.9 per cent hatched and 61.5 per cent produced fledged young.

SPECIES PRODUCTIVITY AND MORTALITY

Many environmental factors determine the mortality of any species from year to year on any area, and there is great variation (Table 46) among species as to where and when in the nesting cycle mortality occurs. For example, the loss for the Red-shouldered Hawk through the failure of eggs to hatch was 19 per cent in 1942 and 36 per cent in 1948. It was 0 and 55 for the Horned Owl, and 22 and 100 for the Marsh Hawk (Plate 33) for the same years. Egg loss for the raptor population was 25 per cent in 1942 and 44 per cent in 1948.

Juvenile mortality of Red-shouldered Hawks was 38 and 17 per cent and that of Cooper's Hawks 0 and 18 per cent in 1942 and 1948, respectively. Juvenile mortality for the entire population was 20 and 12 per cent in the two years. In Superior Township the greatest loss in the raptor population occurred each year in the egg stage. Nestling mortality (except on the Western Area, Table 47) was much less than egg loss.

The stability of the population densities indicates that, though annual fluctuations in the productivity of any species occur (Table 46), in an extensive area and in the course of a period of several years these fluctuations tend to be mutually compensatory. In Superior Township the Horned Owl showed a variation in productivity from 50 to 27 per cent, but data on 39 nests in the same general area indicate that in this region the Horned Owl averages annually about 1.1 young per pair of adults (Table 48). The Marsh Hawks showed the greatest variation in productivity, 70 per cent to 0. The Cooper's Hawk maintained the highest productivity of any of the more numerous raptors, 69 and 64 per cent. The predominant Red-shouldered Hawk evinced the least variability in its productivity, which was 51 and 54 per cent. This was partially in consequence of the larger sample of the Red-shouldered, but may also have been a result of this bird's adaptiveness to the environment.

It is evident that, with the possible exception of the Marsh Hawk, the more abundant species in the Michigan population produced sufficient progeny each year to increase the population greatly if no mortality occurred after the young left the nest. In 1942 the collective raptor population produced an average of 2 young, and in 1948, 1.6 young, per pair of adults. Thus, in spite of a differential nesting success of 11 per cent in the two years, and in spite of a high mortality in some species, survival to the flying stage was great for the raptor populations of both 1942 and 1948.

Table 47 shows that productivity on the Western Area was high in all species except the Swainson's Hawk. This species, which was in direct competition with the more aggressive Red-tailed Hawk, was forced to use inferior nest sites, a situation which may have affected its productivity.

Cooper's Hawks and the Sharp-shinned Hawk showed the highest productivity, but since there were only three pairs it would not appear

TABLE 46

MORTALITY AND NESTING SUCCESS OF RAPTOR POPULATIONS IN SUPERIOR TOWNSHIP

	Egg Deficit								Juvenile Mortality				Nesting Success from Eggs Laid			
	Incomplete clutches* and failure to lay eggs				Egg destruction, infertility and failure to hatch				Mortality in nest after hatching				Fledglings			
	1942		1948		1942		1948		1942		1948		1942		1948	
Species	No.	Per Cent	No.	Per Cent	No.	Per Cent	No.	Per Cent	No.	Per Cent	No.	Per Cent	No.	Per Cent	No.	Per Cent
Great Horned Owl	4	33	3	21	0	0	6	55	4	50	2	40	4	50	3	27
Long-eared Owl	0	0	..	..	0	0	..	..	0	0	..	..	4	100	..	..
Barn Owl	0	0	..	..	6	100	20	33	0	0	..	..	0	0	..	..
Screech Owl	13	20	15	20	18	35	20	33	0	0	0	0	34	65	40	67
Red-shouldered Hawk	29	26	16	22	15	19	20	36	25	38	6	17	41	51	30	54
Red-tailed Hawk	2	50	5	33	0	0	4	40	1	50	2	33	1	50	4	40
Cooper's Hawk	19	42	12	30	8	31	6	21	2	0	4	18	18	69	18	64
Marsh Hawk	19	45	21	39	5	22	33	100	2	11	0	0	16	70	0	0
Sparrow Hawk	1	10	3	15	1	11	6	35	0	0	0	0	8	89	11	65
Total	87	29	75	26	53	25	95	44	32	20	14	11	126	60	106	49

* Computed from potential productivity based on maximum egg number observed.

TABLE 47

MORTALITY AND NESTING SUCCESS OF RAPTOR POPULATIONS ON THE 12-SQUARE-MILE AREA—
MOOSE, WYOMING
1947

| Species | Egg Deficit | | | | Juvenile Mortality | | Nesting Success from Eggs Laid | |
| | Incomplete clutches and failure to lay eggs* | | Egg destruction, infertility and failure to hatch | | Mortality in nest after hatching | | Fledglings | |
	No. eggs	Per cent	No. eggs	Per cent	No. Young	Per cent	No. Young	Per cent
Great Horned Owl	3	25	1	11	0	0	8	89
Long-eared Owl	1	7	1	7	0	0	13	93
Screech Owl	0	0	3	25	0	0	9	75
Red-tailed Hawk	25	52	2	9	4	19	17	74
Swainson's Hawk	3	30	2	29	2	40	3	43
Cooper's Hawk	0	0	0	0	0	0	5	100
Sharp-shinned Hawk	1	12	0	0	0	0	7	100
Sparrow Hawk	11	20	1	2	1	2	42	95
Prairie Falcon	0	0	0	0	1	20	4	80
Raven	4	19	2	12	5	33	10	59
Total	48	25	12	8	13	10	118	83

* Computed from potential productivity based on maximum egg number observed.

TABLE 48

Species	No. Pairs	No. Fledglings Produced	Average Number of Fledglings Per Pair Adults
Great Horned Owl	39	41	1.1
Red-tailed Hawk	22	20	.9
Red-shouldered Hawk	61	111	1.8
Cooper's Hawk	22	48	2.2
Marsh Hawk	19	24	1.3
Total	163	244	1.5

The productivity is computed from successful and unsuccessful nests and non-nesters.

Sixteen records were supplied by W. H. Lawrence and R. L. Patterson.

significant had not this same success been exhibited in the Michigan Cooper's Hawk population and at other nests observed by us in the course of 18 years.

The Sparrow Hawk exhibited a high productivity on both Michigan and Wyoming areas. The nesting success of the Long-eared Owl (Plates 43 and 44) was high on the Western Area, although it nested within Horned Owl ranges (Map 15).

The average success of raptors to fledgling stage for three study years on the two areas, including all the nesting pairs (174), was 62 per cent, or two young per pair of adults.

NEST SUCCESS

Nest success was computed from all paired birds on the areas, including all with active nests. This yields a more nearly accurate figure of raptor nest success than does success determined from active nests only.

Nests destroyed were, in 1942, 18, being 31.6 per cent, and in 1948, 28, being 45.1 per cent. On the Western Area eight nests, or 19 per cent, were completely destroyed. Thus 34 per cent of 161 active nests met destruction, leaving a nest success of 66 per cent. It is of interest that the nest success of these collective raptor populations was similar to that (68 per cent) observed for the Marsh Hawk by Hammond and Henry (1949).

In addition to nests destroyed by one agent or another (Table 49), 7.5 per cent of the paired birds in the population did not produce young because they failed to lay eggs. Some of these made no attempt to nest. If we include this loss in potential productivity with the actual nest loss, we find that the success of all possible nesters was 61.5 per cent.

DETERMINATION OF AVERAGE SUCCESS OF RAPTORS IN FLEDGLING YOUNG

A sample of fledgling success for five of the major raptor species was secured by combining the records from Superior Township with those obtained on the Michigan Check Area. The average success is computed from successful and unsuccessful nests and non-nesting pairs and therefore the figures are not biased by exclusion of early nest destruction and desertion and failure of pairs to nest.

The data of Table 48 show that in the Ann Arbor region the Horned Owl, Marsh Hawk and Red-tailed Hawk produce approximately one fledgling per pair of adults; the Red-shouldered and Cooper's Hawks, two fledglings. The high Marsh Hawk mortality of 100 per cent in 1948 (Table 46) is not typical and when included in a sample of 19 nests (Table 48) lowers the average production.

Hammond and Henry (1949) found the average Marsh Hawk fledgling survival from 15 successful broods to be 3.33. They did not determine the average number of fledglings produced per pair of adults for the 60 nests under observation. From their data, it appears that approximately one young was fledged per pair of adults. The relatively few observations of older broods probably biased the figure.

Stewart (1949), in studying 52 Red-shouldered Hawk nests, found an average of 2.7 young at a period when growth was half completed but does not provide data to permit the determination of fledgling success.

The average number of fledglings produced per pair of adults is in itself a far more usable and significant figure than the number of young that hatch or survive per successful nest. Average fledgling survival allows comparison of raptor productivity with the productivity of their prey species and this relation of numbers is fundamental to a proper evaluation and understanding of the mechanics and function of predation. Such data should be obtained, so that in time an accurate average fledgling success for each raptor species can be determined. The data recorded in Table 48 are a start in this direction.

KINDS OF ENVIRONMENTAL RESISTANCE

Mortality, which may occur at any time during the nesting period, varies with the species and fluctuates from year to year. The causes of mortality are sometimes very complex and frequently the so-called "causative factor" is only the final reaction of a chain of circumstances or related activities. Table 49 lists the observed causes of mortality and the number and per cent of nests destroyed.

Man without question is the greatest destroyer, being directly or indirectly responsible for nearly half the destruction of raptor nests. Not all the mortality due to man is intentional, although shooting of nesting birds accounts for a large proportion of it. Plowing, lumbering and other activities frequently destroy nests, cause desertion, or are instrumental in starting a series of events that result in nest destruction.

In Michigan, raccoons (Plate 15) and oppossums, which ranked next to man as causes of mortality, confined their destruction largely to nestlings. Most of the destruction from these mammals occurred after the observers had made numerous visits.

Infertility is difficult to detect, as other things may cause destruction before this has been definitely determined. The records show, however, that infertility accounts for a significant part of failure to attain the reproductive potential; possibly a greater part than that indicated in Table 49.

Destruction of the nest of a Great Horned Owl by Crows in 1942 is worthy of being reported in detail. After weighing a freshly hatched owlet and a pipping egg on a cold day in mid-March, one of the authors descended the nesting-tree and concealed himself several hundred yards away to await the female owl's return. Within 25 minutes she flew back to the nest with a dozen Crows chasing her. Satisfied that the young owlets would not freeze to death, the observer arose to leave. The owl spotted the movement and immediately departed, with more than half the Crows in pursuit. The remaining Crows did not detect him, so he continued to watch the nest. Four Crows flew at once to the edge of the nest and then to a tree near by. Nervous and wary, they jumped from limb to limb, looking at the nest and cawing loudly. One Crow, bolder than the others, flew three times to the edge of the nest, only to leave immediately on landing. The three other Crows cawed loudly, as though giving encouragement, but made no move to fly to the nest. It appeared that the Crows had a definite project in mind and were summoning courage to carry it out. Finally the bold one flew to the nest, bent down and picked at one of the owlets, flew off, returned again, and quickly and deftly threw an owlet out of the nest with a flip of his bill. As the Crow returned to repeat the performance, the observer interfered. The owlet dropped 40 feet, but was unharmed. There was a bruise on the tip of the wing where the Crow had seized it. Afraid that the performance would be repeated if he replaced the owlet, the observer carried it to the laboratory and fed it egg yolk and sparrow liver before returning it to the nest late in the evening of the same day.

Several days later he returned and took measurements of both owlets, but on his departure the Crows again gave chase to the adult owl. He suspected that the Crows might repeat the earlier performance, so he returned to the nest the next morning and found both owlets dead at the base of the three, each with a bruise on the wing tip as evidence of how it came to its fate. The Crows made no attempt to eat the owlets.

This incident probably would not have occured had man not been involved. The eggs of another pair of Horned Owls eventually were destroyed by Crows when sugar tapping kept the incubating owls away from the nest for long periods.

The Horned Owl is itself no respecter of weaker predatory birds. Two nests of the Red-shouldered Hawk and one of the Red-tailed Hawk were destroyed by Horned Owls. In one instance fledgling Red-shouldered Hawks were killed and partially eaten on the nest. Several Horned Owl feathers were the only clue. Fox squirrels destroyed the eggs of two pairs of Screech Owls and appropriated the nest hollows.

Flooding accounted for the desertion of one Marsh Hawk nest, and a farm dog, following the observer's trail, was suspected of causing early desertion of another. Mink took the eggs at one nest and plowing took a toll of four other nests. The latter were recorded as destroyed by man (Table 49).

Only one nest, in which two nestling Swainson's Hawks died, failed to produce young as a result of disease. The course of the malady was apparently rapid, as the birds were not emaciated. In each the gall bladder was greatly enlarged and the stomach was foul and discolored. The cause of the disease was not determined.

Unknown causes, including desertion, accounted for the failure of twelve nests. Mortality also resulted from eggs or young falling out of

TABLE 49

CAUSES OF COMPLETE NEST FAILURE
SUPERIOR TOWNSHIP, MICHIGAN—MOOSE, WYOMING

Causes of Failure	Number of Nests that Failed[*]			Total Number	Losses Percent of Total Nests
	1942	1948	1947		
Man	7	12	3	22	13.7
Raccoon and Oppossum	3	4	..	7	4.3
Infertility	1	2	..	3	1.9
Crow	1	..	..	1	0.6
Great Horned Owl	1	1	1	3	1.9
Fox Squirrel	1	1	..	2	1.2
Flooding	..	1	..	1	0.6
Dog	..	1	..	1	0.6
Mink	..	1	..	1	0.6
Disease	..	..	1	1	0.6
Unknown Causes Including Desertion	4	5	3	12	7.5
Total Nests that Failed	18	28	8	54	33.5
Total Active Nests	57	62	42	161	..
Loss due to Failure to Lay Eggs, Including cases where there was no nesting attempt,	6	4	3	13	7.5
Total nests and paired birds ..	63	66	45	174	..
Total nest failure					38.5

[*] 1942 and 1948 in Michigan, 1947 in Wyoming.

the nest, from exposure of nestlings to cold rain, from competition among a brood resulting in the death of a runt or younger member, from long exposure to intense heat and sunlight, and from exposure due to lack of brooding under adverse conditions.

ADULT MORTALITY

In 1942 there was a mortality of four adult nesting raptors—one Horned Owl, one Red-shouldered Hawk, and two Screech Owls. The first two were shot and the Screech Owls were killed by a Horned Owl and a Cooper's Hawk. In addition, one immature Red-tailed Hawk was found dead on a road. The mortality recorded thus represented 3.8 per cent of the raptor population.

In 1948 adult mortality was again four—two Marsh Hawks, one Sparrow Hawk, and one Horned Owl. All were shot. This was an adult mortality of 2.7 per cent.

Adult mortality on the Western Area in 1947 was three raptors— one Red-tailed Hawk, one Swainson's Hawk, and one Short-eared Owl. The hawks were shot, the owl was killed and eaten by a Red-tailed Hawk. Mortality for this nesting population was 3.3 per cent

The average adult mortality for all three nesting populations during three years was three per cent.

The winter mortality recorded in Superior Township in 1942 was 10 birds, or approximately six per cent of the winter population (Table 16). In 1948 it was three birds or approximately five per cent.

It was difficult to obtain accurate mortality figures during the fall migration and during the small-game hunting season (Plate 67), but undoubtedly a somewhat higher mortality occurs then. We estimate that the annual mortality was about 12 per cent.

SEX RATIO OF NESTLINGS

The sex of the young raptors was determined by taking weights and measurements at various intervals during the period of growth (Plates 60 & 61). Body weight, feather growth, foot dimensions and diameter of tarsus were found to be the most useful criteria in determining sex. In all raptors studied, the female is heavier and larger than the male and this difference can frequently be measured at 5 to 6 days of age.

Of the 225 young birds of various species whose sex was determined, 121 were females and 104 were males. During the breeding season the mortality of adult females exceeded that of males. This greater mortality of adult females may have some bearing on the larger number of females produced. The smaller size of male raptors is, however, a decided disadvantage during the nestling period, when they must compete directly with the larger females (Plates 10 and 59). A higher mortality of male nestlings may offset the heavier mortality of adult females that appears to occur later.

JUVENILE MORTALITY

Mortality of young after departure from the nest is difficult to determine quantitatively. Field observations indicate that mortality is fairly high during this transition period from complete dependence on parents to independence. The first month after leaving the nest appears to be the most critical period. Hickey (1949) figures that there is a 25 per cent mortality of fledgling Marsh Hawks between the time the young leave the nest and August 1st. Young birds of prey usually remain in the nesting vicinity and hunt in the surrounding area, and frequently return to the nest at night. At this stage they are easily shot and sometimes entire broods are destroyed.

Starvation kills an appreciable number and malnutrition is probably indirectly responsible for a still greater mortality. Young raptors are dependent at first on their parents for food and later undergo a training period during which their instincts to hunt and kill are patterned after and conditioned specifically by those of their parents. Apparently in some adults the instinct to care for the young declines at this stage, and the young are more or less left to shift for themselves, thus decreasing their chance of survival.

Along highways, a toll of young, inexperienced birds is taken by automobiles. The Sparrow Hawk, Screech Owl, and Horned Owl seem particularly vulnerable.

Banding (Plate 33) generally yields few data on birds that fail, as a result of starvation or other normal causes, to survive the first month, but provides a proportionately high return with respect to those that are shot or trapped after they have severed parental contact. For this reason early post-fledging loss computed from banding returns will be lower than actual mortality.

In 1942, 58 (46 per cent) of the 126 hawks and owls produced in Superior Township were banded. In the course of the following six months, a period of learning and of adjustment to the environment, which may be distinguished as the postnesting season, there were four returns, all from birds shot or trapped, indicating a mortality of 7 per cent from these causes. Applying this proportion (Lincoln, 1930) to the seasonal crop of 126 raptors, a mortality figure of nine birds for the postnesting period is obtained. This figure represents that portion of the postnesting mortality in which man's activities were the direct cause of death.

Forty-five per cent of all deaths known through banding returns occurred in the first six months following the end of the nesting season.

Hickey (1949) states that, "Of 102 birds (Marsh Hawks) shot in their first year of life (August 1st to July 31st) and reported with definite recovery dates, 80 per cent were apparently shot in the first five months." Thus there is little doubt but that postnesting season mortality is relatively high, even though the full extent of it cannot be expressed here in quantitative terms.

The average fledgling success of 62 per cent, or two young per pair of adults, in the raptor populations studied, justifies the conclusion that raptor productivity to the flying stage is generally high. Precise data on mortality of birds of the year are not available, but observation and banding indicate that probably it is relatively high during the summer and fall months, perhaps reaching a peak during migration. If an annual adult mortality of 12 per cent be assumed, then, since the nesting populations tended to maintain a stable density level year after year and showed an average productivity of 1 young for every adult, there appears to be an average annual sub-adult mortality of approximately 88 per cent. These figures are presented to show approximate proportions rather than exact percentages, since they are subject to wide error from lack of quantitative data.

Barring concerted persecution by man, predatory bird populations appear capable of maintaining fairly constant nesting populations in suitable habitats year after year.

SEASONAL CHANGES IN POPULATION DENSITY

A raptor density of 4.4 and 1.6 per square mile was found in Superior Township in the winters of 1942 and 1948, respectively. During the nesting periods of these years there was a density of 3.6 and 4.0 adult birds per square mile. The population densities of raptors were stabilized during each winter period, but at levels related to current prey densities.

The nesting densities likewise were stabilized, but at the spring and summer carrying capacity of the land for raptors.

It is interesting that under such favorable conditions as existed in 1942, when meadow mice were abundant, the winter raptor density exceeded the highest nesting density (Table 16).

On the Wyoming area, the raptor density in the winter of 1947 was very low, probably less than one bird per square mile. The nesting density on the same area was 7.6 birds per square mile in 1947 and again in 1948. The low raptor pressure on prey species during the severe winter months may largely account for the heavy nesting density of raptors.

At the close of the nesting seasons of 1942 and 1948 in Superior Township, the crop of young hawks and owls increased the population density to 6.7 and 6.6 raptors per square mile, respectively. On the Western Area the population density at the close of the nesting season averaged 17.4 birds per square mile. These densities are an indication of the hunting pressure exerted by raptors over a large land area.

POPULATION ESTIMATES

Error may arise in projecting population data from small areas to larger ones. Nevertheless, such estimates are valuable and seem to be the only practical device to gain a grasp of population dynamics on a regional or continental scale.

MAP 16. Hardwood forests of southern Michigan

Raptor Population Estimates for Washtenaw County and
for the Hardwood Forest Region of Southern Michigan

The combination of small woodlots and open farm land characteristic of Superior Township and the Ann Arbor area likewise is characteristic of Washtenaw County. Woodlots compose 11 per cent of Superior Township and approximately 13 per cent of the Check Area. Washtenaw County is 13 per cent woodlot (U. S. Census of Agr., 1945).

Because the 36-square-mile check area compared closely with Superior Township in raptor population density and because extensive observations of nesting raptors throughout Southern Michigan further supported the intensive population studies, we feel justified in projecting the population density of Superior Township to the county. Since Washtenaw County contains 716 square miles, projection of intensive local data indicates that the county has an average nesting population of 1,253 pairs of hawks and owls and an average postnesting population of 4,500 bird predators.

The hardwood forest region of southern Michigan (Map 16) includes 29 counties, containing 18,064 square miles. Woodland composes an average of 17 per cent of this area (computed from data of U. S. Census of Agr., 1945). Although conditions for nesting raptors in these 29 counties may vary much, a population estimate for this hardwood forest area should prove reasonably accurate with respect to the total, if not for species composition. General observations of raptors throughout southern Michigan indicated that such an estimate is justified. We estimate for the 29 counties an average nesting population of 31,600 pairs and an average postnesting population of 113,800 birds of prey.

SUMMARY AND CONCLUSIONS

Raptor nesting population densities on the areas studied vary from 1.8 pairs of hawks and owls per square mile in certain cultivated areas to 3.8 pairs per square mile in semi-wilderness areas. These densities are much increased by reproduction, so that by mid-summer they have doubled and raptor pressure on prey forms has increased correspondingly. The composition and population density of these collective raptor populations exhibit remarkable stability from year to year, and therefore the survival of young must approximate closely the death rate of adults. The productivity of a collective raptor population to the fledgling stage is generally high, averaging two young per pair of adults. Annual losses appear to be approximately 12 per cent of the adult population and 88 per cent of the juvenile population. In other words, hawks and owls were subject to decimating forces sufficient to keep the nesting population levels constant. Raptors probably will not normally increase to the point where man need institute widespread control.

This was true in areas of intensive land use as well as in semi-wilderness areas. The investigations here reported on provide little evidence to support those who, advocating concerted control of hawks and owls, use the argument that man has so altered his environment that natural checks no longer operate on raptors and that artificial control is necessary to curb their numbers.

It may be concluded that a collective nesting raptor population tends to keep in biotic harmony with its environment.

Let us, moreover, keep in mind for future reference that the data presented here indicate a stability of raptor-prey relations year after year that strongly supports the vaguely defined "balance of nature theory." It suggests that predation operates as a stabilizing or regulating force—a balance of nature process—both under wilderness conditions, where the members of the fauna have relatively few drastic ecological adjustments to make, and on intensively utilized land, where the fauna is altered and its members are adjusted to rapid man-made changes. We must think of man as a member of the fauna and realize that in spite of the drastic changes he makes in the environment or the predatory pressure he himself normally exerts, the wildlife complex can respond to these changes with adjustments that eventually will tend to recreate a biotic equilibrium in which predation, with other decimating forces, and the counter force of reproduction operate to determine the population levels of prey animals. Gabrielson (1941) stated, "Local readjustments in accordance with varying conditions are often desirable to maintain a balance, though this can never be perfect; the observed average stability of animal populations emphasizes this point." We may state further that the measured stability of collective nesting raptor populations indicates a stability of total prey populations throughout extensive land areas and that the fundamental stability with change that this complex of raptor-prey relationships illustrates is most certainly the result of and involves a balance of forces, a harmonious synchronization of predator-prey activities, a strong tendency of natural biotic forces to maintain a condition of equalized pressure. (See also page 184.)

We may liken this condition of faunal adjustment to that of a climax plant community—the ultimate adjustment of plants to soil and climatic conditions. We may consider it in wilderness areas as a climax condition of collective populations which the fauna has attained in the course of time. In agricultural areas it is more nearly a sub-climax condition that will continue as long as existing land-use practices are not greatly altered. Perhaps the balance of nature becomes more of an understandable concept if we visualize it as we do plant succession—a dynamic process of development toward a more or less stabilized condition that can be maintained over a long period—a condition that involves both flora and fauna. It seems plausible that when we can express quantitatively the condition of collective animal populations and the mutual relations and effects of their members as accurately as we can now define plant communities or the various stages of plant succession, we shall be able to designate stages approaching or being deflected from the climax condition or "balance of nature" in various geographic areas, whether they be of a wilderness nature or ranch or farm land.

CHAPTER 11

Nesting Season Ranges of Hawks and Owls

F OR AN understanding and evaluation of spring and summer predation and its effects it is necessary to know the extent of the area
hunted by a pair of raptors, how long and how intensively they hunt it,
what its geographic relation to the ranges of the other raptors in a
collective population is, and just what effect the spatial limitations
inherent in ranges have upon the application of raptor pressure on a
large land area.

The influx of hawks in spring migration was followed by a gradual
break-up of winter ranges, a movement of some winter residents out of
the Township, a spatial adjustment of other hawks to the changing
population, and then the gradual stabilizing of a nesting pattern. This
pattern, represented by the distribution of nests, territories, and home
ranges, existed as a mould or form from the year before and that year
was a heritage from a previous one. It was not a new annual arrangement, yet it had a degree of flexibility (Chapter 10).

Each season's nesting pattern developed as the raptors defended
their nesting territories of previous years or established new ones when
former nest sites were found occupied. Horned Owls established home
ranges by taking their pick of the previous season's hawk and Crow
nests. Some Red-tailed Hawks also nested within a range in which they

had wintered and had previously nested. Female Sparrow Hawks joined males that had spent the winter on former nesting ranges.

Such activity as soaring, courtship, and territorial defense was typical of the nesting population. Its intensity in the case of each species varied in relation to the chronology of nesting events. In the early stages, warm spring-like days actuated activities related to reproduction and cold wintry days caused activity to revert to that typical of the wintering population. As old territories were reestablished and new ones established and recognized, there developed a pattern of movement which resulted in nesting ranges.

DEFINITION OF NESTING RANGES

A nesting season range is an area over which a pair of hawks moves in performing the activities associated with the nesting cycle. The living requirements of a nesting pair are met within this area. It cannot rightly be called a territory as defined by Noble (1939) and supported by Nice (1943)—that is, "any defended area." The extremities of many hawk ranges are not defended and frequently are mutual hunting grounds that are used by several individuals of the same species at the same time. The nesting territory is generally a much smaller area surrounding a nest and is actively defended. At times, however, the nesting territory and the nesting range are identical. This occurs when populations of a species are high and ranges correspondingly small. The Red-shouldered Hawk ranges (Maps 17 and 18) are in many cases identical with nesting territories. The term *home range* (Burt, 1940) is not fully applicable to migratory individuals or species, since they nest and winter in different areas, often widely separated, but it does apply to year-round residents, such as the Horned Owl, Screech Owl, and Red-tailed Hawk.

METHOD OF DETERMINING RANGES

Hawk ranges were determined by plotting sight records of paired and single hawks. A few owl ranges were plotted both from sight records and field sign, such as roosting spots, pellet accumulations, and calls. The location of all predator nests on the study areas (Maps 13, 14, and 15) showed the spatial relationship of one raptor to another and thus greatly facilitated range plottings. The ranges are bounded by lines drawn through perimeter observations centering around and associated with a particular nest. Thus the nesting ranges (Maps 17 and 18) are observed ranges. The observations determining a range (Tables 50 and 51) are only those made at a distance from the nest. Repeated observations of raptors in the vicinity of a nest were not counted, as they could easily be accumulated to give an erroneous impression of the number of sight records that delimited a range. Most observations represent the sighting of a known bird at a definite spot. When, however, the beginning and the end of a flight increased the range boundaries, each was plotted and was counted as a distinct observation.

CHARACTERISTICS OF RANGES

Size of Ranges

Numerous hawk and owl ranges were plotted, and the areas and diameters of these ranges are presented in Tables 50 and 51 and on Maps 17 and 18. Because observations were insufficient in some cases, the ranges of the birds concerned were not determined. A great number of observations are not necessary, however, to delimit a hawk range rather accurately. The range of a pair of Sparrow Hawks was determined by 122 observations (Table 52), yet the area of this range was comparable with those of ranges 25, 26, and 27 (Table 52), each of which was considered to be accurately mapped with fewer observations.

Though many factors influence the size of individual ranges and though there is much individual variation, nevertheless the computation of average areas and diameters yields useful figures. The following averages have been calculated from Tables 50, 51, and 52.

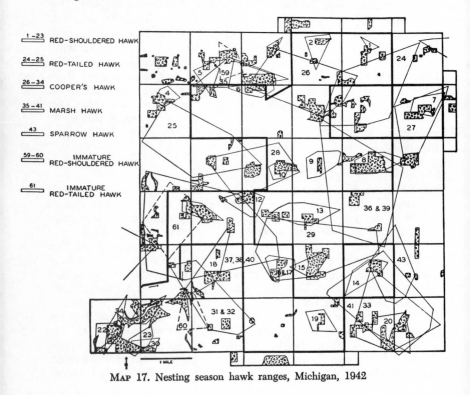

MAP 17. Nesting season hawk ranges, Michigan, 1942

The Prey

PLATE XL. Tracks in the snow were useful indications of the relative abundance of fox squirrels. A count of the number of squirrels seen in a given time on a measured area was made in this woodlot.

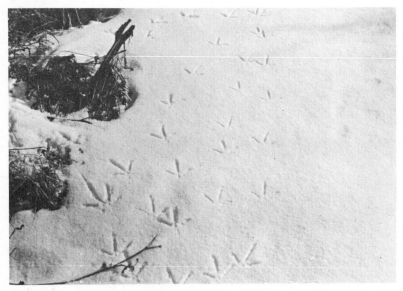

PLATE XLI. Daily movements of pheasants were often indicated by tracks in the snow. Tracks were especially helpful in finding isolated cock pheasants.

PLATE XLII. Above—Osseous remains of six meadow mice in a Short-eared Owl pellet. Below—Little more than a few bones and teeth, including incisors, remain to indicate the number of individual prey animals represented in Marsh Hawk pellets.

PLATE XLIII. Winter pellets of hawks and owls. Left to right, upper row—Red-tailed Hawk, Marsh Hawk, Goshawk, Cooper's Hawk, Sharp-shinned Hawk, Prairie Falcon, and Sparrow Hawk. Left to right, lower row—Horned Owl, Great Gray Owl, Long-eared Owl, Short-eared Owl, Barn Owl, and Screech Owl.

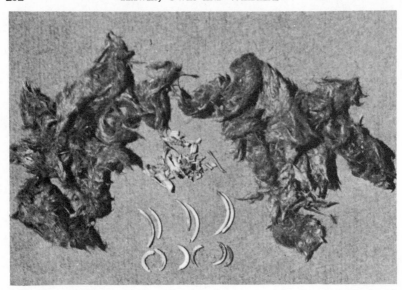

PLATE XLIV. The number of individual prey animals present in Marsh Hawk and Buteo pellets was determined by sorting and pairing incisors. Three meadow mice are represented here.

PLATE XLV. Horned Owl pellet. Dissected bone fragments show that the owl consumed four meadow mice and one rat to form this pellet.

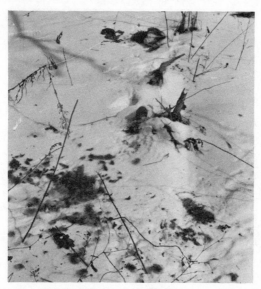

PLATE XLVI. Rabbit killed by mature Red-tailed Hawk. This kill was made during the spring transition period, when mortality of large prey species increases.

PLATE XLVII. Female pheasant killed by a Goshawk. The food of the Cooper's Hawk was largely small birds and was determined through discovery of such kills. More feathers are scattered about kills made by hawks than about those made by owls. Hawks pluck their prey before eating it; owls usually do less plucking or none.

PLATE XLVIII. Young snowshoe rabbit. Immaturity increases the vulnerability of prey forms. Being at low population levels at the time of this study, snowshoe rabbits on the Western Area did not form a significant part of the raptors' diet.

PLATE XLIX. Uinta ground squirrel This species was preyed on heavily by western raptors, but was invulnerable to them for about nine months while it was in estivation and hibernation.

PLATE L. Adult Ruffed Grouse. On the Wyoming study area these birds maintained a rather constant population level over a period of years.

PLATE LI. Richardson's Grouse chick. The young of all grouse are vulnerable to raptors.

In southern Michigan the average nesting range of the Red-shouldered Hawk was found to be about a quarter of a square mile, with a maximum diameter of slightly less than one mile; that of the Red-tailed Hawk an area of one and one-half square miles, with a maximum diameter of two miles; that of the Cooper's Hawk three-quarters of a square mile, with a maximum diameter of nearly one and one-half miles; that of the Marsh Hawk one square mile, with a maximum diameter of one and one-half miles; that of the Sparrow Hawk one-half a square mile, with a maximum diameter of one mile; that of an immature Red-shouldered Hawk an area of one-quarter of a square mile, with a maximum diameter of nearly a mile; that of an immature Red-tailed Hawk an area of nearly a square mile, with a maximum diameter of one and one-half miles. In Jackson Hole, Wyoming, the ranges of the Red-tailed Hawk averaged three-quarters of a square mile in area and one and one-half miles in maximum diameter; those of the Swainson's Hawk one square mile in area and one and one-half miles in greatest diameter; those of the Sparrow Hawk three-quarters of a square mile in area and one and one-half miles in greatest diameter; those of the Horned Owl one square mile in area and one and one-half miles in greatest diameter; those of the Long-eared Owl one-quarter of a square mile in area and three-quarters of a mile in greatest diameter; those of the Raven three and one-half square miles in area and three miles in greatest diameter. These nesting ranges may be compared with the hunting ranges determined in winter (Chapter 2).

FACTORS INFLUENCING SIZE OF RANGES

The specific characteristics of raptors play a part in determining range sizes; for example, the ranges of Prairie Falcons are larger than those of Sparrow Hawks and those of Golden Eagles generally larger than those of Prairie Falcons. In general, the larger birds of a genus have more extensive ranges. Topography, distribution, and type of vegetation, as well as distribution and density of prey, affect range sizes, shape and pattern. The sizes and shapes of the Horned Owl ranges 34 and 36 (Map 19) were influenced by the abruptly rising terrain which separated them. The Cooper's Hawk ranges, 19 (Map 19), 30, and 34 (Map 17), are small because these birds nested in or near dense mixed stands of hardwood and conifers that furnished ample prey close at hand. The number of species nesting in any area also plays a part in delimiting range sizes. The small sizes of the Red-shouldered Hawk ranges are due partially to the dense nesting population, and it would appear that the smaller sizes of the western Red-tailed Hawk ranges as compared with the ranges of the eastern birds are also, in part, a result of a denser nesting population.

TABLE 50

NESTING SEASON HAWK RANGES, 1942, SUPERIOR TOWNSHIP, MICHIGAN

Range No.	Observed area in sq. miles	Max. diameter of range in miles	Inclusive dates of observations	No. of days	No. of observations	Remarks
RED-SHOULDERED HAWK						
1	.03	.4	4/8 −6/15	69	3	
2	.17	.7	3/19−6/17	91	9	
3	.03	.4	4/2 −6/15	75	6	
4	.11	.5	2/27−6/17	111	7	
5	.15	.6	4/8 −5/25	48	10	Still in territory on Aug. 3
° 6	.42	1.0	3/19−6/4	78	14	
7	.19	.7	3/19−6/17	91	16	
8	.28	1.1	2/21−5/30	99	12	Still in territory on Sept. 3.
° 9	.25	.8	3/2 −9/3	186	18	Still in territory on Sept. 3.
°10	.10	.5	4/1 −6/14	75	16	
11	.15	.8	3/19−6/29	103	11	
12	.24	.8	2/14−6/22	129	14	
°13	.37	1.3	2/28−5/19	81	23	
14	.25	1.0	3/2 −5/23	83	13	
15	.17	.7	4/7 −6/19	74	8	
16 & 17	.18	.6	3/24−7/1	100	15	Renesting pair; range includes both nests.
18	.37	1.5	4/16−5/28	43	11	
19	.16	.7	2/28−5/13	75	6	
20	.35	1.0	2/28−7/20	143	11	
21	.16	.9	3/20−6/6	79	12	
22	.20	.7	3/20−7/17	119	11	
23	.15	.7	3/24−7/16	115	12	
RED-TAILED HAWK						
°24	1.65	2.5	3/2 −8/20	172	16	Same nest used in 1948
25	1.79	2.4	4/1 −8/20	142	12	Pair laid no eggs
COOPER'S HAWK						
°26	1.55	2.2	3/2 −6/4	95	16	
27	.76	1.7	3/23−6/22	92	7	
28	1.11	1.8	3/31−5/18	49	12	
29	2.05	2.5	3/24−6/19	88	15	Built 2 nests, no eggs
30	.12	.7	3/22−5/1	41	11	
31 & 32	1.03	2.0	3/31−6/29	91	10	Range of 2 pairs— no eggs
°33	1.43	1.7	3/23−6/14	84	12	
34	.16	1.3	4/11−5/17	37	8	

Note: Asterisk indicates ranges most accurately determined.

TABLE 50 (continued)

NESTING SEASON HAWK RANGES, 1942, SUPERIOR TOWNSHIP, MICHIGAN

Range No.	Observed area in sq. miles	Max. diameter of range in miles	Inclusive dates of observations	No. of days	No. of observations	Remarks
MARSH HAWK						
35	.55	1.8	3/21–5/8	49	10	
36 & 39	2.46	2.7	3/19–7/22	126	37	2 pairs
37, 38, & 40	3.89	3.0	4/10–7/7	89	54	Range of 3 pairs
41	2.14	2.1	4/11–6/30	81	22	Still in range on Aug. 2.
SPARROW HAWK						
*43	.79	1.5	3/24–7/22	121	26	♂ Wintered within nesting range. Still in range on Aug. 3.
IMMATURE RED-SHOULDERED HAWK						
59	.21	.7	4/24–5/23	30	6	Single bird
60	.30	.8	4/13–6/24	73	6	Single bird
IMMATURE RED-TAILED HAWK						
*61	.89	1.4	4/14–5/29	46	13	Single bird

Red-tail No. 24 wintered and nested in area in 1942—Same nest used by wintering and nesting pair in 1948. See Map 17.

Range Pattern

From a study of the nesting season ranges (Maps 17-22), we can draw conclusions as to the nature and causes of the range patterns exhibited.

In Superior Township, the ranges of hawks of the same species did not generally overlap. This is true of all species but is less pronounced in the case of the Marsh Hawks. Where Marsh Hawks nested reasonably close together, range peripheries overlapped to some extent, forming mutual hunting grounds. Ranges of a number of pairs were plotted together because of the difficulty of definitely separating the pairs. Hawks of the same genus (the Red-tailed and Red-shouldered Hawks) did not have overlapping ranges, nor did the ranges of immature birds occupy areas utilized by mature nesting hawks of the same species. Where occasional overlaps are shown, they are the result of encroachment, often repulsed by active defense. This is applicable particularly to the Red-shouldered Hawks, whose small ranges were usually identical with the nesting territories. Overlap also occurred and was tolerated when territorial defense subsided as a result of nest destruction, nest desertion, inability of a pair to nest, or the termination of the nesting period.

In contrast to the pattern of isolated ranges of birds of the same species or genus, the ranges of hawks of different genera overlapped. Nest sites were defended against hawks of different species, but ranges usually were not. The variation in hunting methods, differences in hab-

TABLE 51

NESTING SEASON HAWK RANGES, 1948, SUPERIOR TOWNSHIP

Range No.	Observed area in sq. miles	Max. diameter of range in miles	Inclusive dates of observations	No. of days	No. of observations	Remarks
RED-SHOULDERED HAWK						
1	.10	.8	4/3 –5/24	51	7	Birds shot at, nest destroyed
2	.32	1.0	4/3 –5/24	51	9	
3	.26	.8	3/25–6/11	79	10	
°4 & 5	.56	1.4	3/25–6/14	82	15	Renest—Range includes both nests
6	.51	1.2	3/31–6/6	68	8	
7	.60	1.5	3/28–6/17	82	21	
8	.47	1.6	4/27–5/20	24	9	
9	.17	.7	4/27–5/14	18	4	
10	.20	.9	3/28–5/14	48	7	Non-nesting pair
11	.13	.7	4/4 –6/6	64	7	
12	.29	1.2	4/17–6/12	57	10	
13	.12	.5	3/28–6/10	75	9	
°14	.17	.6	3/20–6/8	81	11	
15	.25	.8	3/20–5/22	34	11	
16	.21	.6	5/3 –6/20	49	7	
17	.22	1.1	3/10–6/12	95	7	
18	.29	1.0	4/7 –6/3	58	6	
19	.17	.8	3/18–6/3	78	8	
RED-TAILED HAWK						
20						Deserted. No data
21	.50	1.3	3/31–4/21	22	5	Deserted by May 9.
°22	1.35	2.2	2/18–7/1	135	17	
23	1.28	2.2	2/30–6/12	114	11	
°24	2.15	2.1	3/28–6/9	74	29	
COOPER'S HAWK						
25	.11	.6	4/24–6/19	57	6	
°26	1.45	2.3	4/4 –6/17	75	12	
27	.98	1.6	3/28–5/27	61	8	
28	.17	1.0	5/27–6/17	22	6	
°29	.37	.9	4/27–6/15	50	7	
30	.93	1.3	4/7 –6/9	64	7	
31	.29	1.3	4/5 –5/24	50	6	
32	.07	.5	3/18–6/15	90	6	
MARSH HAWK						
33	.43	1.2	3/25–5/16	53	16	
°34	1.15	1.7	3/30–6/17	80	25	
35	.38	1.1	4/24–6/12	50	15	
36	1.44	1.9	5/20–6/12	24	8	
37	.58	1.6	4/30–5/19	20	9	
38	.87	1.4	4/10–5/27	48	16	

Note: Asterisk indicates ranges most accurately determined.

TABLE 51 (continued)

Nesting Season Hawk Ranges, 1948, Superior Township

Range No.	Observed area in sq. miles	Max. diameter of range in miles	Inclusive dates of observations	No. of days	No. of observations	Remarks
39	.46	1.0	3/28–5/15	49	6	} 10 Renester
40			5/26–6/19	25	4	
41	1.85	2.4	4/17–6/19	63	19	
*42	.86	1.5	3/25–6/9	77	34	
SPARROW HAWK						
43	.09	.5	3/25–5/1	37	7	Female shot
*44	.83	1.4	4/11–6/15	66	27	Male wintered within nesting range in 1949
45	.73	1.3	4/11–5/9	29	13	Male wintered in nesting range in 1948 and 1949
46	.08	.6	4/7 –6/17	72	7	
IMMATURE RED-SHOULDERED HAWK						
68	.26	.9	4/20–5/14	25	8	2 birds
IMMATURE RED-TAILED HAWK						
69	1.71	2.4	4/17–6/12	57	19	2 birds
70	.66	1.5	4/21–6/12	53	20	2 birds
71	.27	.9	5/14–6/9	27	5	1 bird
AMERICAN ROUGH-LEGGED HAWK						
72	1.01	1.8	3/31–4/25	26	15	2 birds
SINGLE SPARROW HAWK						
73	.44	1.2	5/26–5/31	6	12	Non-nester

Note: See Map 18.

itat commonly hunted, the chronological separation of nesting activities between species, (Tables 37 and 38), and the variations in nest site requirements explain the overlapping ranges and dearth of inter-specific strife.

On a basis of the evidence gathered in southern Michigan, it would be reasonable to conclude that the territorial demands of hawks of the same species precluded overlap and that thus the resultant isolated nesting-range pattern expresses an intolerance arising from definite spatial requirements within a species. On the other hand, it is evident that the distribution of nesting sites or woodlots might automatically produce a pattern of isolated ranges. This is even more evident when it is realized that the predominant nesting species (the Red-shouldered Hawk) hunted largely in the woodlots. These contained kettle holes, favorite hunting grounds, which made it unnecessary for this hawk to range far in search of food. The other hawk species, fewer in numbers, had, because of the greater distances between their nests, less likelihood of overlapping. The Marsh Hawks, whose nest sites were not dependent upon isolated woodlots, did overlap ranges when nesting close together. This overlap occurred in mutual hunting areas and not in the vicinity of nest sites. It would appear, then, that the isolation of ranges might result largely from the arrangement of vegetation and the distribution of nesting and hunting areas or from a combination of both and not from any inherent specific spatial requirement or intra-specific intolerance. In order to throw more light on this subject, we shall examine the data gathered in the West on an area characterized by a heterogeneous distribution of vegetation. Here we see (Map 20) that Buteo ranges tended to be distinct, yet overlapped to a much greater degree than the ranges plotted in Superior Township. Not only did ranges of Red-tailed Hawks overlap, but they overlapped those of Swainson's Hawks also. Overlap rarely occurred near nest sites, but rather at range peripheries and in

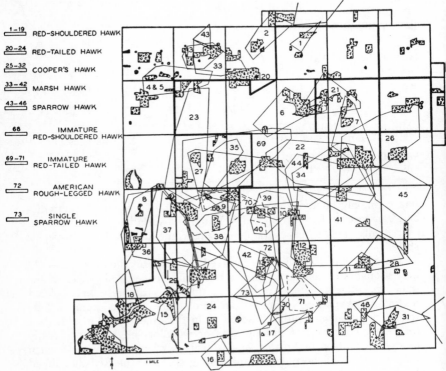

MAP 18. Nesting season hawk ranges, Michigan, 1948

TABLE 52

NESTING SEASON HAWK RANGES, 1947, MOOSE, WYOMING

Range No.	Observed area in sq. miles	Max. diameter of range in miles	No. of observations	Remarks
RED-TAILED HAWK				
1	.90	1.7	17	
* 2	1.01	1.8	22	
3	.71	1.2	16	
4	.50	1.1	7	
5	1.29	1.9	14	
6	.82	1.6	13	
9	.30	1.0	9	
10 & 11	1.49	1.8	25	2 pairs
12	.32	1.1	6	
SWAINSON'S HAWK				
*13	1.65	3.0	23	
14	1.56	2.1	13	
15	.52	1.0	21	
16	.28	.9	7	Deserted
17	.75	1.2	18	
COOPER'S HAWK				
18	.79	1.4	16	
SHARP-SHINNED HAWK				
19	.51	.9	8	
20	.26	1.5	6	
GOSHAWK				
21	.82	1.7	7	Ranged into study area
SPARROW HAWK				
22	1.13	2.0	20	
23	.53	1.2	11	
24	.16	.7	9	
*25	.64	1.2	18	
*26	.43	1.2	27	
*27	.72	1.2	41	
28	.23	.8	10	
29	1.11	2.0	19	
30	1.93	2.3	19	
31	1.06	1.6	26	
32	.65	1.3	20	
PRAIRIE FALCON				
*33	9.96	4.4	21	
GREAT HORNED OWL				
34	.62	1.8	10	
35	.45	1.4	13	
*36	1.10	1.6	10	
37	1.11	1.6	15	

Note: Asterisk indicates ranges most accurately determined.

TABLE 52 (continued)

NESTING SEASON HAWK RANGES, 1947, MOOSE, WYOMING

Range No.	Observed area in sq. miles	Max. diameter of range in miles	No. of observations	Remarks
LONG-EARED OWL				
38	.13	.7	5	
39	.41	1.0	4	
40	.10	.5	5	
RAVEN				
45	4.06	3.6	24	
*46	2.63	2.8	25	
*47	4.17	3.3	34	
IMMATURE SWAINSON'S HAWK				
48	.06	.4	6	Single bird
GREAT GRAY OWL				
49	1.00	1.4	39	3 non-nesting birds Range of pair observed off area from May 1 to July 16.
SPARROW HAWK (A)				
*	.42	2.1	122	Outside study area

favorite hunting areas. The ranges of Red-tails (Map 20) overlapped so consistently that a separate range for each pair could not be plotted accurately. It is possible that more overlapping of ranges in hunting territories occurred than is shown: overlapping ranges are more difficult to plot than are isolated ones, as it is not always possible to determine to which nest an observed bird belongs.

Sparrow Hawk ranges (Map 21) show a tendency toward isolation, but when nests were close together (25 and 26) there was much overlap. Overlap also arose at range peripheries, and in such hunting areas hawks were observed to hunt together in harmony.

Horned Owls, Long-eared Owls, Cooper's Hawks, and Sharp-shinned Hawks did not hunt within the ranges of other members of the same species. As in Superior Township, this could have resulted from a spatial arrangement of nests that made the distances between them too great to allow an appreciable movement of one pair within the range of another. It is possible that in the case of the Horned Owl certain spatial requirements limited the number nesting on the area, for neither food nor suitable nesting sites appeared to be lacking.

The Prairie Falcon (Plate 27) illustrates the fact that certain species have definite spatial requirements. One pair of these falcons nested on

Blacktail Butte in 1947 (Map 21—No. 33) and only one pair has been known to nest on this Butte during other years for which records were kept. In 1947 the falcon's nest was on a cliff at the north end of the butte. In 1948 and 1949 it was on a cliff at the south end, close to where a Raven nest was located in 1947. Though three possible sites were available, only one pair of falcons nested in this area; presumably they kept other pairs away. This spatial requirement of nesting Prairie Falcons has been observed at numerous other locations by the authors, but on some long continuous cliffs the falcons may nest closer together than usual. Nevertheless each nesting pair will maintain a hunting range distinct from those of other nesting pairs. This likewise is true of the Duck Hawk (Plate 25). It is evident that spatial requirements and a dearth of acceptable nest sites will tend to limit the nesting population of both Prairie Falcons and Duck Hawks in any area.

Raven ranges were usually, but not always, distinct (Map 22). Where overlap occurred near nest sites it was the result of territorial aggression

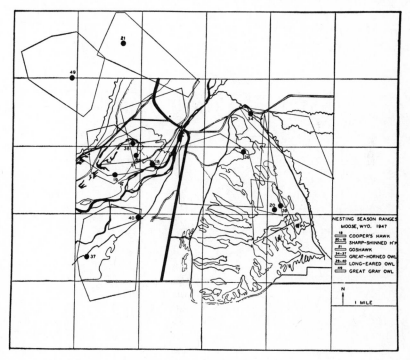

MAP 19. Nesting season ranges, Western Area, 1947

and was met with forceful and persistent defense. Nesting territory was defended against other Ravens and against hawks, but the entire range was not.

We are justified in concluding that the ranges of most raptors are distinct from those of neighboring pairs of the same species and that this separation is partially due to spatial requirements that vary with circumstances and with species. We may also conclude that overlapping at range peripheries of birds of the same species will normally occur and the degree to which this is exhibited is influenced by the distribution of nest sites and hunting areas and by the abundance of food. Where nest sites are distributed rather uniformly and where nesting territories and hunting areas are compact units, as are those of Red-shouldered Hawks in Michigan, ranges are usually small and isolated. Where nesting territories are distinct from hunting areas, as exemplified by the concentration of nests on Blacktail Butte and along the Snake River bottom with hunting areas on the surrounding sagebrush benches, the hunting areas are used commonly by many birds of the same species. Ranges overlap in these hunting areas, but not in the vicinity of nests.

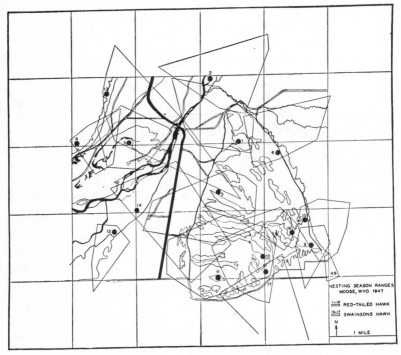

MAP 20. Nesting season ranges, Western Area, 1947

RANGES OF IMMATURE HAWKS

On both study areas, immature hawks and owls (young of the previous year) were a very small proportion of the spring population. As nesting patterns became fixed, immature hawks and owls were harassed constantly and buffeted from one territory to another. They not only were driven from territories by birds of their own species but also by those of other species and by Crows and Ravens. Eventually they settled down in an ecological niche not taken over by other nesting raptors or at least not by those of the same species. Once settled, they developed a range that often was limited by those of their neighbors. If this range contained hunting grounds as well as suitable nest sites, it was quite possible for the immatures to nest there in following years. Observations and banding indicated that this may have occurred. The ranges of immatures are likely to be somewhat smaller than those of mature birds. This is because they must be content with available niches, which are at times small, and also because the ranges often represent movement of one bird rather than two.

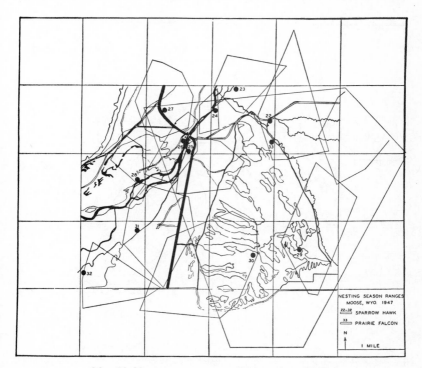

MAP 21. Nesting season ranges, Western Area, 1947

TERRITORIALITY

Contained within hawk nesting ranges are both hunting and nesting territories. Nesting territories are most actively defended early in the season, when territorial adjustments are being made. Defense may be divided into two types—that exhibited when man is not the disturbing factor and that shown when man contributes to, or is, the disturbance. Defense may vary from vocal protests to blows delivered. In general, the frequency and intensity of territorial defense declines as the season progresses. During incubation, hawks become more secretive and are less likely to reveal their nests by unncessary defense. It also appears that upon laying they definitely have established their right to the territory and consequently are slower to defend it, while other hawks are less likely to challenge. Defense against an actual intrusion on the nesting territory by man or other potential enemy usually is greatest during the first two weeks following hatching. Hawks will defend their territories against birds of another species much more actively if man is invading their territories at the same time. For example, at such times Prairie Falcons (Plate 27) will stoop on and attack Ravens with whom they

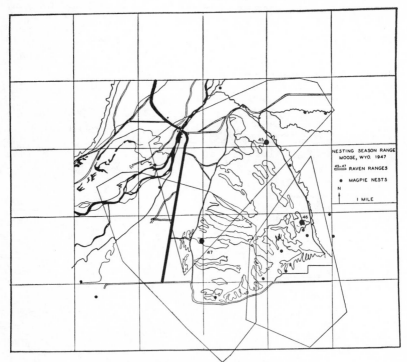

MAP 22. Nesting season ranges, Western Area, 1947

previously have had no quarrels. Under these conditions, Duck Hawks will strike neighboring Horned Owls when the latter are flushed from their nests. Defense is more intense against an outsider than against an encroaching neighbor. Typical of this was the case of a Golden Eagle that traversed Blacktail Butte and was attacked by every bird whose territory it passed over. The attacks were so severe that though the eagle several times attempted to alight, it was driven on only to have another hawk or Raven continue the pursuit. Similar examples could be cited for other species.

Defense expressions vary with the advance of the nesting season. Early in the season the entire nesting territory is defended; subsequently territorial defense wanes, but the nest itself is defended more actively, especially if young are present. After the young have left, an area where the young and adults are hunting may be defended but no defense made at the nest. Defense of nests and young against members of the same species is determined and forceful, but it does not occur nearly as frequently as might be imagined, since an adjustment (illustrated by the ranges) has already been attained as a result of past nesting patterns, which minimizes such conflicts through mutual recognition of territories. We suspect that most of the territorial conflict is caused by new members of the community that are adjusting themselves to the rather definitely established territorial pattern or by pairs that have had their former territory usurped by an earlier nester, such as the Horned Owl.

Conflicts often are minimized by chronological differences in defense activity. Young and adult Long-eared Owls (Plate 37) have been observed in the very heart of a Horned Owl territory, but this has occurred after the young Horned Owls were well advanced toward caring for themselves (Figure 7 and Map 19). A young Great Gray Owl (Plate 62), purposely released within the nesting territory of a Horned Owl after the latter's young were fledged, was allowed to establish a limited range without interference.

There is extreme variation in the extent to which birds of the same or different species will defend their nests and young from man (Plate 29). These are individual differences and vary from no defense other than a protesting scream as the hawk departs to downright suicidal attacks (Plate 56). At one time or another we have been attacked and struck, with either talons or wingtips, by every raptor species treated in this study, and wounds inflicted by Goshawks, Duck Hawks, and Horned Owls have required medical attention.

Besides attacking, many owls have staged elaborate wounded-bird displays, accompanied by a fantastic assortment of utterances and antics (Bent, 1937). The Horned Owl and the Long-eared and Short-eared Owls (Plates 32 and 36) have been observed most often to display this type of behavior.

Up to this point we have been discussing nesting territories, each of which is a defended area within a range. Often very localized hunting territories exist within a range. These are small areas that are more consistently hunted than others and are defended. In the case of the Red-shouldered Hawk, it is often a kettle hole. For the Cooper's Hawk, it may be a small bushy woodlot. For at least a short time, a hunting ground generally exists near the nest if sufficient prey is present. As a general rule, while one hawk incubates, the other hunts away from the nest. When the young have hatched, one bird either broods or perches close by the nest while the other hunts a distant hunting ground. The parent at the nest will hunt the surrounding area while standing guard. As the young develop and appetites increase, both parents may range farther from the nest and each may go to a separate hunting ground. These distant hunting grounds generally are not defended. There is a tendency for the hunting range to increase in size as the season advances.

VERTICAL TERRITORIALITY AND ITS EFFECT ON RANGE OVERLAP

Nesting territories are three-dimensional. Hawks soar in the upper parts of their territories, and when reestablishing them prior to nesting will drive away intruders that soar not far above them. In Superior Township, soaring was most conspicuous from noon until evening. At such hours in spring, territories were observed to extend vertically 1,000 feet or more. On the Wyoming study area nests were not, as in Michigan, on one altitudinal level, since the elevation varied with the topography. Vertical territories over Blacktail Butte exhibited a staircase-like pattern, the ceilings for territories at the base of the butte being lower than territorial ceilings mid-way up the slope and these in turn lower than those at the top of the butte. As a result, a hawk soaring or flying to hunting grounds in the valley occasionally would pass high above the territory of another hawk of the same species. When this occurred at a sufficiently great height, territorial defense was not exhibited. If, however, as occasionally happened, a hawk or a Raven flew low over the territory of another member of the same species, defense was immediate and intense. This staircase-like pattern of the territory ceilings produced a two-dimensional range overlap (the birds from above passing unchallenged over the ranges of lower nesting pairs). This phenomenon was responsible partially for some of the apparent range overlap shown on Maps 20 and 21.

DURATION OF RANGES

A nesting range may be initiated in spring with the return of a nesting pair, or it may be a continuation of a winter range. Where territorial disputes occur, permanent ranges do not develop until ownership is established. The adults and young in most cases tend to remain within the nesting range until fall migration. With the departure of young from the nest, however, ranges and territories become less distinct and merge into one another to a much greater degree than earlier. The duration of nesting ranges as shown by Tables 50-52 is merely the length of time these ranges were observed. Many ranges existed as spatial units until the fall migration. Late summer and fall appear to be the periods of greatest instability, with the break-up of nesting ranges and the slow formation of fall and winter ranges by both residents and migrants.

Some pairs of both hawks and owls remain within a limited home range not only for one year but for several. In Superior Township this was noticeable particularly with such species as the Red-tailed Hawk, Horned Owl, and Screech Owl. Male Sparrow Hawks nested in the areas in which they wintered and a few Cooper's Hawks did likewise. The nesting Red-tailed Hawks of Ranges 22 and 24 (Map 18) wintered in Range 2 and 4, respectively, as shown on Map 3; many such examples could be cited.

THE ROLE OF RANGES

The nesting ranges exhibited by a population of raptors are the result of adjustment of that population to the environment—that is to their own population densities, to nest site locations, to topography, vegetative distribution, hunting areas, and prey abundance. As such, these ranges will within limits tend to vary with environmental changes or, when different areas are involved, with environmental differences (Plate 26). To some extent, ranges are an expression of activity, and it can be clearly seen how the activity of any single bird is curtailed by a complexity of environmental forces. It is literally true that the activity of any pair, and even of any single bird, affects some other one to the extent that a population is always automatically undergoing adjustment to its members' activities and is constantly responding in many ways to the dynamic forces operating on it. The tendency toward isolation of nesting ranges of hawks of the same species gives a certain amount of freedom from these adjustments at a time of year when breeding activities demand full attention.

THE RELATION OF RANGES TO HAWK NAVIGATION

In view of the work of Yeagley (1947) on pigeon homing and his emphasis upon a method of navigation that does not depend upon learned landmarks, it is interesting to note that adult and immature hawks remain within very limited areas throughout the breeding season and thus have no opportunity to learn the landmarks of any large area. As the time of migration nears, they may extend their local movements somewhat. The adults may, however, leave an area almost overnight. In spring these previous nesters return, apparently without confusion, not only to their former nesting woodlots but often to the same nests.

Young hawks establishing a range for the first time are buffeted about by established nesters, but even at this time the area over which they move is not extensive and often is not that in which they were raised. This seems to suggest that adult hawks return in migration to the same pin-point area year after year without having ever hunted or flown over an extensive region surrounding their nests. If landmarks play an appreciable part in hawk navigation, they are most likely landmarks along the migration flight and not recognizable ones over a large homing area.

THE RELATION OF RANGES AND HUNTING TERRITORIES TO PREDATION INTENSITY

In the discussion of hunting territories it was stated that small intensively hunted areas exist within ranges. On the other hand, some distant parts of a range are seldom if ever visited. There are also blank spaces (inter-range areas), as indicated by the maps, into which hawks or owls seldom if ever move. This helps to explain the varying degree of predation pressure observed by different workers on small areas or on localized populations, such as Bob-white or Pheasant broods. Broods outside the normal ranges of such raptors as the Cooper's Hawk, Horned Owl, and Red-tailed Hawk will not be preyed upon by these birds; broods within their ranges will suffer loss, which will be greatest if the prey lives within the raptors' daily hunting grounds. Broods raised within the boundaries of overlapping ranges of these three raptors may suffer unusually high losses from predation. Localized predation pressure on any prey population is thus affected by the range pattern of a nesting population of bird predators.

CHAPTER 12

Spring and Summer Food Habits

THE tedious task of dissecting a hawk pellet or examining nest food items and comparing the fur, feathers, or portions of bone, teeth, bill or claws found therein with laboratory collections to determine what a hawk has eaten may seem to the layman not only a foolish waste of time, but also a very disagreeable chore. On the contrary, such work has a detective-like fascination—the unfolding of a dramatic story completely concealed except for a little tell-tale evidence. On an area where such ecological data as the number and movements of hawks, the density and distribution of prey species, weather conditions, vegetative cover and other factors are known, food identification becomes a major clue, revealing minute details of previous events.

For the moment let us imagine that we are 50 feet above the ground, clinging to the rough bark of a white oak tree with one arm, while we examine and remove the food contents of a Red-shouldered Hawk's nest. The three young hawks eye us fixedly as we pick up the head of a meadow mouse, the hind leg and tail of a rat and three large pinching claws of crawfish. Then, digging in our climbing spurs and securing ourselves with the flip rope, we free both hands to work the fresh food out of the bulging crops of the nestling hawks. Each resists this undignified treatment, but in a few minutes we have the remains of a green frog and a leopard frog. We record the food data at once, descend the tree and then if time permits we reconstruct preceding happenings. We suspect the meadow mouse was captured a quarter of a mile from the nest, where we have seen this hawk perched above matted orchard grass where meadow mouse runways are conspicuous; the rat, a frequenter of barnyards, was probably killed within shotgun range of the farmer's house; while the crawfish surely came from the only kettle hole lying within the hunting range of this pair of hawks. The green frog probably inhabited the same kettle and the leopard frog might have been captured anywhere in the pasture, where its kind are abundant.

On another visit, the remains of a young cottontail give us the first inkling that young rabbits were out and about. Then on a final visit we find only the soft, long-quilled, half-developed feathers of the young hawks, and, still adhering to the flaky oak bark, the guard hairs of a raccoon. And so the story of what the hawks eat ends with a record of what the raccoon eats.

The food story of this nest is interesting, but it tells us little about predation until we combine this information with similar data gathered from all other hawk and owl nests on our study area during the same season.

Let us examine in greater detail how the food habits data of a raptor population were gathered and what they show when interpreted with reference to measured ecological factors.

The spring and summer food habits of each species of nesting raptor in Superior Township and in the Western Area are expressed in per cent of individual prey items found in a food sample for each species. As in winter, the size of the food samples varied with collecting conditions. Consequently, the diets of some species are reported more accurately than are others. In spite of a dearth of food data for the Marsh Hawk and Screech Owl, the tables of raptor diets are considered to be reasonably representative of the food habits of the collective raptor population inhabiting each study area.

The food taken by the raptors in Superior Township during spring and summer will be compared with the fall and winter diets (Chap. 6) and with the diet of the nesting raptor population of the Western Area. From these comparisons general conclusions will be drawn.

METHOD OF COLLECTING FOOD DATA

Food data were collected from a large number of nests, for it was believed that data thus obtained would represent more truly the diet of a species than would an equal number collected from only a few nests.

Methods of studying food habits varied with the raptor species. Items found at nests were recorded, food was squeezed from the crops of nestlings (Errington, 1932), and fledglings of most species were tethered (Plate 58). When little direct food evidence could be recovered at the nest, pellets were analyzed. To minimize error, all food remains and pellets were removed from the nests at each visit. When young birds were tethered near the nest by jesses, their mortality was remarkably low. Tethering all occupants of a nest proved most effective; otherwise, the tethered ones were neglected by the parents in favor of those that were not confined.

In comparing food data of two or more years it is important to gather the data regularly and at comparable periods throughout the nesting season since the food of any raptor changes to an important extent between the beginning and the close of the season. A preponderance of data collected during a short period will prejudice an account of the seasonal diet. Diets of the Cooper's Hawks illustrate this point. The much higher representation of Ring-necked Pheasant in the 1948 diet is due almost entirely to the fact that raptor fledglings were tethered and their food determined through July, when young pheasants become highly vulnerable. In 1942 no hawks were tethered during this time. Tethering the young of late nesting species and individuals yielded data on early and midsummer food habits (Plate 58). Had it been possible to sample the food habits of the collective raptor populations throughout the entire summer, the percentage of each species in the diet (Table 58) would have been more correctly reported. Since, on the other hand, there is no evidence that the basic diet recorded would have shown any fundamental changes, it will be treated as representing the spring and summer food of the raptors.

Pellet analysis is a poor method of determining quantitatively the food of nestling hawks, for such pellets contain very little osseous material. Prey species can be identified readily from most nest pellets, but it is frequently impossible to determine the number of individuals concerned. Therefore pellets were utilized consistently only when necessary, as in the case of the Red-shouldered Hawk and Sparrow Hawk, and then both the frequency of occurrence of species in the pellet and the number of prey items determined by osseous remains were recorded. (Tables 65, 66, 72 and 82). Contents of some other pellets were tallied if they contained prey items that were different from the non-pellet items identified at the same nest.

Pellets of young owls generally contained sufficient osseous material to yield good quantitative food data, and pellet data were utilized for all owls. Care was taken to prevent duplicating food remains recorded at the nest with prey determined from pellet analysis.

IDENTIFICATION OF FOOD ITEMS

Much of the food recovered at the nests could be identified on the spot and recorded. Doubtful items were collected and compared with specimens. Fledglings and immature birds were especially difficult to identify. Many small birds could not be identified to species and were accordingly lumped under the prey groups, "Small and medium-sized birds." No attempt was made to identify insects.

NESTING SEASON FOOD OF RAPTORS
IN SUPERIOR TOWNSHIP, MICHIGAN

The food data recorded in Tables 65 through 75 show the food habits of raptor populations on a definite area during the breeding season of hawks and owls. The food data for each nest have been tabulated separately to show individual variation in hunting and feeding. Each nest has a number used consistently throughout this study so that nest location, hunting range, productivity, and food habits for any nest can be compared (See Maps 13, 14, 17 and 18).

Red-shouldered Hawk Diet

Three hundred twenty-six individual prey items were recorded at 20 Red-shouldered Hawk nests in 1942, and 247 items at nine nests in 1948. The per cent of each food item is recorded in Tables 65 and 66.

The major prey species of the Red-shouldered Hawk during the two years were: meadow mice, 31.9 and 25.1 per cent; small and medium-sized birds, 20.2 and 29.1 per cent; snakes, 13.8 and 11.3 per cent; frogs, 8.6 and 16.2 per cent; and crawfish, 14.1 and 9.3 per cent. Since frogs did not appear in the pellets, their percentage representation in the diet was computed from non-pellet evidence only and thus the recorded proportion of this food in the diet is probably low (Plate 18).

The wood frog, cricket frog, leopard frog, green frog, and bull frog were identified specifically. The leopard frog, wood frog, and green frog were important items in the diet. All snakes recorded were of the garter snake genus.

The small and medium-sized birds identified to species were: Eastern Bluebird, Bobolink, Cardinal, Downy Woodpecker, Flicker, Common Goldfinch, Indigo Bunting, Meadowlark, Red-wing, Song Sparrow, Starling, Vesper Sparrow and Wood Thrush.

During both years, sparrows as a group were most frequently represented in the diet, followed by the Bobolink and Starling. Meadowlarks and Bluebirds were well represented in the diet in 1942. In that year all food items representing game birds and mammals were young of the year with the exception of two adult fox squirrels. The same general situation was evident in 1948. In both years a high proportion of the small and medium-sized birds were immatures and fledglings.

The percentages of major food groups are similar for the two nesting seasons (1942 and 1948), with game mammals composing 5.7 and 3.6 per cent; small rodents, shrews, and moles, 36.1 and 30.3 per cent; birds, 21.4 and 29.1 per cent; and lower vertebrates and invertebrates, 36.5 and 36.8 per cent, respectively.

The data in Tables 65 and 66 indicate that there is probably little variation in this diet from year to year in this region. This in turn suggests an average stability of collective prey populations. (See pages 333 and 334.)

Red-tailed Hawk Diet

Sixty-one individual prey items were recorded at the only Red-tailed Hawk nest in the township in 1942, and 150 items were recorded at two nests in 1948. There is much variation in the diets recorded for these three nests (Tables 67 and 68), but meadow mice and small and medium-sized birds were the food items that appeared most frequently at all nests.

The occurrence of six Ring-necked Pheasants in the diet in 1942 (Map 13, Nest 24) and the 11 muskrats and 7 fox squirrels recorded at nest 22 (Map 14) in 1948 clearly illustrate individual diet differences. Pheasants were abundant in the vicinity of the nest studied in 1942 and were preyed on rather heavily. Nest number 22 (Map 14) was situated at the edge of a large marsh, and thus muskrats suffered proportionately heavy losses. The high percentage of large mammals taken by this pair of birds, however, cannot be explained solely by the availability of large prey in their hunting territory. These hawks were exceptionally fierce and defiant, and their aggressive natures were reflected in the type of food they preyed on. That spirited, aggressive individual hawks can be trained to capture larger and stronger quarry than milder-spirited birds of the same species has been demonstrated time and again by falconers. Such individuals deviate from the general food preferences of a population, and their hunting activities may have a significant effect on local prey populations, but such individualism is not an important factor in the dynamics of predation.

Game mammals were 13.2 and 18.7 per cent of the Red-tailed Hawk diet during 1942 and 1948, respectively—substantially higher than the proportion of these mammals in the diet of the Red-shouldered Hawk. Game birds were also more heavily preyed on by the Red-tailed Hawk, while the lower vertebrates and invertebrates played a minor role in its diet.

Thus, though the Red-tailed and Red-shouldered Hawks competed for the food staples (meadow mice, and small and medium-sized birds) they did not compete in all food categories. The Red-tailed Hawk in southern Michigan meets little competition from the Red-shouldered Hawk for game birds and mammals and does not compete seriously with it for invertebrates and cold-blooded vertebrates.

Among the game species recorded in 1948, three out of four pheasants, six out of seven rabbits, two out of seven fox squirrels and five out of 12 muskrats were young. There appears to be a tendency for the Red-tailed Hawk to take a higher proportion of adult game animals than does the Red-shouldered Hawk. This may be viewed as consistent with the larger size, more powerful feet, and generally more aggressive nature of the Red-tailed Hawk.

The following small and medium-sized birds were identified as food of Red-tailed Hawks: Starling, Flicker, Red-wing, Domestic Pigeon, Bluebird, Wood Thrush, and several species of Sparrows.

Cooper's Hawk Diet

Two hundred and twenty individual prey items were recorded at six nests in 1942, and 296 items at five nests in 1948.

Mammals played a minor role in the diet of the Cooper's Hawks, being 6.8 and 11.8 per cent of the diet during 1942 and 1948, respectively. Small and medium-sized birds were 87.7 and 72.0 per cent, respectively, in the two years.

Of the 193 small birds recorded in 1942, 73 were not specifically identified. Of the 120 identified, all but the Yellow-breasted Chat were recorded again in 1948, and since the records for the latter year show the proportion of each species of bird (sparrows and warblers excepted) in the diet of Cooper's Hawk, they will be discussed.

The 137 small birds that could be identified in the 1948 diet included the following species in the order of the number of times they were taken. Many sparrows which could not be specifically identified are placed in one group:

Sparrows—36	Blue Jay—4	Wren—1
Bobolink—19	Yellow-billed Cuckoo—4	Rose-breasted
Robin—11	Towhee—4	Grosbeak—1
Meadowlark—8	Wood Thrush—3	Red-breasted
Starling—8	Tufted Titmouse—2	Nuthatch—1
Flicker—7	Cowbird—2	Mourning Dove—1
Wood Warblers—7	Common Goldfinch—2	Domestic Pigeon—1
Cardinal—6	Hermit Thrush—1	Downy Woodpecker—1
Red-wing—5	Oven-bird—1	Sora Rail—1

Pheasants were preyed on significantly in both years. A higher percentage of pheasants recorded for the Cooper's Hawks in 1948 (Table 70) as compared with 1942 is not indicative of difference of feeding preference in the two years. The discrepancy in the data arose because the young Cooper's Hawks (Nests 29 and 30, Map 14) were tethered for longer periods in 1948 and most of the pheasant data were recorded from these birds in late July. A comparison for the two years can best be made from the number of pheasants recorded at the nests. In 1942, nine pheasants were recorded, and in 1948 only four of the 44 tallied were recorded at the nest. There is no doubt but that the occurrence of pheasants in the diet of the 1942 birds would have been equivalent to that recorded in 1948 had the young hawks been tethered during late July. At this time, haying operations and the cutting of oats and wheat significantly reduce the protective cover for pheasants. This reduction in cover over a large land area increases substantially the vulnerability of young pheasants to predation. The half-grown young become particularly vulnerable to Cooper's Hawks at this time, as is shown by data in Table 70 and further verified by data recorded during the summer of 1949.

The two major food groups are not unlike for the two years (Tables 69 and 70), with mammals representing 6.8 and 11.8 per cent of the diet and birds 93.2 and 88.2 per cent, respectively.

The 14 pheasants recorded in 1948 whose age could be definitely determined were all young birds. A large proportion of the small birds preyed on during both years were fledglings.

Marsh Hawk Diet

Records of food of the Marsh Hawk (Plate 33) are set forth in Table 71, but the data are too incomplete to be considered representative of the species in the township. The complete destruction of all Marsh Hawk nests before hatching in 1948 precluded gathering any quantitative data for this species in that year. Field observations verified a diet high in meadow mice for both years. These mammals continued to form the staple food of the Marsh Hawks during the periods of courtship and nest building before the spring vegetation furnished them protective cover. As cover increased the dwindling meadow mouse population became less vulnerable and frogs and small birds became major food items. Frogs, though an important item in the diet, do not appear in Table 71. Such quantitative data as were obtained, plus field observation, indicate that meadow mice, small and medium-sized birds, and frogs were fed upon in the order mentioned.

Sparrow Hawk Diet

No data on food of the Sparrow Hawk were obtained in 1942. Data gathered in 1948 (Table 72) indicate a diet high in meadow mice and small birds. Grasshoppers and other insects are important items in the diet which are not fully represented in the Table.

Sparrow Hawks were observed regularly hunting meadow mice and were frequently seen carrying them to the nests. There is little doubt that meadow mice were the major food item of these falcons during both study years. For additional information about food of Sparrow Hawks, see Table 82.

Horned Owl Diet

Ninety-nine individual food items were recorded at two nests in 1942 and 161 items at three nests in 1948. The data in Tables 73 and 74 show the varied diet characteristic of this owl.

Errington and Hamerstrom (1940) interpreted the extremely varied Horned Owl diet as predation governed by opportunism, which means that the Horned Owl preys on whatever comes its way. There is no doubt that this owl (Plate 38) is a very efficient hunter and that it can and does prey on a range of prey wider than that of any other North American raptor.

The wider the range of prey that a raptor species can take, the more difficult it is to correlate the prey consumed with the density of prey populations available over an extensive area. In 1942 mammals were 40.5 per cent and birds 59.6 per cent of the Horned Owl diet; in 1948 mammals were 62.1 per cent and birds 29.7 per cent. This change in diet did not reflect any known changes in the density of bird and mammal populations in the township.

The Horned Owl diet (Tables 73 and 74) is an excellent illustration of a situation where prey risks modify the effect of prey densities to such an extent that the diet of a single raptor does not show a consistent correlation with prey densities over an extensive land area. Nevertheless, the food recorded at each nest does roughly reflect the prey densities within the hunting range of each pair of owls as modified by such conditions affecting risk as dispersion, cover protection, and specific hunting area.

Small birds recorded in the diet include Meadowlark, Blue Jay, Red-headed Woodpecker, Downy Woodpecker, Robin, Starling, Red-wing, and Rose-breasted Grosbeak. No food data were gathered after May 15, and it is not until after this date that the small bird population curve climbs rapidly to the July peak. Therefore, this prey group undoubtedly forms a more significant portion of the Horned Owl diet as summer progresses.

In 1948 all Pheasants recorded were adults in the ratio of 16 hens to 5 cocks. This ratio does not differ importantly from the known winter sex ratio of 4 hen pheasants to 1 cock.

All of the pheasant mortality recorded in Tables 73 and 74 is that of adult birds from the residual winter population. They were 20.2 and 13.0 per cent of the diet in the nesting season in 1942 and 1948, respectively. The increased activity of pheasants at the time of dispersion from protective winter roosting areas to crowing territories, when protective cover is at a minimum, made the Ring-necked Pheasant more vulnerable to the Horned Owl at this period. It is significant that this increased vulnerability of adult Ring-necked Pheasants (Plate 38) occurs at a time when other prey populations are at or near a seasonal low. Errington and Hamerstrom (1940) found similarly heavy predation on Ring-necked Pheasants by Horned Owls in the early spring months.

Barn Owl, Long-eared Owl, and Screech Owl Diets

Table 75 contains records of food of the Barn Owl (Plate 35), Long-eared Owl, and Screech Owl. Meadow mice occurred most frequently in the diet of all three species. Wallace (1948), in his study of the Barn Owl in Michigan, found that, "In ordinary times close to 80 per cent of the Barn Owl's diet is of Microtus (meadow mouse), and that during periods of abundance of these rodents, the catch may rise to 90 per cent or more or drop correspondingly during periods of low availability."

It is significant that the proportion of small birds in the diet of the nesting period of the Screech Owl and the Long-eared Owl shows increase over that found in their winter diets (Tables 26 and 28).

MORTALITY OF ADULT RING-NECKED PHEASANTS

The average winter Pheasant population in Study Areas A and B was 226 birds in 1942 and 222 in 1948, with a sex ratio of four females to one male during both winter periods. Winter mortality was insignificant prior to March 15.

A census of crowing cocks was made on these same areas in April. In 1942, 41 crowing cocks were tallied and in 1948, 43 were tallied. Assuming the winter sex ratio of four to one to be applicable to the spring population, we compute a total pheasant population of 205 and 215 birds for these areas during the height of the crowing period (April 15). Because of the homogeneous nature of the township, movement into and off the areas can be considered mutually compensatory.

Thus, in the short period of 30 to 45 days there was an apparent loss of 21 birds in 1942 and of seven birds in 1948, or an average loss of 6.3 per cent of the two wintering populations. Observations and raptor food data indicated that most of this loss was due to predation by Horned Owls (Plate 38), Cooper's Hawks, and foxes.

The Horned Owl food data (Tables 73 and 74) show that throughout the township 20 and 21 adult pheasants were taken by Horned Owls from February 20 to May 15 during 1942 and 1948, respectively. Pheasant remains are so conspicuous at the owls' nests and perches that they are not readily overlooked. Thus the food data indicate that the Horned Owls alone accounted for an average of approximately 2 per cent of the over-wintering pheasant population of the township during a period of three months. When we consider that the number killed by Cooper's Hawks was also substantial (not measured quantitatively) and that the Red-tailed Hawks took adult pheasants during this period (Tables 67 and 68), we can estimate the average minimum adult pheasant mortality from raptors in the spring of the year to have been in the neighborhood of 4 to 5 per cent.

FOOD OF RAPTORS IN THE NESTING SEASON IN THE WESTERN STUDY AREA, WYOMING

Tables 76 through 82 show the food habits of the raptor species on the Western Study Area in the nesting season of 1947. Percentage of diet in terms of individuals taken is stated in each case.

Red-tailed Hawks, the predominant species, exerted greatest pressure on the Uinta ground squirrel and meadow mouse. These mammals were, respectively, 41.8 and 33.3 per cent of their diet. The second most common nesters, the Sparrow Hawks, preyed most heavily on meadow mice, which were 57.3 per cent of their diet.

The bird hawks exerted greatest pressure on the small bird group; in this respect their diet was similar to that of the Cooper's Hawk in Michigan. The diet of the western bird hawks as a group was higher in mammals. They were 60.5 per cent of the Goshawk diet (Plate 28), 26.5 per cent of the Cooper's Hawk diet, and 6.5 per cent of the Sharp-shinned Hawk diet.

Meadow mice and pocket gophers were 64.5 and 23.5 per cent, respectively, of the Horned Owl diet. Birds were only 3.7 per cent of the diet, in contrast with 59.6 and 29.7 per cent, respectively, for the years 1942 and 1948 in the diet of the Michigan Horned Owls. Had the young Horned Owls been tethered, birds would undoubtedly have formed a larger proportion of the diet, since this prey group becomes relatively more vulnerable at the time when young Horned Owls are on the wing; the general feeding trend would not, however, have been significantly altered.

Both the Long-eared Owl and the Great Gray Owl (Plate 37) exerted greatest pressure on the meadow mouse and pocket gopher. It is significant that the owls exerted much greater pressure on the pocket gopher, a species that is chiefly nocturnal, than did the hawks. Again we re-emphasize this dovetailing of pressure that enables a collective raptor population to exert a proportional pressure on all prey forms.

COMPARISON OF THE RAPTOR DIET IN 1942 WITH THE DIET IN 1948 IN SUPERIOR TOWNSHIP

A comparison of the diet of the collective raptor population in 1942 with that in 1948 (Tables 53 and 58) shows significant similarities in the percentages of the various prey species and groups utilized as food during the two nesting seasons. A comparison of the diet of the various raptor species in the two years (Tables 65 through 82) also shows little variation.

This indicates that there was little variation in the spring population densities of the various prey species during the two years in spite of much variation in the fall and winter populations of the preceding years (Chapter 5).

When we consider the probability that season after season small birds, meadow mice, snakes, frogs and crawfish will be the most abundant prey available during the nesting season, and that top-heavy winter prey populations will tend to be reduced proportionately by spring through the action of wintering raptors and other forces of the environment, it is logical to conclude that with minor variations the diet of the collective raptor population in Superior Township in the nesting season will be basically the same, year after year. The food and population data support this conclusion.

The concept is one of a stable nesting raptor population supported, season after season, on a basically similar diet. This could be the case only if the nesting raptors tended to balance prey populations by exerting greatest pressure on the most numerous prey species. It appears that pressure on a species or prey group does not approach the level

TABLE 53

COMPARISON OF THE SPRING AND SUMMER FOOD OF NESTING RAPTORS IN SUPERIOR TOWNSHIP, 1942 AND 1948

| Predator Species | Mice, Shrews, Rats, Ground Squirrels, Moles, Chipmunk, Weasels, Red Squirrel | | | | Snakes, Frogs, Crawfish | | | | Rabbit, Muskrat, Fox Squirrel, Pheasant, Quail, small and med.-sized birds, Crow | | | | Total No. of Individuals | |
| | No. of individuals | | Per cent diet | | No. of individuals | | Per cent diet | | No. of individuals | | Per cent diet | | | |
	1942	1948	1942	1948	1942	1948	1942	1948	1942	1948	1942	1948	1942	1948
Red-shouldered Hawk	118	75	36.2	30.4	119	91	36.5	36.8	89	81	27.3	32.8	326	247
Red-tailed Hawk	28	94	45.9	62.7	5	2	8.2	1.4	28	54	45.9	36.0	61	150
Cooper's Hawk	10	22	4.6	7.4	0	0	0	0	210	274	95.5	92.6	220	296
Marsh Hawk	32	..	62.7	..	0	..	0	..	19	..	37.3	..	51	..
Sparrow Hawk	..	41	..	75.9	..	0	..	0	..	13	..	24.1	..	54
Great Horned Owl	32	90	32.3	55.9	0	13	0	8.1	67	58	67.7	36.0	99	161
Long-eared Owl	65	..	86.7	..	0	..	0	..	10	..	13.3	..	75	..
Barn Owl	40	..	100.0	..	0	..	0	..	0	..	0	..	40	..
Screech Owl	30	..	52.6	..	16	..	..	..	11	..	47.4	..	57	..
Totals	355	322	38.2	35.5	140	106	15.1	11.7	434	480	46.7	52.8	929	908

1837=Total No. of Individuals
36.8%=Average per cent of mice, rats, etc., for two-year period
13.4%=Average per cent of snakes, frogs, and crawfish for two-year period
49.9%=Average per cent of rabbits, pheasants, small birds, etc., for two-year period.

where there is danger of even local extinction, but apparently it can have a controlling effect in limiting population increases.

COMPARISON OF THE WINTER AND SPRING DIETS OF THE RAPTOR POPULATIONS IN SUPERIOR TOWNSHIP

Figures 8 and 9 graphically illustrate the difference between the winter and spring raptor diets in Superior Township. Even when we include small and medium-sized birds in the Game Food group, we find that only 4.5 per cent of the diet of the entire winter raptor population during 1942 and 1948 consisted of game food considered beneficial by man. On the other hand, 50 per cent of the food of the raptor population falls in this category during the nesting season. Only about 13 per cent, however, is truly game species, the remainder being small birds. For an analysis of the seasonal diets see Table 54.

Mice are the staple food during the cold months, with small and medium-sized birds assuming this role during the warm months. Fox squirrels, rabbits, Pheasants and Bob-whites are all more heavily preyed on in spring and summer than in fall and winter. Meadow mice are a major prey species throughout the year. White-footed mice are taken in greater numbers in cold weather than in warm weather (Table 54).

The winter diet of the raptors leaves no doubt as to the beneficial nature of their activities during cold weather.

The spring and summer diet, when considered in the light of man's interests, would appear to cancel out some of the good effects. When it is realized, however, that in Superior Township, season after season, a numerically stable nesting raptor population supports itself on a diet roughly proportional to prey densities (to be proved later), it is evident that the collective raptor population must be in equilibrium with the collective prey population.

COMPARISON OF THE RAPTOR DIET IN MICHIGAN WITH THAT IN WYOMING

A comparison of the nesting season food of the raptors in Superior Township (Table 53) with the diet of the raptors in the Wyoming area (Tables 55 and 59) shows significant differences.

In the Western Area 82.5 per cent of the diet of the entire raptor population consisted of mammals, 17.2 per cent was birds and only 0.3 per cent was cold-blooded vertebrates (Table 59). This is in contrast to the diet of the raptors in Michigan, where a two-year diet averages: mammals, 42.3 per cent; birds, 44.1 per cent; and cold-blooded vertebrates and invertebrates, 13.4 per cent. Figures 8, 9 and 10 show that the diet of the nesting raptor population in Wyoming more closely parallels the winter than the spring and summer diet of the Michigan raptors.

The difference in the species of raptors present only partially accounts for the tremendous difference in diets recorded in Tables 53

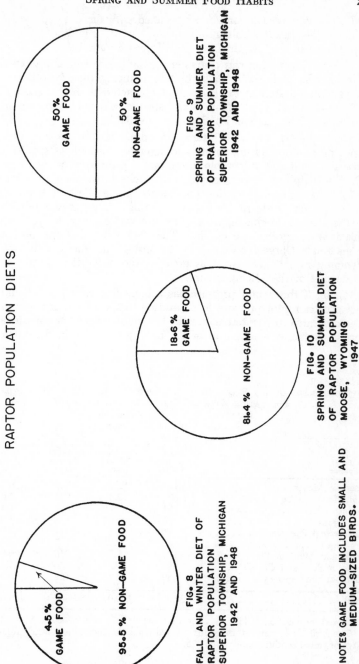

RAPTOR POPULATION DIETS

FIG. 9
SPRING AND SUMMER DIET
OF RAPTOR POPULATION
SUPERIOR TOWNSHIP, MICHIGAN
1942 AND 1948

50% GAME FOOD

50% NON-GAME FOOD

FIG. 10
SPRING AND SUMMER DIET
OF RAPTOR POPULATION
MOOSE, WYOMING
1947

18.6% GAME FOOD

81.4% NON-GAME FOOD

FIG. 8
FALL AND WINTER DIET OF
RAPTOR POPULATION
SUPERIOR TOWNSHIP, MICHIGAN
1942 AND 1948

4.5% GAME FOOD

95.5% NON-GAME FOOD

NOTE: GAME FOOD INCLUDES SMALL AND MEDIUM-SIZED BIRDS.

and 55. It is, rather, a condition associated with the prey species available and their relative vulnerabilities. For variations in diet of the raptor populations in the two areas consult Tables 58 and 59.

The Michigan area was characterized by heavy winter predation on rodents, which played a major part in reducing their populations by spring. The Western Area, because of characteristic heavy snow cover and the hibernation of many rodent species, did not support a heavy winter raptor population. Food was not available.

Consequently, in the Western Area predation on rodents assumed major proportions during spring and summer, when these species were most abundant and vulnerable to raptors. If a function of raptor predation is to help keep rodent populations in check, we should expect that spring and summer predation on rodents in the Western Area, where there was very little raptor pressure during winter, would be notably greater than in Michigan, where pressure was exerted the year round. This is exactly what we do find. The data show that the proportion of mammals (largely rodents) in the spring and summer diet of the Wyoming raptors was twice as great as the proportion of this group in the diet of the Michigan raptors.

Again, if this raptor pressure has a balancing effect, we should expect greater raptor pressure to be exerted in spring and summer in the Wyoming area to counterbalance the lack of significant raptor predation during winter. A comparison of the nesting densities in the Michigan

TABLE 54

COMPARISON OF THE FALL AND WINTER DIET OF THE COLLECTIVE RAPTOR
POPULATION WITH THE SPRING AND SUMMER DIET
(Data for 1942 and 1948 Combined)
Superior Township, Michigan

Prey Species	Fall & Winter—1942-48		Spring & Summer—1942-48	
	No. prey indiv.	Per cent diet	No. prey indiv.	Per cent diet
Meadow Mouse	11,167	82.7	546	29.7
White-footed Mice	1,670	12.4	55	3.0
Rabbit	91	.7	45	2.4
Fox Squirrel	5	.04	34	1.9
Ring-necked Pheasant	20[1] +	.1[1] +	108[2] +	5.9[2] +
Bob-white	1[1] +	..[1] +	5	.3
Small and Medium Sized Birds	165	1.2	676	36.8
Frogs, Snakes, and Crawfish	22	.2	246	13.4
All other prey species	359	2.7	122	6.6
Total	13,500	100.04	1,837	100.0

[1] Biased by omission of kills by Cooper's Hawk.
[2] The addition of some 40 individuals would make this figure more nearly representative of actual conditions. See page 278.

TABLE 55

Spring and Summer Food of Nesting Bird Predators In Area
Moose, Wyoming
1947

| | Non-Game Food | | Game Food | | |
| | Small rodents, shrews, weasels | | Rabbits, grouse, ducks, small and medium-sized birds | | |
Predator species	Number of prey individuals	Per cent diet	Number of prey individuals	Per cent diet	Total number of prey individuals
Red-tailed Hawk	170	89.9	19	10.1	189
Swainson's Hawk	30	81.1	7	18.9	37
Sparrow Hawk	178	80.9	42	19.1	220
Prairie Falcon	80	80.0	20	20.0	100
Cooper's Hawk	16	26.7	44	73.3	60
Sharp-shinned Hawk ..	3	6.5	43	93.5	46
Goshawk	46	51.7	43	48.3	89
Great Horned Owl ...	338	93.6	23	6.4	361
Long-eared Owl	127	98.4	2	1.6	129
Great Gray Owl	82	98.9	1	1.2	83
Totals	1,070	81.4	244	18.6	1,314

and Wyoming areas shows that in the Wyoming area, where raptor predation was largely confined to spring and summer months, there were 3.8 hawk and owl nests per square mile, whereas in Michigan, where raptor pressure was exerted the year round, there were only 1.8 pairs of nesting hawks and owls per square mile (Tables 39, 40, and 41). This indicates that raptor pressure in the nesting season in southern Michigan was half of the similar pressure in the Wyoming study area at that season.

In both areas, as we shall soon see, the raptor population tended to be in proportion to the vulnerable food, with the heaviest predation on the most abundant prey species and groups. In the Western Area collective pressure on the densest group, the rodents, during a period of six or seven months, has a balancing effect on prey populations, while in the Michigan area a similar effect resulted from continuous year-round pressure on two major prey groups, rodents and small birds.

In both areas, predation by mammalian predators was occurring and field observation and scat analysis indicated that, similarly, greatest pressure was exerted by these predators on the densest prey populations. Could a comparable quantitative study also have been simultaneously conducted on a collective mammal predator population, the entire predation phenomenon would show a still more effective application of pressure, which is instrumental in keeping in balance the species concerned.

CHAPTER 13

The Function of Raptor Predation

OUR DATA and our interpretations of them have led to the point where we must ask the questions: What are the functions of raptor predation as a biological force? What does the continual proportionate killing of prey by raptors accomplish besides the satisfying of basic food drives? How is raptor pressure exerted?

Before we can explore more fully the function of raptor predation in Nature we must examine further the mechanics of this phenomenon. To do this will involve a discussion of the densities and productivity of prey populations in Superior Township and a comparison of the raptor-prey density relation in spring and summer with that in fall and winter. In other words, we must answer the question: Does the density relation shown to have been a fundamental principle of predation in fall and winter also apply to predation during other seasons (Chap. 7)? If so, of what value is this fact?

AVAILABILITY AND RELATIVE DENSITIES
OF MAJOR PREY POPULATIONS

Great and significant changes occur in prey populations as winter gives way to spring. These changes are well known and obvious, but their relation to the ecology of raptor predation has not been clearly presented. Instrumental in producing these changes are migration, cessation of hibernation, and reproduction.

The spring migration, resulting in the establishment of thousands of small birds in Superior Township, markedly altered the composition and relative densities of the collective prey population. The emergence of frogs, snakes, and crawfish from hibernation was a change almost equally significant. These prey species, invulnerable to winter raptors, suddenly, often almost overnight, became highly vulnerable to the nesting raptor population. The reproductive efforts of all prey species increased the available prey population and greatly altered the ratio of young to adults. Thus total and relative prey densities changed between winter and summer, and with this change occurred changes in vulnerability to raptors.

There is a dynamic difference between winter prey populations and those of the breeding season. Winter population curves gradually descend from a high fall peak to a low in spring, with little or no reproduction to offset losses. During the breeding season, prey populations increase through reproduction, while being simultaneously retarded by environmental resistance forces, including predation. At this time reproduction usually does, and in this case did, outweigh the collective forces of environmental resistance. Thus from low residual winter prey populations there is a steady increase during spring and summer to a peak in the adult and juvenile prey populations (Table 56). This peak is attained at different times by different prey species and the maximum collective prey population is attained some time in late summer or early fall.

This steady increment with simultaneous losses makes it difficult to establish even rough estimates of actual densities of prey available to raptors. Potential densities based on average reproductive rates can be calculated and are helpful, but the eggs or tiny young of most prey species are not highly vulnerable to predation by hawks and owls, and by the time young are abroad and vulnerable to avian predators their numbers are far below calculated potentials. Also, since the resistance of the environment is greater for some species than for others, the average number of young of each species reaching an independent stage is not in direct proportion to the average number of young produced. Thus a calculation of prey available to raptors on a basis of either an average or a potential productivity will contain errors.

In spite of this difficulty in computing theoretical population densities, a relative abundance rating for major prey species can be determined from data at hand (Table 60).

By using the average number of litters or broods per year and the average number of eggs per nest or young per litter, we can calculate a theoretical average productivity of a pair for a breeding season.

It is evident from Table 57 that under ideal conditions the meadow mouse will produce more young in a breeding season than will any other prey species with which we are here concerned. From this it might appear that meadow mice should have been the most abundant prey in Superior Township during the spring and summer. This, however, was not the case, as meadow mice and all other prey species, with the exception of the small-bird group, reached a period of minimum population density in spring (Table 56) prior to building up their

TABLE 56
PERIODS OF MAXIMUM AND MINIMUM PREY NUMBERS

Species	Period of Minimum Yearly Numbers	Period of Maximum Yearly Numbers
Meadow Mouse	March	Oct. & Nov.
White-footed Mice	May	Oct. & Nov.
Fox Squirrel	May	August & Sept.
Ring-necked Pheasant	May	July & August
Bob-white Quail	May	July & August
Cottontail Rabbit	May	July & August
Small Birds	Jan. & Feb.	July

TABLE 57
PRODUCTION OF YOUNG BY PREY SPECIES

Species	Broods per Year	Young per Brood
Meadow Mouse	7 to 8	5
White-footed Mice	4 +	4 +
*Rabbit	3 +	5 +
*Fox Squirrel	1 +	3
*Ring-necked Pheasant	1	10
*Bob-white Quail	1	13
Song Sparrow	1 to 4	4 +
	Average 3 + Nesting Attempts	

* After Allen, D. A. (1943).
Meadow Mouse and White-footed Mice data taken from Hamilton (1937;1941) and Burt (1940) respectively.
Song Sparrow after Nice, M. M. (1937).

densities through reproduction. Small birds, on the other hand, moved into the area in large numbers and settled in all habitats. Over extensive areas their initial breeding population far exceeded that of the meadow mouse. With this numerical advantage, they remained more abundant than meadow mice during April, May, June, and mid-July. As the season progressed the meadow mouse population increased with continual acceleration. The young of early litters produced offspring, and by late summer meadow mice surpassed small birds in population density.

Small birds

Young (1949), working in the vicinity of Madison, Wisconsin, found 94 breeding pairs of birds on a five-acre tract, a density of 18.8 pairs per acre. There was a nest density of 164 nests, or 32.8 per acre. Passerine species averaged 49.8 per cent successful in nesting attempts. He did not compute total productivity for the study area. There was, however, a 47-per-cent nesting success of 121 active nests of six species. Computations from Young's data show that for the nests of six different

species there was an average of three fledglings per successful nest or
1.4 fledglings per nest in which at least one egg was laid.

From his description it seems certain that the high population
densities existing on this tract were not representative of more extensive areas. Nevertheless, it illustrates the very high population densities of small birds that can build up during the breeding season. Nice
(1937), in comparing 9 studies of passerine birds, showed that there
was a percentage of success of some 41 to 46 per cent of eggs and nests,
concluding that this would appear to be normal in temperate North
America.

It was obvious from general observation and from the number of
passerine nests located at random throughout Superior Township that
the population density of small birds was very high. John L. George
(1952) studied small-bird populations on a 171-acre farm in Section 14, Superior Township, throughout the 1948 breeding season
(Map 1). A preliminary inspection of his data showed an adult small-
bird population averaging three birds per acre in April, rising to a
potential maximum of 13 birds per acre in July (egg mortality omitted).
This farm is reasonably typical of the farm land (woods and open
country) in the township. It we project these data to the township, we
find that at the height of the breeding season an approximate population
of some 300,000 small birds was potentially available to raptors and that
even in April, when the population was at a minimum, some 69,000 adults
were available. The actual number available to raptors was much less
than the calculated 300,000, since young birds do not become significantly
vulnerable to most hawks and owls until just prior to completion of
fledging. Nevertheless, it appears certain that small birds were the
most abundant prey group available during the spring and early
summer, exceeding even the meadow mouse.

Meadow Mice

Hamilton (1937) considered the meadow mouse to be the most
rapid breeder of all known mammals, producing an average of 5.07
young per litter.

Bailey (1924) recorded 17 litters a year for a captive female, and
Hamilton (1937) noted nine litters in eight months from a captive
meadow mouse and suspected that eight to ten litters are produced
in "mouse years" and not more than five or six litters in years of scarcity.
This remarkable reproductive ability is matched by heavy and continuous mortality.

Hamilton found breeding to be well under way by early April, with
complete cessation during most winters. In March and April litters
occur with increasing rapidity, and throughout the summer the population increases, but independent immature meadow mice never equal the
breeding population. Since young meadow mice become independent

of the female at two to three weeks, it appears from Hamilton's data that a tremendous mortality occurs during the period of dependency and very soon thereafter.

We can conclude that though there is a high reproductive potential, only a fraction of this potential normally reaches an independent status, and the population of independent immature meadow mice does not increase suddenly but grows slowly as the breeding season progresses.

Blair (1940), working in Michigan, found meadow mouse density in moist grasslands to be at a low in late May. Hamilton's data (1941) would indicate a seasonal low occurring in late April. In Superior Township breeding was general in early April and at that time meadow mouse populations reached their seasonal low. The populations rose steadily after the first litters were abroad, approximately May 1, and attained a peak some time in October or November.

We estimated a population of approximately 300,000 meadow mice for Superior Township in the fall of 1941. This represents the peak population recorded for the area. A much lower population existed in 1947. During both winters raptor pressure, as well as other decimating forces, greatly reduced these meadow mouse populations, leaving relatively low numbers to initiate the breeding cycle in spring. In spite of rapid reproduction, a meadow mouse population (exclusive of dependent young) normally does not rise rapidly until late summer and fall, and we can conclude that populations existing in Superior Township during April, May, June, and mid-July, when raptor food data were collected, were far below the fall levels. This being the case, meadow mice ranked second in population density to small birds. In September, with the departure of the passerines and the rapid expansion of the meadow mouse population, these mice again ranked first in population density.

Frogs, Snakes, and Crawfish

So little is known about the population densities of frogs, snakes and crawfish that the density rating in Table 33 is questionable. If all species of frogs were included, the population probably would exceed that of small birds. Only those recorded as staple items in the raptor diet (leopard frog, wood frog, and green frog) however, will be included. The tree toads, spring peepers and swamp cricket frogs, although present in large numbers, seldom appear to be taken by hawks and owls, probably because of their small size.

The leopard frog, green frog, and wood frog are prey species that exhibit high densities locally, but because they are confined to specific habitats of limited extent the populations on any large area are not nearly as great as their local densities might lead one to believe.

Some idea of the local density of leopard frogs was obtained when two men systematically cruised an acre of favorable habitat for one hour. They caught 27 frogs and tallied 13 others that were seen but escaped capture. It was thought that these 40 individuals were very nearly the entire population of leopard frogs on this one-acre plot.

White-footed Mice

Burt (1940) found that the population curve for the eastern wood mouse in southern Michigan is lowest in May, before the first young are abroad, and rises rapidly until the latter part of June or July, reaching a peak in October and November, when the young-of-the-spring females appear. The curve then descends gradually to its lowest point in the following May.

It is apparent from Burt's population curve (Fig. 11) that the population of white-footed mice is much lower during the spring and summer months than during the fall and winter. We can assume that the population of this species in Superior Township would show a similar curve, and thus the number of maturing young and adults available to raptors during the breeding season of hawks and owls would be much less than the population computed for the fall period (Table 34). Though this curve is intended to show the actual population, it is based on the results of trapping and there is reason to think that, in addition to the trappable animals that it represents, a substantial portion of the potential increment to the breeding stock is, while still too young to be trapped, available to raptors and preyed on by them. Therefore, the total number of mice present and available to raptors during the breeding season is greater than Burt's curve would indicate. If, on the other hand, we calculate the average number of offspring from a single pair of mice (Burt, 1940) or from a residual spring population, the potential

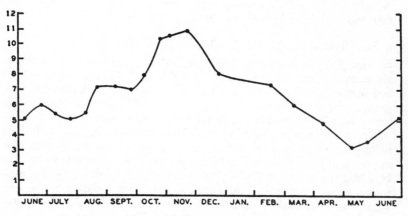

FIGURE 11. Population curve for white-footed mice

number available is much more than the population curve of Fig. 11 would indicate. Only the young that reach the stage where they are abroad can be considered potential prey for hawks and owls, and this number must lie somewhere between Burt's curve and a curve drawn for the average number of young per litter.

Other Prey Species

The populations of Pheasants, Bob-whites, rabbits and fox squirrels increased a great deal above their winter populations. Allen (1943) found that an average population of fox squirrels in an average year produced about four young per female, which normally resulted in a fall age ratio of about 60 young to 40 adults. English (1945) found fall pheasant populations to be about double the winter populations. The cottontail rabbit generally has at least three litters a year, with an average of more than five young per litter. Haugen (1943) found that only about one young cottontail rabbit in five survived until fall, so that the total annual accession to the spring breeding stock is only a little more than 150 per cent of it. Of 15 young, 12 die before becoming adult. No doubt some of these deaths are caused by raptor predation. The reproductive increment of Ring-necked Pheasants, Bob-whites, rabbits, and fox squirrels did not alter the relative density ratings of these species to one another or with respect to other prey (Table 57). The population of each reaches a low some time in May and rises gradually to a peak in late summer or fall.

We can conclude that the most abundant prey in Superior Township during spring and summer were, in the order of their abundance: small birds; meadow mice; frogs, snakes, and crawfish as a group; white-footed mice; game birds; rabbits; and fox squirrels. With prey density ratings as recorded in Table 60, we are in a position to review the food data to see if there is a relation between predation and relative prey densities and to determine causes of prey vulnerability.

PREDATION AS RELATED TO PREY DENSITIES

The diet of the nesting raptor population of Superior Township is set forth in Table 58. During both 1942 and 1948 small birds ranked first in the diet, meadow mice a close second and the cold-blooded prey groups (frogs, snakes, and crawfish) third. The diet of this nesting raptor population is in harmony with our earlier conclusion that a collective winter raptor population takes its prey, in general, in proportion to the prey population densities. In other words, the changes in both raptor and prey densities and composition that occurred during the breeding season did not alter the basic density relation. As Table 60 shows, small birds, with the highest density or population rating (1), were most frequently represented in the raptor diet and were subject to predation from all the avian predators. The meadow mouse, with

TABLE 58

FOOD DATA SUMMARY FOR THE COLLECTIVE NESTING RAPTOR POPULATION
SUPERIOR TOWNSHIP, MICHIGAN
1942 and 1948
February to August

Prey Species	No. of prey individuals		Per cent of Diet		Per cent of Diet	
	1942	1948	1942		1948	
Meadow Mouse	258	288	27.8		31.7	
White-footed Mice	54	1	5.8	37.3	.1	34.4
Other mice, rats, shrews, and moles	34	24	3.7		2.6	
Thirteen-striped Ground Squirrel	3	4				
Chipmunk	0	5	.9	.9	1.0	1.0
Weasel	1	0				
Red Squirrel	5	0				
Woodchuck	1	0				
Fox Squirrel	17	17	1.8		1.9	
Muskrat	3	19		4.6	2.1	6.5
Rabbit	22	23	2.4		2.5	
Raccoon	0	1				
Ring-necked Pheasant	39	69	4.2		7.6	
Bob-white	2	3				
Chicken	2	1				
Duck	0	1				
Rails	2	2		5.7		8.9
Crow	7	4				
American Bittern	0	1				
Screech Owl	2	0				
Small and medium sized birds	337	339	36.3	36.3	37.3	37.3
Frogs	28	40	3.0		4.4	
Snakes	50	30	5.4	15.1	3.3	11.6
Crawfish	62	36	6.7		3.9	
TOTAL	929	908		99.9		99.7

the second highest population or a prey density rating of 2, was represented second in the diet and likewise was subject to predation pressure from all of the raptor population. The exact prey density rating of the cold-blooded prey group may be questionable; but the fact that it ranked third in the raptor diet, while chiefly vulnerable to a much smaller number of birds (Table 60) than were either small birds or meadow mice, is in itself an indication that it was very abundant.

The presence of other species and groups in the raptor diet (Table 58) also shows that predation was proportional roughly to relative prey densities. White-footed mice in 1948 are the only exception to this trend

(Table 60). The low proportion of white-footed mice in the raptor diet can be traced to their scarcity in the Horned Owl diet (Table 74). The proportionately heavier winter predation on white-footed mice in 1948 (Table 33), which resulted from the relatively low meadow mouse population of that year, may have reduced the population densities of white-footed mice to such an extent that they were relatively invulnerable by early spring. Food habits data, however, strongly indicate that prey density governs avian predation in spring and summer as well as in fall and winter.

The data gathered in the Western Area follow much the same pattern. Table 59 shows that meadow mice were most frequently represented in the raptor diet and were followed by small birds, pocket gophers and Uinta ground squirrels in that order and that the population densities of these species parallel their frequency of occurrence in the diet. It is of interest that Olaus Murie (1935) found meadow mice and pocket gophers to rate first and second, respectively, in the year-long diet of the coyote in Jackson Hole. This suggests that relative population densities have a major part in all effects of predator on prey and vice versa.

Meadow mice were by far the most abundant prey on the Western Area, having attained a peak in their cycle during the fall and winter of 1946. In the spring of 1947 the woody vegetation of extensive areas was heavily girdled (Plate 6).

A 1/10-acre plot in one of the more heavily damaged areas along the Snake River showed 300 girdled stems of soapberry out of 336. Forty cottonwood saplings were not girdled. Thus 300 stems out of 376 were girdled by the wintering meadow mice.

Another plot of the same size in an aspen grove on Blacktail Butte showed 185 out of 213 aspens girdled. Only saplings one inch or less in diameter were girdled.

On a third plot, 20 out of 29 clumps of bitterbrush and 15 out of 29 aspen saplings were girdled. Observations throughout the study area showed that the meadow mouse population was generally high and environmental destruction widespread. These mice formed nearly 50 per cent of the diet of the collective raptor population (Table 59).

High meadow mouse populations were general throughout the Jackson Hole valley. In his narrative report of January to April, 1947, A. Nelson described conditions on the Federal Elk Refuge: "Meadow mice the past two years increased by thousands and even with the use of traps and poisoned grain in our yard, we were unable to catch up with the increasing numbers. Hundreds lived beneath the snow banks in our yard the past winter, where they burrowed a mass of furrows in the dry lawn grass. Suddenly, early in March, the mice began dying in large numbers and, as the snow banks melted in the spring, as high as thirty were found scattered within a four-square-yard area, and in nearly

TABLE 59

FOOD DATA SUMMARY FOR THE COLLECTIVE NESTING RAPTOR POPULATION
MOOSE, WYOMING
1947
FEBRUARY TO AUGUST

Prey Species	Total Number Represented in Pellets	Percent Total Represented in Pellets		
Meadow Mouse	607		46.2	
White-footed Mice	53	4.0		
Other Mice	37	2.8		
Shrews	5	.4 } 7.5		
Weasels	4	.3		
Ground Squirrel	149		11.3 } 82.5	
Pocket Gopher	159		12.1	
Red Squirrel	20	1.6		
Chipmunk	11	.8		
Marmot	8	.6		
Jack Rabbit	8	.6 } 5.4		
Snowshoe Rabbit	13	1.0		
Unidentified mammals	11	.8		
Ruffed Grouse	12	.9		
Sage Grouse	4	.3 } 1.4		
Mallard Duck	2	.2		17.2
Small and medium-sized birds	205		15.6	
Short-eared Owl	1	.2		
Sparrow Hawk	1			
Fish	3	.3		.3
Snake	1			
Total	1314	100.0		

TABLE 60

THE RELATION OF THE PREDATION BY A COLLECTIVE RAPTOR POPULATION TO
RELATIVE PREY DENSITIES
SUPERIOR TOWNSHIP, MICHIGAN

Prey Species or Groups	Prey Density Rating		Vulnerability Rating		Percentage of prey in the diet of the collective raptor population		Number of raptors to which prey species or groups were vulnerable	
	1942	1948	1942	1948	1942	1948	1942	1948
Small Birds	1	1	1	1	36.3	37.3	133	148
Meadow Mouse	2	2	2	2	27.8	31.7	133	148
Frogs, Snakes, Crawfish*	3	3	3	3	15.1	11.6	46	38
White-footed Mice	4	4	4	7	5.8	.1	39	47
Game Birds	5	5	5	4	4.4	7.9	96	106
Rabbits	6	6	6	5	2.4	2.5	96	106
Fox Squirrel	7	7	7	6	1.8	1.9	82	88

*Frogs include green, wood, and leopard frogs only.

every nest built of grass on the surface of the ground from one to three dead mice were found. A few mice remain, but during their peak, and as the snows melted, we found they had damaged a number of the small poplar trees planted in the yard last fall. From the 75 poplar trees planted, the mice under the snow girdled the bark from 28. The bark on some of the trees has been girdled a foot in height. Rose bushes were also damaged."

Relation of Predation on Grouse to their Population Densities

The prey recorded at individual raptor nests also shows the density relation. This, perhaps, is most clearly illustrated by the predation on Ruffed Grouse, Sage Grouse, and Richardson's Grouse (Plates 50 and 51).

A drumming census of Ruffed Grouse (Plate 50) revealed an average density of 3.4 males per square mile or seven birds per square mile on a basis of a one-to-one sex ratio. The heaviest concentrations were along the Snake River bottom, where one square mile supported 17 drumming grouse. Evidence of predation on Ruffed Grouse by the Horned Owls was greatest at nest 35, where the grouse were most numerous. As Table 81 indicates, grouse were more frequently represented in the diet of the Horned Owls nesting in the vicinity of highest grouse concentrations.

Strutting ground censuses of Sage Grouse yielded a maximum count of 70 males on two display grounds. Observations of dispersal patterns indicated that about half of the birds came from in or near the study area. On the basis of an even sex ratio, approximately 70 or 80 birds inhabited the study area, with an average density of about six birds per square mile. Only one Horned Owl nest was within hunting range of the Sage Grouse strutting ground and one Sage Grouse was identified at the nest of this owl. Three others were recovered at the nests of Red-tailed Hawks whose hunting ranges were in the vicinity of greatest Sage Grouse concentration.

The density of all grouse was low in comparison with that of other prey species and, as we would suspect, they formed an insignificant portion (1.2 per cent) of the total raptor diet. It appears, however, that raptors accounted for a minimum of seven per cent of the adult grouse population (all species) prior to and during the nesting season.

CONDITIONS AFFECTING PREY RISKS

As has been shown earlier, the relation of predation in winter to prey densities is constantly modified by conditions affecting prey risks. The result is prey vulnerability.

The conditions affecting prey risks are so numerous and so inter-related that only those having special significance in determining prey vulnerability during spring and summer months will be discussed. For clarity it is necessary to treat each such condition individually; it should, however, be remembered that they are intricately related to one another and to prey population densities. For the earlier discussion of conditions affecting risks the reader is referred to Fig. 6 and pages 176-180.

Protective Cover

In late winter and early spring, protective cover was at a minimum for most prey in Superior Township. This tended to counterbalance lowered prey densities so that prey populations still remained vulnerable to heavy predation. The months of March and April were critical. Table 16 shows that during each year a transient raptor population equal to or exceeding both the winter and the adult breeding population exerted pressure on prey populations exposed by a dearth of protective cover. With the emergence of new ground cover in early May, meadow mice, the staple winter food, became less vulnerable to all raptors (Table 54). As the vegetation grew during spring and early summer, all prey species enjoyed greater protection, but since their population densities increased through reproduction, they remained vulnerable to predators in spite of increased habitat security. Thus the vulnerability of the collective prey population did not decrease significantly with an increase in protective cover for at no time did low collective prey densities coincide with excellent cover to lower prey vulnerability.

Movement, Activity and Habits

Movement and activity increase the conspicuousness of a prey species, which in turn usually increases its vulnerability to predation. For example, populations of passerine birds are especially active and conspicuous during the periods of territorial selection and courtship. With the advance of the nesting season, activity increases with family responsibilities. It is safe to state that, in general, a total prey population shows greater activity and movement during the reproductive season and in this respect then runs a greater risk from raptors than at any other time.

Age of Prey Species

The young and immature of all prey species are, in general, more vulnerable than adults (Plate 48). Small birds, which ordinarily are not highly vulnerable to many raptor species, are very vulnerable when young, as is shown by Tables 65 through 75. The heavy predation on small birds during spring and summer in Superior Township is due partly to the greater risk that immaturity imposes on a population. The age factor perhaps is second only to prey density in determining prey vulnerability during the warm months. Immaturity and reproductive increment are difficult, however, to evaluate separately and thus the effect of these two factors in determining vulnerability cannot be separated.

Dispersion

The role that dispersion plays in increasing vulnerability of a species is exemplified by the Ring-necked Pheasant. In late March and early April these birds leave the winter coverts and spread over a wide area for nesting. This dispersion, accompanied by increased daily activity related to the breeding season, while ground cover is at a minimum, makes them more conspicuous and takes them into the ranges of new individual raptors. As a result their vulnerability increases, and it is during this period that they frequently are preyed upon.

Predator Activity

The raptors showed a definite tendency to take larger and more powerful prey during the nesting season than during winter (Table 54, Figs. 8 and 9). This tendency was evident for all species. It does not necessarily indicate that they consciously select larger in preference to smaller prey to meet the food requirements of their young, but rather that certain physiological changes associated with the reproductive processes alter their temperament so that they exhibit greater aggressiveness and courage. This occurs when food demands to meet the need of both young and adults are at a maximum. This change in temperament is reflected in the defense of nest and young as well as in the type of food taken. A pair of Red-tailed Hawks (Nest 22, Table 68) that fed almost exclusively on meadow mice during the winter and on three occasions were observed to exhibit complete indifference to cotton-tail rabbits and fox squirrels, nevertheless preyed heavily on rabbits, muskrats, and fox squirrels during the nesting season. This occurred within the same general hunting range. Other examples could be cited. There is sufficient evidence to indicate that a change in temperament exemplified by aggressive behavior plays a part in determining food habits.

These conditions affecting prey risks, and other less significant ones, all operated to modify the density-relation already discussed.

The degree of their influence upon vulnerability varied much with circumstances. At times, as has been pointed out, one or more of these conditions may be so important as to mask the density-dependent relations. In the broad view of predation, however, they are, regardless of season, secondary in importance to prey density in determining what raptors feed upon.

THE APPLICATION OF RAPTOR PRESSURE AND ITS LIMITING EFFECT ON PREY POPULATIONS

Superior Township

We have seen that raptor pressure tends to be applied proportionately to the densest prey populations and that mobility (migration) of raptors enables them to apply this pressure most effectively during winter. We shall now see that the time element also is significant in predation and must be considered when evaluating predation as a force that takes part in balancing prey populations.

Taking a broad chronological view of predation in Superior Township, we are confronted with these facts:

1. The Horned Owls' diet during the first month of the nesting season was essentially a winter diet.

2. Winter prey continued to be affected by an unrelenting raptor pressure (Table 16) until after the middle of March, when the arrival of early passerine migrants supplemented prey that had wintered. Thus early spring was a critical period for all the winter prey species.

3. By the middle of March the breeding raptor population had settled and nesting territories had been established.

4. In mid-March cock Ring-necked Pheasants established crowing areas and attracted harems. The lack of cover, the dispersion of the birds, and their increased activity rendered them more vulnerable to the Horned Owl and Cooper's Hawk than at any other time.

5. By the first of April, raptor pressure on warm-blooded prey was relieved in part by the emergence from hibernation of snakes, frogs, and crawfish. The predominant Red-shouldered Hawk population shifted from a meadow mouse diet to one in which these species formed a significant part.

6. In late April and early May, the first young meadow mice and rabbits were abroad.

7. Not until May did small birds become the most generally vulnerable prey, equalling and surpassing meadow mice as the most readily available source of food.

8. Predation on young Pheasants, fox squirrels and rabbits began in May and increased gradually as their progeny increased in number.

9. From the first of June to mid-July the young of small birds became increasingly abundant and the major source of raptor food.

10. The first litters of white-footed mice were not abroad until about the first of June and did not become an important food item until fall and winter.

11. A large migration of small birds took place in late August and in September, and many hawks moved with them. As summer residents left Superior Township other birds arrived from the north. Hawks migrating from farther north encountered peak seasonal rodent populations in Superior Township in October and November.

12. In winter, meadow mice absorbed most of the raptor pressure and were the major food of the raptor population.

13. Although the degree and duration of pressure exerted on any prey species varied with phenological events, the total raptor pressure exerted on the collective prey population was unrelenting.

The Western Area

A similar phenological situation existed at Moose, Wyoming. The breeding raptor population began exerting pressure on rodents soon after the winter snow blanket receded. This continued while rodents came out of hibernation in early May and reached a maximum when the rodent population was at its annual peak. Because of the early hibernation of the Uinta ground squirrel (Plate 49) and some other rodents, the heaviest rodent densities occurred in late summer, before the breeding season of hawks and owls was completed. Thus on the Wyoming area it was the *breeding raptor population* that exerted the maximum pressure on the peak rodent densities, whereas on the Michigan area it was the *fall and winter raptor population* that did so. In both cases, substantial reduction occurred in the rodent populations, as is evident from the food analysis samples (Tables 58 and 59).

How Raptor Pressure is Applied

Limiting pressure by a breeding raptor population is applied to prey populations in three ways: *First, by application of pressure at critical periods; second, by application of maximum pressure at periods of maximum prey density; and third, by continuous proportionate pressure on a collective prey population.*

The winter prey in Superior Township continued to support raptor pressure unrelieved until after the middle of March, at which time early passerine migrants arrived to augment the prey population and draw pressure away from the resident species that were then at their lowest density levels. It was not, however, until the first and second weeks in April, when frogs and snakes emerged in numbers, that any great shift of raptor pressure from the winter prey occurred. Thus, until early April the diet of the transient and breeding raptor populations (Table 16) was essentially a winter diet.

No raptors except the Horned Owl started nesting until many of the summer prey species were available; nevertheless, most adult raptors were on the area exerting pressure by the middle of March (Tables 37 and 38). Prior to early April the winter and transient raptors (Table 16) took part in maintaining a continual pressure on the winter prey. In most years raptor pressure from early March to early April greatly exceeded that exerted in winter.

In 1942 this pressure equalled that applied in the previous winter (159 raptors), but in 1948 the increase was 150 per cent (Table 16). Thus a very heavy pressure was applied to the winter prey species in early spring. Since these prey populations had survived the rigors of winter, they cannot in a true sense be considered surplus. They were essentially adult populations of potential breeders and the heavy mortality at this time drastically reduced the productive possibilities. When we further consider that cover was at a minimum and their activity was approaching a maximum, there is little doubt but that the generally increased raptor pressure exerted in early spring on minimum populations may produce a critical situation for a prey species.

We may conclude that during this critical period of early spring, raptors in Superior Township significantly reduced the breeding stock of prey and that the meadow mouse population definitely was limited by predation.

As the breeding season progresses this pressure on the increasing prey populations is continued by the simultaneously increasing raptor population. Food analysis shows that various prey species became prominent in the raptor diet as their young appeared. By the first of August, when the raptors attained maximum density (Tables 39 and 40), the collective prey population was at or near a peak. Therefore the maximum amount of raptor pressure came when prey was most abundant and could best support it. In late fall and early winter, when small rodents attained peak populations, migrating raptors exerted the pressure and, as has been shown, this pressure is greatest where the small rodent populations are highest.

Thus the annual picture of predation in Superior Township is one of continual application of raptor pressure (Table 16) characterized by:

1. A fall and winter pressure that tends to be proportional to the prey densities and is exerted largely on peak small rodent populations.

2. A heavy early spring pressure that is applied on minimum prey populations when cover is at a minimum.

3. A pressure exerted by adult nesting raptors that is not distributed each season in accordance with the density of prey species, but as determined by the past nesting pattern of the raptors (Chap. 10). Development of the latter has no doubt been affected by average prey population densities.

4. A maximum pressure exerted at the period of maximum prey density.

5. A spatial distribution of raptors that tends to distribute hunting pressure in proportion to immediate prey vulnerability and permits a maximum hunting pressure to be exerted.

6. Both diurnal and nocturnal pressure applied in all major habitats and on a wide range of prey species.

7. The application at all times of this pressure on the various prey populations in proportion to their respective densities.

8. The continuous application of pressure, exerting a steady regulatory influence.

These adjustments of a raptor population to a prey population are important in limiting the numbers of prey animals (Fig. 12).

Figure 12 illustrates how regulatory predation pressure is continuously applied, and, by operating on all components of a collective prey population, tends to balance each specific population at a density level with a more or less fixed relation to those of all others. In the broad perspective, it is erroneous to think of raptors preying on one prey population and then shifting to another as the more abundant prey declines. It is rather a case in which a population of raptors is adapted and adjusts itself to exert continuous pressure on each component of a prey population so that a proportionately greater number of individuals is taken from the more numerous prey species.

PREDATION AS A LIMITING FACTOR

It is important that there be no confusion as to just what is meant by the statements that predation tends to limit prey populations. Much misunderstanding has arisen from failure to make clear the author's meaning when using the terms limiting and depressing. In this discussion limiting and depressing mean that predation has a measurable and significant effect in keeping down population levels. The term, "the limiting factor," will be used as defined by Leopold (1937), "the factor that outweighs all the others in the extent to which it pulls down the unimpeded increase." It might be more nearly correct to say, the factor that appears to outweigh all others. So intricately interrelated are all factors that it is often impossible to single out a major one. It is like saying that the human heart is more important to the body than is the system of veins and arteries.

Predation is a powerful force tending to establish and maintain equilibrium among animal populations. It is to be expected that this force, which operates simultaneously and in conjunction with other regulating forces of the environment, cannot continually exert the same pressure. At times predation will appear insignificant, and other forces will be or appear to be controlling. When, however, we consider the basic mechanics of raptor predation and the continual application of its pressure in nature, there is little doubt that it can be and frequently is the limiting factor in controlling a certain kind of prey and at all times has a regulatory effect, tending to maintain an equilbrium among the varied elements of prey populations (Fig. 12). This is true on areas where man has greatly altered the original environment as well as where he has more or less left it untouched.

A raptor population is like an individual raptor species in that it is an evolutionary product. Its numbers and species have become adapted to fit the immediate environment, the numbers and kinds of prey species present being influential in guiding this process. It is, however, much more flexible than is an individual raptor or a population of a single raptor species because it responds to changed environmental conditions by change in numbers and composition.

Concrete evidence has been submitted to show that raptor predation was the factor limiting the increase of meadow mice. This was accomplished by the great reduction of their numbers during fall and winter

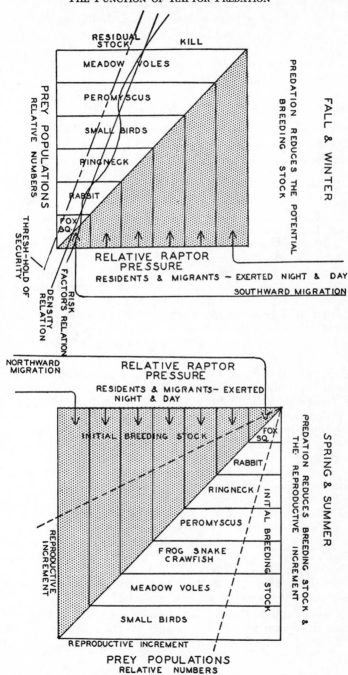

Figure 12. Diagram of predation dynamics

and by the critical pressure exerted on them during early spring. Likewise, raptor predation has been proved to be critical with respect to all the winter prey populations of Superior Township during the period of minimum population density, when it is frequently the factor limiting all available prey species.

THE RELATION OF PREDATION TO OTHER REGULATORY FORCES OF THE ENVIRONMENT

Population levels are regulated by the relations between environmental forces and the internal reproductive force. The evolutionary process has produced a delicate yet extremely complex balance between the reproductive potential of animal life and the forces or processes of the environment that resist population increases. No single force, such as predation, intra-specific self-limiting mechanisms (Errington, 1946), habitat limitations, disease, or emigration, will be controlling at all times. It would appear that a combination of such forces is necessary to keep populations in check. It is reasonable to suppose that there has been simultaneous evolution of these forces that regulate levels of animal life and tend to prevent overpopulation and of their counterbalance, high reproductive rates (compensatory breeding included). In any event these basic yet apparently opposing forces of procreation and destruction are intricately related. Predation as a population depressant may well be constructive in its destructiveness and effective either alone or in combination with other regulatory forces.

It appears that the role of predation as a population depressant has been underestimated. Errington (1943) stated, "Losses from predation and intraspecific attack increased about in proportion to the decrease of the others and vice versa, with intraspecific intolerances being dominantly operative in the event of unusually low losses through other agencies. Predation then by minks, and probably by all or most of the predators upon young muskrats of the north-central region, may often be correctly regarded more as an indicator of vulnerability than as a primary depressant of productivity." He then concluded that "intraspecific self-limiting mechanisms in conjunction with habitat limitations seem primarily to govern the general population status of muskrats in the north-central region. Even when locally nearly annihilative, predation rarely showed evidence of functioning as a true population depressant, insofar as it usually only took the place of some other mortality factor and as, in its absence, some other factor, particularly intraspecific strife, tended in its turn to become sufficiently operative to compensate for decreased predator pressure."

It would appear that Errington underestimated the importance of predation because he saw it replaced by other regulating forces, of which he believed intraspecific strife to be dominant. He stated (1943),

"Intraspecific strife usually seems to be dominantly operative and thus the ultimate check on population levels within limits of habitability of the environment." In his summary of predation and vertebrate populations (1946), Errington went so far as to conclude that "Predation looks ineffective as a limiting factor to the extent that intraspecific self-limiting mechanisms basically determined the population levels maintained by the prey." It is doubtful if we can assign a permanently controlling dominant role to any one of these basic regulating forces, which is apparently what Errington (1946) has attempted to do, and yet he recognized the fact that any one of them may be ineffective at any given time or place. He has pointed out (1943) that the failure of one of these forces to operate is offset by the functioning of some other one. It is likewise questionable if we can say accurately, as Errington (1943) has indicated, that a large proportion of the victims of predation are doomed anyway, regardless of the presence or absence of predators, and therefore predation is ineffective in controlling population levels. There is no question that a certain proportion of an annual population is doomed. The important thing is not that the individuals constituting that proportion are doomed, but how they die, and that by their removal nature contrives to maintain population levels in harmony with the environment. Predation is extremely effective as a force operating in conjunction with other resistance forces to bring about this harmony. Numerous cases in which the forces of predation, intraspecific self-limiting mechanisms, and habitat limitations exerting tremendous pressure have all been necessary to check an expanding population could be cited; therefore we cannot assign a basic role to any one. It seems clear, however, that the continual pressure, governed by relative densities, that is exerted by a population of predators on prey populations is a force so powerful and so accurately meshed with all of life that it cannot be dismissed as ineffective. We almost surely shall fail to see the function of predation if we approach it from the standpoint of predation on a single prey species or predation by a single predator species, but we need only to visualize an animal community with predation eliminated to grasp immediately how important this force is and how intricate and widespread its ramifications.

We have seen that the pressure exerted by a raptor population tends to be in proportion to the relative densities of the various prey species. Thus the raptorial pressure on the prey population of an extensive area tends to depress the various species more or less simultaneously toward the threshold of security (Fig. 12). Observations of predation by a single raptor species at times will show marked pressure on one or a few prey species and will at other times show little or no pressure on the same species. In evaluating such data we should keep in mind the fact that they are only part of the pertinent information. On such a basis we could, according to the partial data at our disposal, judge predation to be either nearly annihilative or ineffective. The mechanics of predation are such that no single species in a multiple prey population can under normal environmental conditions draw sufficient predation pressure to keep its population level dangerously depressed. This we should naturally expect, and this fact cannot be used as an argument against the effectiveness of predation as a controlling or regulating factor. The killing of a prey animal by a predator does not necessarily mean a lower prey population than would have existed had the act of predation not occurred; but the continual killing of prey animals in a prey population means a continuous proportional loss of animals, which tends to keep the population within limits. The argument that other regulatory forces would become operative in the absence of predation or that intercompensation (Errington, 1946) would offset the forces tending to lower populations has no bearing on the role of predation as a force regulating and at times limiting populations. The same could be argued for any other limiting factor.

As a matter of fact, if predation frequently can act as a limiting factor, and if as we have shown it becomes most effective in late winter and early spring reducing the over-wintering populations (breeding stock) then we could expect to find a reproductive mechanism evolved to counterbalance it and other depressants simultaneously operative. Compensatory breeding would appear to be such a mechanism. Just as predation or intraspecific strife are responses to high density levels, so compensatory breeding is a response to low density levels. This response is exhibited when spring to fall prey population gains show an inverse ratio to spring breeding densities. It does not indicate the ineffectiveness of predation but rather the complexity of population producing and destroying mechanisms.

It should be noted that food supply, for example, is sometimes a factor limiting populations and when this factor has a strong depressing effect, predation naturally will tend to regulate prey levels only within the limits imposed by the food supply. In such cases, predation cannot be considered the limiting factor.

Although predation can be the limiting factor, we should perhaps have a truer concept of it if it were thought of not in terms of when and how it may assume this role, but rather as a regulatory force continually operating to lower prey increase in proportion to prey density and to do this before more drastic but less steadily functioning forces become effective. These other forces seldom, if ever, affect the total prey population simultaneously, but are confined to specific prey only. Disease may strike one prey, while food shortage may regulate another. In contrast, predation strikes all components of the collective prey simultaneously and continually.

In evaluating the relative importance of predation as a natural regulator of animal populations, it must be considered with reference to all other biotic and physical forces operating simultaneously to effect control. In some situations, physical conditions—temperature, snow depth, moisture—impose a high order of control, while in others biotic conditions, such as food supply, cover or disease, dominate. The fact that seldom, if ever, is any single regulating force independent of others makes it difficult or impossible to determine the relative importance of each.

Nicholson (1933) differentiated between decimating and controlling factors and formulated the idea that only those mortality factors whose pressure automatically increases with increase in the size of the population could be considered as truly controlling a population. Nicholson termed such a factor a density-dependent mortality factor. Whether or not we accept Nicholson's definition of a controlling force, we must recognize the significance of his concept. It is not alone the number of animals that a mortality factor or force accounts for that determines its controlling effect, but the way in which it operates. This becomes even more significant when we think in terms of collective rather than specific populations. Our data show predation to be a force controlling prey populations because it tends to reduce them in proportion to the relative numbers of each included species—it provides continuous control, regulated by relative densities.

As a suppressive force, predation does not reduce populations to very low numbers, as do epizootics or starvation. Moreover, the number of any single prey species accounted for may be far less than the number killed by a hard winter or a wet spring. Sudden drastic reduction in numbers temporarily releases a population from the pressure of control forces, thus allowing population densities to be again increased. This creates a type of control characterized by excessive fluctuations in numbers. Where predation is dominantly operative, however, control is characterized by continuous and proportionate reduction that tends to keep population levels near a mean. The fact that predation operates as a steadily functioning force throughout the seasons and year after year, in spite of continually changing physical and biotic conditions of the environment, gives it a great advantage, in comparison with regulators that operate intermittently or only under special conditions, in harmoniously regulating animal populations with one another and with the rest of their environments.

Some workers have advanced the thought that predation may tend to maintain a balance under wilderness or primitive conditions, but that where man has altered the land complex, a "balance of nature" no longer exists and predation then loses its regulatory function. Our data tend strongly to refute this hypothesis, since the force of predation operated similarly to maintain balance both in an area of intensive land use and in an almost unaltered wilderness region. It would appear that on any area of land, animal populations tend toward stability of interrelationships and that predation in greater or less degree plays an important role in establishing and maintaining such a state of balance.

CHAPTER 14

Predation's Annual Toll

W E HAVE attempted to comprehend the meaning of a complex life pattern and our interpretations have presented a view of how raptor predation operates and how it may function as a population regulator. This has afforded a glimpse of processes and relationships that have been evolving for many thousands of years between raptors and their prey. A question that now presents itself is this: How many prey animals actually are destroyed by raptors, and can we express this predation in Superior Township in terms of numbers of mice, pheasants, rabbits and other major prey annually killed? If so, will such figures refute or confirm the statements and conclusions we have made concerning raptor predation? A quantitative evaluation of the approximate numbers of major prey killed by the hawks and owls may be expected to emphasize a new method of approach and broaden our search for the basic facts.

In order to calculate the prey that raptor populations annually took off Superior Township we must have the following information:

1. The number of raptors and the length of time present.
2. An adequate sample of the raptor food.
3. The relative abundance of major prey species.
4. The average weights of major prey species available to the raptors.
5. The quantitative food requirements of the raptors.

We have already presented data on the first three requirements. Our task is to fulfill the last two, and then interrelate all our information in such a way that we can calculate the approximate numbers of prey species killed. In brief, our procedure will be to determine raptor-days by multiplying the average numbers of hawks and owls in Superior Township by the number of days they were known to hunt the area. This will give us our first expression of raptor pressure. By dividing the raptors into weight classes and multiplying the number of raptors in each class times the average weights of the various raptor species, we obtain total raptor weight as a second expression of raptor pressure. From feeding experiments, we shall determine average daily rations for the raptor species and for the raptor weight classes. This information, when multiplied by numbers of raptor-days, gives us raptor pressure expressed in terms of grams of food consumed by the raptor populations. On a basis of known prey weights, we can then determine the weight of raptor food represented in the food samples and calculate the percentage by weight of each prey species in the raptor diets. With this accomplished, we shall then compute the total number of grams of the various prey animals consumed by the raptor populations. Then, in the final step, we shall convert grams of prey animals to approximate numbers of the various prey animals consumed. We must recognize, however, that there are minor variables which we have been unable to measure, as well as some shortages in the data that we have measured. Thus, we can hardly expect a simple foolproof calculation and the end product will not be exact values but approximate ones. The reader is urged to consult tables 83-100 in the appendix.

DETERMINATION OF RAPTOR FOOD REQUIREMENTS

Quantitative Feeding Experiments

From 1939 to 1942, 29 birds of prey, representing 11 species, were fed to obtain quantitative data on food requirements (Tables 83, 84 and 85). Captive birds were used in the experiments (Plates 64, 65, 66). All were equipped with jesses and leash, (Plate 66), placed on perches, and handled and cared for with the falconer's techniques (Craighead, 1938). The birds were tame and well adjusted to captivity. They were as close to the wild state physiologically and psychologically as it was possible to keep them and still have them under control.

Some investigators have questioned the use of captive birds in determining the food requirements of wild birds. In order to minimize the differences between the wild and captive states, the raptors were handled individually each day to keep them tame. Some were flown periodically to give them exercise (Plate 64) and others were completely free for varying periods of time (one for as long as 127 days), returning regularly to be fed and weighed. A docile captive bird far more closely approaches its wild counterpart in its activity, food requirements and behavior than does a wild "unmanned" bird kept captive in a cage. We do not know just how closely the food requirements presented in Tables 83 and 84 approach the requirements of wild birds. Experience indicates, however, that the feeding of captive birds is a practical and fairly accurate method for obtaining this information. Results may be somewhat biased by a diet consisting largely of raw beef and by lack of constant exercise. Nevertheless, it is believed that the variable factors in the feeding experiments are compensatory to a degree that would tend to make the recorded food requirements approximate closely the true state.

Feeding Method

The hawks and owls were kept out of doors, exposed to weather, and they were normally fed once a day at a regularly scheduled time. They were allowed to take only such food as they could consume within a 30- to 40-minute feeding. Uneaten food was then collected. The daily ration of each raptor was weighed to the nearest 1/10 gram. Each hawk and owl was weighed every four or five days to record gain or loss in weight. The daily rations of the birds were regulated so as to hold the raptors at rather constant body weights throughout the feeding experiments. Where there was a slight gain or loss, the bird was brought back to the starting weight before the feeding was terminated. Maximum and minimum temperatures were recorded daily. During the fall and winter periods, the feeding experiments were conducted with individual birds for as long as 118 days and in spring and summer for a maximum of 86 days (Tables 83 and 84). The data indicated that for well-adjusted birds, four weeks of feeding will yield an average seasonal ration for each species studied that is sufficiently accurate for the purpose in hand.

The staple ration was raw lean beef, occasionally supplemented with ground "green" bone. Rodents, birds and other natural prey were added to the beef diet two or three times a week and in some cases more frequently. This rounded out the dietary requirements and afforded roughage, such as fur, bone, and feathers, which was ejected in the form of pellets. The birds remained healthy on this diet. One Duck Hawk used in the experiment remained healthy and vigorous for more than seven years (Plate 65). Three of the experimental birds were kept for a year or longer and comparative data obtained from them for the cold and warm seasons (Table 85).

Feeding Data

A summary of the feeding data is presented in Tables 83 and 84. It should be noted that regardless of species, the percentage of average body weight eaten per day is correlated closely with the weight of the raptors and with air temperature. The only major exception appears to be the Great Gray Owl (Plate 62), which consumed only 7.4 per cent of its body weight at an average temperature of 14 degrees Fahrenheit.

This bird is more densely feathered than the Horned Owl, and it is quite possible that the added insulating effect accounts for the relatively low food consumption in relation to body weight and temperature. A lowered metabolism at low temperatures may also be involved. The Barred Owl likewise did not follow the general trend for the medium-weight class of raptors.

Table 85 shows the differences in percentage of body weight consumed during cold and warm weather. Here raptors have been grouped into weight classes; the large raptors ranging from 800 to 1,200 grams, the medium raptors from 200 to 800 and the small raptors from 100 to 200. This has been done to facilitate the application of the feeding data. It proved to be impracticable to feed the same individuals, or even the same species, during both summer and winter periods in all cases and thus the data are not strictly comparable, except for those that are starred (Table 85). The correlation of food consumed with raptor weight and average temperature is sufficiently constant, however, to justify a comparison of weight classes, even though the species treated are different.

The data show that in the relatively colder weather of fall and winter the large raptors ate 3 per cent more of their average body weights than raptors of this weight class consumed during the warmer weather of spring and summer. The medium-sized and small-sized raptor classes ate about 3 and 8 per cent more, respectively, in cold weather than in warm weather.

APPLICATION OF FEEDING DATA

Raptor Pressure Expressed as Raptor-Days

An objective of the feeding experiments was to determine the amount of food required by raptors to maintain their body weights, so that this information could be applied to compute the amount of food necessary to maintain the known raptor populations inhabiting a definite land area, Superior Township.

Table 16 shows the year-round raptor populations in Superior Township during the study years. These data for 1941-42 and 1947-48 are converted to raptor-days in Tables 86 and 87. During this time there were 98, 85, and 14 hunting-days during the fall, winter and spring transition periods, respectively. The average number of hawks and owls of each species times the number of hunting-days on the area, yields raptor-days for the fall and winter periods and is one expression of the raptor pressure. This value is 32,471 and 13,317 raptor-days for 1941-42 and 1947-48, respectively (Table 86).

Raptor-days for the spring and summer periods were determined in a similar manner. In this case, however, the number of hunting-days varied for each species (Table 87). The spring period for all raptors was considered to begin on March 18 (the end of the spring transition period) and to extend, in the case of each species, to the average hatching date.

The summer period extends from the average hatching date for each species to September 1, the beginning of the fall period. Thus the spring raptor pressure is exerted only by adults while the pressure in summer is the result of predation by both adults and juveniles (Table 87). For example, the average hatching date for the Red-shouldered Hawks for both study years was May 12. Thus the spring period, during which adults selected territories, built nests, and laid and incubated eggs, was 55 days. The summer period, during which the adults fed their young in the nest and then both adults and juveniles remained within the hunting ranges, was 112 days (Table 87). The hunting-days for fall, winter, spring transition, spring and summer constitute a year. Thus some Red-shouldered Hawks were present in Superior Township throughout the year and exerted pressure by hunting 197 days in fall, winter and spring transition and 167 days in spring and summer.*

In the spring of 1942 there were 40 Red-shouldered Hawks hunting for 55 days or 2,200 Red-shouldered Hawk days. In summer there were 77 exerting pressure for 112 days or 8,624 raptor-days. This makes a total of 10,824 Red-shouldered Hawk days for the spring and summer periods of 1942 (Table 87). In the fall and winter of 1942 there was a total of 813 Red-shouldered Hawk days (Table 86). Thus the Red-shouldered Hawk population for 1941-42 produced a hunting pressure totalling 11,637 raptor-days. In a similar manner raptor days were computed for each raptor species (Tables 86 and 87).

All nestlings that attained the fledgling stage were considered to represent raptor pressure, even though they did not capture prey but were fed prey killed by the adults. For example, a one-day-old Red-shouldered Hawk that survived to leave the nest was considered to represent one raptor-day for every day in the nest and for every day after leaving the nest until September 1. Raptors that hatched but did not survive to fledge were not converted to raptor-days or considered as raptor pressure.

The next step is to convert raptor-days to raptor pressure expressed in grams of raptor weight and then to derive from this a statement of grams of food consumed.

Raptor Pressure Expressed as Raptor Weight

Tables 88 and 89 express raptor pressure in terms of raptor weight. Information in this form was obtained through computing the average number of raptors of each species in three weight classes (large, medium, and small) that periodic censuses showed were present in Superior Township during the fall and winter periods. The average weight of each raptor species (Table 99) was then multiplied by the average number of individuals present to give total specific weight and the sum of these weights was obtained for each class and for the col-

*Calculations were made for 364 days instead of a complete year.

lective population. Tables 88 and 89 give weights for each raptor species, for the weight classes and for the collective populations on the area. In 1941-42 the raptor pressure through the fall and winter amounted to 112,094 grams of raptor weight and in 1947-48 to 59,271 grams. The average weights of the large raptor groups on the area were 1,211 grams and 1,318 grams, respectively, in the two study periods. For the medium and small weight classes they were 469 and 539 grams, and 159 and 160 grams, respectively. These figures were obtained by dividing the total raptor weight of each weight class by the average number of raptors in each class (Tables 88 and 92).

In the spring and summer periods the raptor pressure was 96,579 grams and 105,703 grams in 1942 and 1948, respectively. The average weights of the large raptor groups were 1,398 grams and 1,318 grams. For the medium and small weight classes they were 554 and 559 grams, and 160 and 159 grams, respectively (Tables 89 and 93).

In Table 89 the numbers of raptors represent averages of the spring and summer populations. As has been pointed out, the summer populations consist of adult birds and fledged juveniles. For the purposes of our calculations the juveniles from the time they hatch are considered equivalent to adults. Thus the total raptor weights in Table 89 are computed by multiplying the average number of raptors (adults and juveniles) times the average weights of adult birds (Table 99). This would appear to yield a biased figure for the total raptor weight, since the juveniles are growing, and those of large raptors approach adult weights only at 8 to 9 weeks of age. It can be justified, however, by the fact that during this period growth replaces weight as a factor in food consumption. Feeding experiments with young birds showed that in general the average ration of young raptors, 1 day to 7 weeks old, was only slightly less than the average spring and summer ration for full-grown raptors of the same species. For example, the average ration for two Red-shouldered Hawks fed from the day of hatching (hatched in incubator) to 51 days of age was 55.8 grams for the male and 62.4 for the female. The ration for full-grown birds during the summer months was 58 grams for the male and 71 for the female (Table 84). Thus the total raptor weights of Table 89 actually are an expression of raptor weight and raptor growth as they relate to food consumption.

This means that during the growth period a young Red-shouldered Hawk consumes an average quantity of food that nearly equals an adult's consumption for the same period of time. The food intake of the growing bird increases from about 4 grams at one day old to a maximum of 100 grams at 3 to 4 weeks, then gradually tapers downward. Growth curves for young Sparrow Hawks (Plates 60 and 61) and Horned Owls further support this. If we accept the fact that the food requirements of a growing raptor during its developmental stage closely approximate the requirements of an adult bird, then the raptor pressure for a nesting population of adults and maturing juveniles can be rather simply calculated by multiplying raptor numbers times average adult weight records as in Table 89 and this can in turn be converted to food consumed by multiplying the adult ration by the number of raptor-days (Table 91). This eliminates the necessity of presenting growth and food curves and unnecessarily complicating our calculations.

Raptor Pressure Expressed as Grams of Food Consumed

Tables 90 and 91 progress one step further by converting the raptor pressure expressed as raptor-days into raptor pressure expressed as grams of food computed to be consumed by the raptor populations. These data were obtained by multiplying the computed average weight of food required per day by a raptor of each weight class by the total number of raptor-days for that weight class, shown in columns one of Tables 90 and 91. These rations required per day by a raptor of each weight class on the study area are computed, as shown in Tables 92 and 93, by applying the percentage of average body weight eaten per day by each raptor weight class (large, medium, and small) in the feeding experiments to the average weight of raptors of the same weight class on the study area (Tables 88 and 89).

Computed rations were sought, instead of utilizing the rations determined through feeding, because compensation then can be made for variations within raptor weight classes in the area. Food consumed is so closely correlated with raptor body weight that the raptor rations for the populations are most accurately expressed in terms of percentage of average body weight eaten. Variation in weights of the raptor weight classes on the study area is shown in Tables 92 and 93. These values in the two study years are different because of differences in the numbers of individuals of the several species making up each weight class; for in each year the average weight of a class was calculated by multiplying the constant average weight for each included species by the number of individuals of that species present, adding the results of these multiplications and dividing the total thus obtained by the total number of individuals present in that class (Tables 88 and 89). If one of the heavier species in the class was more numerous, this tended to increase the average weight for the class. As an example, the average weight of the large raptors was 1,211 grams in 1941-42 and 1,318 grams in 1947-48. This difference is due largely to the increase in Horned Owl numbers in the latter year. The computed rations correct for this.

It was computed that the raptor population consumed about 2,964,-392 grams of food in the fall and winter of 1941-42 and 1,182,633 grams in the corresponding periods of 1947-48. During the spring and summer periods of 1942 and 1948 the populations are calculated to have consumed 2,037,687 grams and 2,058,909 grams, respectively.

With these figures as a basis, the next step is to compute the approximate number of the various prey animals consumed by the raptor populations. We can do this by projecting our food data samples. Within the limitations of our data, this will give an expression of predatory pressure in terms of animals killed. The numbers of the assorted prey animals computed to have been consumed, as well as the data used to determine these, are presented in Tables 94 and 95.

CALCULATION OF PREY ANIMALS CONSUMED

The numbers of prey animals computed to have been consumed by the raptor populations during the fall and winter periods (Table 94) were calculated in the following manner: With an exception discussed below, the average weight of each prey species (Table 100) was multiplied by the number of that species in the food samples of Column 3, Table 94. This gave the total food weight in grams of each prey species or group represented in the food samples (Column 4). The exception is that in the case of large prey, whose average weight exceeded the maximum a raptor could consume at one feeding, the average maximum raptor ration, instead of the average prey weight, was used in the calculations. This latter figure (recorded to the nearest decagram) is an average of the maximum daily rations eaten by the Red-tailed Hawks and Horned Owls. It does not represent the maximum that one of these raptors can eat by gorging. This exception was made because there was evidence that large prey were not utilized completely by raptors but were frequently abandoned by them following an initial maximum feeding and were not revisited. Some were consumed by mammalian predators before the raptor could return for a second meal. Where prey was abundant, raptors seldom returned to a cold kill. There is no doubt, however, but that in a number of cases a large prey does furnish a large raptor with several meals and conversely it is not infrequent that a kill is made, the raptor disturbed, and little or no food consumed. The two situations may well be compensatory to the extent that on the average and over an extended period of time one maximum meal per large prey most closely represents the utilization of such prey by raptors during the cold months.

The food samples of Column 3 include not only data obtained from pellet analysis but also kills observed in the field. For example, we have included small bird kills for the Cooper's Hawks in order to have the samples more nearly complete (compare with Table 33). To make these food samples still more representative of the entire raptor populations, it was necessary to weight the diet of the Horned Owl. In 1941-42 the Horned Owl pellets represented 5.9 per cent of all raptor pellets collected and the owls themselves represented 6.7 per cent of the entire raptor population. In 1947-48, however, Horned Owl pellets represented 37.9 per cent of all raptor pellets, while the Horned Owl population was only 15.4 per cent of the total raptor population. There was thus a preponderance of food data for these owls in 1947-48 and furthermore these data included most of the large prey forms recorded for the entire raptor population. For example, of 47 rabbits and 16 pheasants recorded in the food sample for the 1947-48 raptor population, 43 and 15, respectively, were accounted for by Horned Owls. It was reasoned that an ideal food sample for a raptor species would be one whose percentage of the total food sample for the population equaled the percentage of that species in the total population. The 1941-42 Horned Owl diet was thus thought to be reasonably representative and required no altering. The 1947-48 diet was corrected by dividing 15.4 per cent (the percentage of the Horned Owls in the total raptor population) by 37.9 per cent (the percentage of Horned Owl pellets to all raptor pellets), which yields 41 per cent. Then 41 per cent of the number of each prey item in the 1947-48 Horned Owl diet alters the sample so that it represents more accurately the Horned Owl diet in the total raptor diet. This reduces the Horned Owl food sample by 59 per cent. The data presented in Column 3 of Table 94 are then as representative a diet for each raptor population as we can offer. Column 5 of Table 94 shows the percentage of each kind of prey in the total weight of food samples. Large prey of course form a higher percentage of the total prey weight than of total prey numbers. This difference may be seen on comparing Column 5 of Table 94 with Table 33. One large prey form, such as a pheasant, is equivalent in raptor food value to approximately seven meadow mice. With such data it is possible to evaluate predation not only in terms of prey numbers but in terms of prey weight. For example, let us consider a hypothetical raptor winter diet that consists of only two prey species— meadow mice and Ring-necked Pheasants. This diet contains 28 meadow mice and two pheasants, which, by numbers, is 93 per cent field mice and 7 per cent pheasant. By prey weight, using the weight of an average maximum raptor ration to represent each pheasant, the diet is 67 and 33 per cent, respectively—the larger prey forms thus assuming greater significance in the diet.

The computed numbers of grams of prey animals consumed by the raptor populations are found in Column 6 and were obtained by applying the percentages of Column 5 to the total grams of food consumed by the populations (Table 90). The computed numbers of prey animals consumed (Column 7) were obtained by dividing the figures in Column 6 by either the average weight of prey (Column 1) or the average maximum raptor ration (Column 2). If the average prey weights were used initially to compute the total prey weight in grams included in the food sample (Column 4), then they were used again to convert the data of Column 6 to the prey numbers recorded in Column 7. The average maximum raptor ration was used in corresponding fashion.

Table 95 shows the computed number of prey animals consumed by the raptor populations during the spring and summer periods. With minor exceptions, the computing procedure is similar to the one we have just reviewed for Table 94. There are some differences in the type of data that should be noted.

A high percentage of all prey taken by raptors during the spring and summer months proved to be juveniles. As would be expected, weights of these immature prey varied greatly during this period. It was found that the average weight of juvenile small-prey forms, corresponding to those found at raptor nests, was normally only about half the average weight of winter adults and the average weight of juvenile large prey was generally about one-third the adult winter weight. Accordingly, in Column 1, Table 95, the average weight of small prey is recorded as one-third the winter weights of Table 100, with the exception of frogs, crawfish, and other items, for which the adult weights were employed. These exceptions were made because, as far as could be determined, the only frogs and crawfish taken by raptors were adults. There was also no logical reason for altering the weight of "other items," since this category encompassed a wide weight range of miscellaneous prey.

In the case of large prey, calculated juvenile weights, rather than the average maximum raptor ration, were employed in computing grams of food consumed. The reason is that while young were being fed, much of the raptor food was carried to them and with few exceptions was completely consumed. Calculations were based on our observations that, contrary to the situation in fall and winter, large prey usually were utilized fully during the warm months. Though this was true in Superior Township, we have elsewhere recorded exceptions.

In order to make the food sample of Column 3, Table 95, more representative, it was slightly altered from that presented in Table 33. The number of frogs in the sample was doubled because a comparison of regurgitated food with items found at the nest indicated that at least half left no identifiable evidence and thus were not recorded. Forty pheasants were added to the Cooper's Hawk diet of 1941-42 to make it comparable to the 1947-48 diet. This has been explained and qualified in Chapter 12. The computed prey kill appears in Column 7 (Table 95).

INTERPRETATION AND EVALUATION OF COMPUTED PREY KILLS
Fall and Winter

To what extent our calculations are correct can be judged to some degree by comparing the computed raptor kill for the cold periods with the estimate of prey populations (Table 34). By calculating the percentage killed out of each major prey population, we can see if our computed kill is reasonable. This we have done, and Table 96 shows the approximate percentages of the major prey populations computed to have been killed by raptors during the fall and winter periods. The figures indicate that predation is responsible for a significant reduction in the prey populations. In 1941-42, raptor predation accounted for 24 per cent of all the major prey species in Superior Township. In 1947-48 it accounted for 20 per cent. This is in harmony with the view, already stated, that predation is effective as a population control. These percentages are based on the prey population projections and the prey numbers computed to have been killed, and are presented chiefly to show relationships.

It should be noted that in the case of the meadow mouse and white-footed mice there was not an increase or a decrease in the percentages of these populations killed to correspond with the size of the populations. There was, however, greater raptor pressure on the numerically larger populations and a greater number of individuals killed. Nicholson (1933) formulated the concept that, ". . . the action of the controlling factor must be governed by the density of the population controlled." He explains that the pressure exerted by this factor must automatically increase with increase in the size of the population. Predation, he points out, varies in intensity with the numbers of the prey and thus exerts a controlling effect. On this basis, Nicholson classifies predation as a "density-dependent mortality factor." It is not clear whether Nicholson means that the automatic increase of predation pressure is such that the percentage of prey taken increases or only whether the total number taken increases. For example, if predation accounted for 20 animals in a population of 100, then, to effect the same extent of control in a population of 200, predation would have to account for 120 animals. In the first case it would account for 20 per cent of the population and in the second case for 60 per cent, but in both instances 80 animals would be left to reproduce. We have no evidence that raptor predation effected this degree of control. The number of prey animals killed by raptors was substantially greater when the populations were at high levels but the percentages killed either did not rise or did not rise sufficiently to maintain the same extent of control. In 1941-42, when the meadow mouse population was at a high of approximately 303,000, the raptors accounted for what we have calculated to be about 26 per cent of the population. In 1947-48, when this prey species was at a much lower population level, approximately 75,000, the raptors accounted for 22 per cent. Approximately 22 per cent of the white-footed mouse population was killed by raptors when the population level was 33,000, and 24 per cent at a population level of about 27,000 (Table 96). On this basis we conclude that raptor predation is density-regulated, and we can say it is density-dependent, thus supporting Nicholson's findings; provided that by a density-dependent mortality factor, Nicholson means a factor or force killing a greater number of animals but not a greater percentage when populations are high than when they are low. This density relation, as we have seen, when applied to the components of a collective prey population, enables predation to function as a precise and powerful regulatory force.

The cottontail rabbit appears to have been the most vulnerable cold-weather prey, and both it and the fox squirrel apparently were as vulnerable at very low population levels as at higher levels. Why this should be is not entirely clear, but probably is intricately related to risk factors and to the relative vulnerabilities of the other major prey species. Their vulnerability (Table 96) was high because they were largely confined to woodlots, 11 per cent of the total land area; and the Horned Owls, their major predator, were similarly confined. The greater chance of contact between raptor and prey tended to compensate for low prey density. We suspect the percentages of game birds taken by raptors represent much of the winter mortality from all causes. The 18 per cent calculated for 1947-48 is a high winter mortality for Ring-necked Pheasants and Bob-whites, but there is no reason to believe that this does not express rather accurately what took place.

The weakest data in our calculations are the food samples. If these fairly represent the food of the populations then the numbers of each prey species computed to have been killed or consumed should be fairly close to the true state. If we assume that the total numbers of prey calculated to have been consumed are approximately correct, we find that the 11,569 individual prey items in the fall and winter food sample for 1941-42 are approximately a 13 per cent sample. The 1,557 prey items of 1947-48 are a little more than a 5 per cent sample. In each year the spring and summer food samples are more than 3 per cent of the total food. Thus we believe that our calculations for the fall and winter of 1941-42 are the most accurate and that in general our fall and winter calculations of prey animals consumed (Table 94) more closely approach the true state than our computations for spring and summer (Table 95).

Spring and Summer

Limitations effective during the warm months must be recognized before we can evaluate our prey-kill calculations. We found it more difficult to obtain a representative food sample in this period. Not only is the physical problem of collecting greater than in winter, but in addition, the following conditions presented difficulties:

1. The number of species preyed upon is greater and thus, to attain significance, a larger sample is required.

2. The young of prey species become very vulnerable when they first venture abroad and then become progressively less vulnerable as they mature.

3. The chronology of nesting raptors is such that some have completed nesting when other species are just beginning. This necessitates a long period of sampling.

4. The young of prey forms producing only one brood or litter appear and are especially vulnerable for relatively short periods of time, while the young of prey species that produce offspring throughout the spring and summer remain vulnerable for a much longer aggregate period. This necessitates regular sampling to obtain adequate samples and avoid disproportionate ones.

5. Food samples can be readily gathered during the nesting period, but it is difficult to obtain data prior to nesting or after departure of fledglings. This tends to bias the data in favor of the spring periods. There is as yet no feasible way to obtain adequate food samples for all raptor species in late summer.

In interpreting our calculations, we must consider that we are utilizing food samples largely representing prey taken by adults during the nesting and early post-nesting season and projecting these for late summer populations of both adults and juveniles. This we recognize as our greatest potential source of error. Field observation indicates juveniles do not take as many large prey forms as do the adults. Also there is evidence that the adults turn to a diet consisting of fewer large prey forms in late summer, when the instinct to feed their young has diminished.

Because of these inherent difficulties we wish to caution that our calculated prey kills for the spring and summer periods express the predation in Superior Township as a projection of the food samples, which are somewhat biased in favor of spring feeding habits. Therefore, the large prey forms probably are overemphasized and the small prey forms, especially rodents, underemphasized. Nevertheless, we believe that it will prove instructive to project the spring and summer food samples as we have done in Table 95, since they are the most nearly complete food data ever accumulated for raptor populations on a definite land area. There is evidence that the calculated kills are close enough to the true state to be significant. Consequently, they should prove useful in helping us to visualize predation in quantitative terms. It is not the figures, however, but the method that we particularly wish to emphasize. We shall interpret Table 95 with this in mind.

The data of Column 7, Table 95, require some explanation and qualification. Our computations indicate a high kill of major game species, with a particularly heavy take of Ring-necked Pheasants. To evaluate the raptor kill properly, we should have data on the productivity and total mortality of the game species in Superior Township. Such data could not be obtained, but we can compute productive potentials from the information at hand and determine the percentage of potential fall populations lost to raptors. Our population data relating to the Ring-necked Pheasant are more accurate than those obtained for the other game species. By calculating the productive potential for this species in the township (English, 1945) we shall establish a basis with which to compare the calculated pheasant loss to raptors. In Table 97, the pheasant mortality from raptors is expressed in terms of the degree of reduction of a theoretically possible pheasant population. The 2,487 pheasants calculated to have been lost to raptors in 1941-42 constitute a 16-per cent destruction of potential birds. English found that the actual fall pheasant populations in Michigan were double the winter populations and that the total loss of the potential fall population was nearly 90 per cent. Thus mortality calculated on a basis of potential birds is generally high and a raptor destruction of about 16 per cent may occur, especially under circumstances where nest destruction is at a minimum and where some re-nesting occurs. The percentage of pheasants produced that is taken by raptors, of course, is appreciably higher.

We believe, however, that our calculated pheasant loss to raptors is higher than what actually occurred. We suspect that a larger food sample, with more data for late summer, would have shown some reduction in the ratio of captured pheasants to other prey and thus a reduction in the total number of pheasants computed to have been consumed. Likewise, more data in late summer in both years would undoubtedly have yielded a food sample proportionately higher in meadow mice, small birds and other small prey forms.

In view of the relatively low rabbit population in the winter of 1941-42, the number of rabbits computed to have been killed the following spring and summer appears at first to be excessive. When we consider, however, that only about one young cottontail rabbit in five survives until fall (Haugen, 1943), then the computed kill by raptors for 1942 is roughly one-third of the mortality to be expected from all causes. In 1948 it was apparently much less than this, perhaps only about one-tenth of the total cottontail mortality.

It is difficult to reconcile the relatively frequent occurrence of fox squirrels in the raptor diets and the computed fox squirrel kill (Column 7) with the small wintering populations and with what we know of the productive potential of these animals (Allen, 1943). It would appear that either the food sample is biased in favor of the fox squirrel or else the winter population estimates are low. We cannot rule out, however, the possibility of abnormal age ratios or of compensatory breeding which would allow a greater than normal increase of fox squirrels. All these factors may have a bearing. In general, however, the calculated kills are within the realm of possibility. In other words, they show, on a basis of our knowledge of the existing prey populations and their reproductive potentials, that the kills computed to have been made in each species or group could have occurred.

There have been few quantitative measurements of mortality from predation within a given prey population. Tinbergen (1946) measured the mortality inflicted on four species of small birds by the European Sparrow Hawk and compared this loss to the total expected mortality for each species. He concluded that Sparrow Hawk predation accounted for at least half the summer mortality of House (English) Sparrows; about a quarter of the mortality of Chaffinches and Great Tits; and a small percentage of that of Coal Tits. The magnitude of the raptor mortality we have calculated is in harmony with the findings of Tinbergen.

It is reasonable to conclude that the number of large prey forms consumed by the raptors in Superior Township during the warm months was significant and indicates generally a higher mortality among game species than most workers have ascribed to raptor predation. Direct evidence of predation on a prey species is generally difficult to obtain and for this reason biologists have usually expressed such mortality at a minimum.

A QUANTITATIVE EVALUATION OF PREDATION

In the light of our quantitative approach, we are now prepared to reconsider some important questions. Do our quantitative data indicate that raptor predation can function as a control by precisely and effectively reducing populations? Do they show that it acts on specific prey populations and on collective prey populations as a regulatory force?

In the winter of 1941-42 raptor predation accounted for approximately 26 per cent of the meadow mice (Table 96). Here raptor predation controlled by drastically reducing a population of high density. There is also evidence that in early spring it temporarily limited further increase by reducing the productive possibilities. During the winter of both years raptor predation accounted for nearly a quarter of the white-footed mouse populations, thus greatly reducing numbers of these species. As we have seen, it accounted for approximately 24 and 20 per cent of two winter prey populations.

A comparison of the percentages of prey species in the total prey kill with the percentages of those species in the total prey population further confirms the earlier conclusion that predation is largely determined by the relative densities of the prey, or, to repeat what was stated previously, that in the case of each major prey species or group the ratio of the total population of that species or group to the total of the collective prey population approximates the ratio of the kill from the species or group by a collective population of raptors to the total kill from the collective prey population by the same raptor population. This relationship is shown in Table 98.

We can conclude that the total weight of food required by a raptor population and the number of prey animals killed during a year are of such magnitude that raptor predation must be recognized as an effective biological control; furthermore, that the way in which raptor predation acts on collective prey throughout the year to effect this prey reduction strongly suggests a precise regulatory force. The degree of effectiveness that such predation may exhibit will vary much at different times and in different places, since it will depend on other population regulators simultaneously operative and the degree to which they are effective. Therefore predation should not be judged solely on how much it may reduce prey populations, but also on the way in which it brings about this reduction—its ability to do a regulated control job. The removal or suppression of any population control eventually will throw a greater burden on intact remaining ones. The question is which ones —disease, predation, emigration, inter- and intra-specific strife, food supply, climate, human activities—do we wish to be controlling. Where man desires and is able to exercise complete control, the other causes of population loss can be suppressed without damage; where he does not wish to control or cannot control them it appears that predation helps to perform the task in a way more precisely regulated than those of other forces. This means that in raptor predation we have at our disposal a basic natural phenomenon that if allowed to function can serve as a useful tool in management.

CHAPTER 15

The Relation of Predation
Phenomena to Wildlife Ecology

WE HAVE seen that over large land areas raptor predation is
continuous, and that in any area it exerts on prey species pressure
that is proportional to their relative densities, and that this pressure
is directed on all components of the collective prey, and varies from
season to season and from year to year. Migration enables the various
raptors to concentrate during fall and winter in areas where major
prey species have attained the highest population densities. Spring
migration alters both the numbers and the composition of a collective
population of raptors as well as the number and composition of some
prey species, thus making continuous and proportional pressure possible.
Raptor reproduction produces an increase of pressure that coincides with
increasing prey densities. Heavy pressure by raptors on prey species
frequently occurs at critical periods in the life cycles of the latter. The
combined pressures are influential in controlling prey populations.

The data and conclusions with respect to raptor predation that have
so far been presented provide a source of factual material that invites
us to relate it to other types of predation and to certain problems of
wildlife ecology.

327

By applying the findings of this detailed study to the broad horizons of ecology, we may obtain a clearer perspective of the ramifications of predation and gain a greater insight as to its relative position in nature; nature which should ultimately be comprehended as an indivisible unit.

BIOLOGICAL VERSUS ECONOMIC ASPECTS OF PREDATION

Perhaps the greatest misconceptions concerning predation have arisen through a misunderstanding of the biology of predation and a confusion of the biological with the economic aspects. No sound economic evaluation of predation can be made without considering the general biological effects. In this book we have dealt only with the biological side. The economic appraisal must be expressed in money values, which derive largely from computing the value of predation as a natural regulating force. An economic evaluation must be determined on a basis of quantitative data comparing collective predator populations and their consumptions of animal life with prey populations and their consumptions and destruction of both animal and vegetable life. This, however, is beyond the scope of the present work. It is sufficient that we realize that the economic aspect has been created by man, but the biology is a product of life itself, and that these two aspects are quite different and not to be confused. Those who have championed widespread predator control cite the saving of valuable poultry and livestock and the increase in game populations accruing from such control. They have failed, in general, to comprehend the mechanics of predation and thus failed to see its biological significance. On the other hand, those favoring no control overlook the fact that predation by raptors and other predators can and does cause economic loss to individuals. The fact that this loss is in accord with the biological functioning of predation does not alter the immediate and important fact that the loss in poultry or livestock may mean the difference between profit and loss—perhaps in some cases the margin necessary to gain a livelihood.

A COMPARISON OF AVIAN WITH MAMMALIAN PREDATION

The basic operation of predation by a collective mammalian population appears to be identical with raptor predation, except that mammals lack the mobility needed to concentrate quickly from distant regions into areas of high prey density. A cursory examination of fox, skunk, mink, and weasel scats picked up in Superior Township during the winter of 1941-42 revealed that, like the raptors, the mammalian predators were feeding largely on meadow mice. The diet of the largest carnivore, the fox, responded to changing vulnerability factors from winter to spring like that of the Horned Owl, and in spring included a greater proportion of larger prey species—Pheasants, chickens, crows, rabbits, and fox squirrels.

On the Western Area the predatory pressure exerted by the raptors was supplemented by similar pressure from such mammalian predators as bears, coyotes, badgers, weasels, mink, marten, and shrews. The results of field observations and of a cursory examination of feces were in harmony, as in Superior Township, with the idea that the relation of mammalian predators to relative prey densities resembled that of raptors. Their specialized hunting and feeding habits would have enabled them to place pressure on some prey forms not highly vulnerable to raptors. The coyote, like the Horned Owl, utilizes a wide variety of food (O. J. Murie, 1935). Thus the combination of all predatory pressure on the area applied in the manner described previously for raptors results in a yet more intricate but more perfectly harmonized force that touches nearly all elements of animal life and may be said to regulate their respective population levels (Craighead, F. C., 1951).

HUMAN PREDATORY PRESSURE

In the management of land or of wildlife species we cannot divorce ourselves from the fact that man is an integral part of the fauna. In considering the effects of predation we must realize that man himself exerts much pressure on all wildlife forms, especially on game species. This pressure, when directed at harvesting pheasants, Bob-whites or rabbits or toward killing meadow mice, pocket gophers or ground squirrels, still more intricately expands and ramifies the force of predation. This pressure that man exerts in semi-wilderness as well as in civilized areas differs somewhat from other predator pressures in that it can be highly selective, need have no correlation with relative prey densities and can and has been so efficiently applied as to be annihilative. If we think of both the harvesting of desirable wildlife species and the destruction of "vermin" as predatory activity—the one often a sport, the other a chore—then we realize that man's control measures should consist of the application of human predatory pressure only to the extent that it dovetails with the sub-human predator pressure to maintain a stability that in the aggregate is beneficial to the entire faunal complex. Obviously, to advocate no control by man is absurd, since his very existence creates predatory pressure. To advocate leaving an area untouched, where man is or has been a member of the fauna, is to urge the creation of an artificial condition. The other extreme is planned wholesale poisoning of both predators and rodents that are natural components of the fauna. An example of the middle ground seems to be the regulated hunting season, in which man scientifically harvests his game species with an eye to management. In this instance he applies his pressure, but under regulations.

This leads us to a definition of control as applied by man. We normally use the term to denote objective measures instigated to reduce the number of rodents or predators, but the planned harvesting of pheasants or deer is also control. Therefore, for the sake of clarity, let us term the former control, since its objective is simply a reduction in numbers, and term the latter management, since its objective is largely to harvest surplus game. Both can be thought of as human predatory pressure.

The ultimate step is for man to learn how to harmonize his own rapidly expanding predatory pressure completely with sub-human predation and other processes that regulate populations. He should accept the fact that his problem is not to eliminate sub-human predation but to dovetail it with his own erratic, greatly altered, greatly biased, pressure on the environment. Predation is a tool for him to use, not an evil force to be destroyed.

PREDATION AND PERIODIC PREY FLUCTUATIONS

In the fall and winter of 1941-42 meadow mice in Superior Township had attained a high population level corresponding to the cyclic crest of more northern species, but not approaching in magnitude their peak fluctuations. This year was considered, generally, to be, throughout northern latitudes, an abundance year for other wildlife species as well. The high meadow mouse population was attributable at least partially to vegetation conditions resulting in an abundance of food and cover for this species. The much lower meadow mouse population of 1947-48, conversely, may be traceable to less favorable conditions of food and cover. Prior to 1941-42 a large number of crop fields were abandoned. These were near-marginal lands that were not sufficiently productive to warrant, under existing economic conditions, their continued use. On these fallow fields plant succession proceeded without interruption by agricultural practices. Annual weeds were present the first fall, and succession progressed through types 4 and 5, favorable meadow mouse habitat (Table 61). By 1941-42, this fallow land supported a dense cover of grass and mixed perennials (Type 6). Some fields abandoned earlier than others were characteristic of type 7, containing invading shrubs typical of the site conditions. These fields, in changing from the grass and mixed perennials to the shrubs, passed their optimum food and cover capacity for meadow mice in the winter of 1941-42. The excessive girdling of the shrubs and large-scale "eating out" of grass cover was evidence of this. Many other fields were in slightly earlier successional stages, providing near-maximum food and cover for meadow mice. From the time the land was abandoned, meadow mouse populations were favored by the plant succession. They were not made excessively vulnerable by fall or spring plowing, and thus relatively high breeding populations were possible. The highest meadow mouse densities were maintained in the fallow fields, but by the fall of 1941 all favorable habitats were populated from the well-distributed centers of higher meadow mouse density.

By 1947-48 the fallow fields that had supported high meadow mouse populations in 1941-42 had either been reclaimed as crop land, because of the stimulus of high prices during World War II, or had so advanced into the shrub stage of succession that they were no longer capable of supplying an equal amount of food and cover for meadow mice. On the other hand, the fields grown up to shrubs were more favorable rabbit habitat and undoubtedly played a part in producing the higher rabbit population of this latter year.

The largest area of high meadow mouse density in the township in 1947-48 was crop land. An extensive timothy field was left standing, thus furnishing abundant food and cover for meadow mice throughout the winter. By spring, the winter population of meadow mice was drastically reduced by raptors and further decimated by spring plowing. Prior to plowing, food was still abundant and was not a factor limiting this population.

It would appear that in 1941-42 plant succession on abandoned farm land was in a stage where vegetative conditions provided abundant food and cover for meadow mice and that this, at least partially, was responsible for the high meadow mouse population. In Superior Township land-use practices determined largely whether such conditions could develop. The conditions influencing land use might possibly be traced back through those of an economic nature to those that were meteorological. It may be significant that the year 1941-42 was a year of high prey populations throughout northern latitudes. The effects of predation in limiting prey populations and its controlling effect on a high, perhaps even cyclic, meadow mouse population, found in this study, give us a foundation of fact upon which to explore the role of predation in the drama of northern periodic prey fluctuations.

Much interest in the possible causes and chain of events associated with periodic fluctuations of animal populations in far northern regions has been aroused. Workers have attempted to explain rodent cycles as consequences of climatic and other causes or of a predator-prey relation. Those advocating a predator-prey relation have generally agreed that predators are not sufficiently numerous to bring about a steep decline in rodent numbers. The affect that predation or lack of it might have on initial population build-up, however, has not been given sufficient consideration. It is during the build-up that we should expect predation to be most effective. We have shown that a function of predation is to prevent wide fluctuation of prey numbers. Yet, in the north, fluctuations are characteristic of predator-prey interactions. Thus it could be reasoned that, in cyclic areas, predation is not as effective a population regulator as elsewhere. If this is a fact, it merits further investigation because of the bearing it might have on the build-up of cyclic populations. The purpose of the following discussion is to explore this thought.

In studying cyclic phenomena we are concerned with three distinct aspects, the causal factors responsible for the rise in population numbers, the type of periodicity (length of cycle) and the causal factors responsible for the decline.

Whatever the initiating cause or causes, the build-up must be brought about by either increased reproductive performance, a reduction in environmental resistance forces or a combination of both. Similarly, the die-off must be due to decreased reproductive performance, an increase in environmental resistance forces or a combination of both. The question is whether known mechanisms account for the rise, decline and periodicity, or a force or forces as yet unknown or at least currently unrecognizable cause them. The relationships of reproductive forces to decimating forces are so complex that we should expect to find an explanation for cyclic phenomena when we have studied the many facets of these opposing forces in greater detail. Butler (1953) reminds us that perhaps too much research has been directed toward finding one cause of cycles rather than seeking numerous interrelated ones. This appears to be sound. Similarly we should seek an explanation through an exhaustive study of known mechanisms before falling back on unknown forces. When we consider that reproductive performance and mechanisms that limit populations are the basic ingredients of cycles and that these are so complex that we cannot explain their functioning by single causes, we should expect to find many interrelated factors contributing to cycles. Cycles involve both predator populations and prey populations the relation between which is predation; yet a persistent approach in the attempt to explain cycles has been the general assumption by some workers that predation is a force too ineffective to be worth serious consideration. Errington's concept (1946) that only excess or surplus prey are taken by predators has supported this line of reasoning. Also this trend of thought has persisted because it has been generally assumed that the prey cycle is the basic one which the predator cycle follows and on which it is dependent. The predator cycle following the prey cycle has then been observed to be ineffective in controlling it.

For example, an increase of predators accompanying rising prey populations has been noted and the inability of the former to control the population increase of the latter has been emphasized. Dymond (1947) stated, "The dependence of the periodicities of a number of predatory mammals and birds on the periodic fluctuations in the populations of their prey is obvious. The dependence of the Canada Lynx (*Lynx canadensis*) nine to ten-year periodicity on corresponding periodicity in the hare (*Lepsus americanus*) is generally admitted." Again he stated, "No one, I believe, questions the direct connection between the four-year periodicity of the Arctic fox (*Alopex lagopus*) and Snowy

Owl (*Nyctea scandiaca*) and the corresponding periodicity of the lemming (*Lemmus trimucronatus*) and (*Dicrostonyx hudsonius*)."

Butler (1953) states, "The good years for lynx, fox, fisher, coyote, and wolf are connected closely with those for rabbit, usually beginning in the middle of the good years for rabbit and continuing for several years after." He shows a similar association between muskrat and mink. In the case of these predators and their prey there is also a close habitat association. We have shown how winter raptor populations tend to be proportional to prey populations (Table 34).

Dymond (1947) stated further, "In the case of carnivorous mammals and birds it is the size of the population of the prey animals that determines the size of their populations and not vice versa." We have demonstrated, however, that although the size of raptor populations in fall and winter is more or less proportional to the food supply, the spring or breeding population of meadow mice had to a large extent been reduced by raptor concentrations. Thus the breeding population of meadow mice was determined to a significant extent by the raptor populations. Should this apply to cyclic populations as well, especially during the initial increase, then we could conclude that the predator cycles are affected by the prey cycles and that the reverse is also true. *Though predation appears to be ineffective in preventing periodic fluctuations, the lack of sufficient predatory pressure appears to be a definite causative agent in such periodic phenomena and to be a factor permitting the great amplitude in the fluctuations of prey populations that is characteristic of far northern regions.* If we consider that under certain conditions predation is an effective balancing force operating on prey populations and that it also can be a factor limiting population increases, its absence or its reduction certainly would affect the increase of prey populations, particularly where low winter predation pressure allows a maximum population of potentially high breeders to reproduce each spring. In this respect, it is interesting to note that from north to south the number of prey species increases, but the populations of any species per unit of area are less. Likewise, marked periodic fluctuations are characteristic of prey species of the far north, but non-periodic fluctuations and slight periodic fluctuations characterize prey populations in the northern United States. Lesser fluctuations of any one species occur in the south. Two main types of periodicity in Canada are generally recognized: the 10-year cycle of the forested country, and the four-year cycle of the Arctic. Butler (1953) points out that there are more factors affecting animal populations in the forested country than there are on the tundra, and he reasons that favorable combinations of regulating or control forces should occur more frequently in the Arctic and perhaps account for the shorter arctic cycle. From north to south the number of species of bird and mammal predators increases and this is particularly true in winter, as a result of southward migration of raptors. Migration reduces the number of raptor species present as well as the number of individuals. Raptor pressure, therefore, is least where pronounced periodic fluctuations of birds and mammals occur. Raptors do concentrate in winter in northern regions of high prey abundance and they may also concentrate locally during the breeding season in response to available food; but total pressure exerted over large areas is less than in more southerly regions. Decrease in risks, resulting from the greater depth of snow cover, hibernation or periodic inactivity, leads to decreased prey vulnerability in the far north. Dymond (1947) has pointed out the relative scarcity of predators in the case of the lemming. He stated that, "'The arctic lemming has comparatively few predators," and lists four species. "Whereas a rodent of comparable size in the Austral region of North America has a host of predators." He gives an incomplete list of 16 mammal and bird predators. If we consider that in these same northern regions the predatory mammals, furbearers, are extensively trapped winter after winter and have been harvested since the beginning of the fur trade, it seems obvious that the mammalian predation pressure has been and still is greatly reduced by man. Butler (1953) shows that in spite of certain drawbacks the fur collection can be used as a direct indication of fur animal abundance. He states, "In spite of these drawbacks all available information indicates that, when the population size increases or decreases, the fur collection also goes up or down in a relationship which is direct but not proportional." The comparatively low predatory pressure in the north can be the key that allows a prey species to start a breeding season at a relatively high population level. Under environmental conditions favorable for propagation, such as occur where plant species providing food and cover are relatively few, but those few are abundant over large areas, predation would be and is ineffective in controlling mounting populations of prey with a high breeding potential, even though predation pressure may be locally heavy or may rise generally with prey population increases. A rising population eventually reaches a point where it destroys or damages its environment, as is illustrated by the destruction of food and cover by meadow mice in Superior Township in the winter of 1942 and on the Western Area during the winter of 1947. In Superior Township, however, predation, not starvation or disease, was the major force that reduced the meadow mouse populations. In higher latitudes insufficient predatory pressure over large land areas frequently allows starvation, disease, and emigration to become major controlling factors. Rowan (1948) stated that rabbit (*Lepus americanus*) damage in the vicinity of Edmonton in 1942 was so severe that "Almost wherever one cared to take a country walk, rabbits were incessantly encountered, small trees were denuded of their bark, while the snow, in winter time and on favored spots, was trampled flat and hard by innumerable rabbits."

Crowding and the malnutrition resulting from reduction in food and cover create optimum conditions for epizoötics. Whatever the immediate cause may be, whether emigration, disease, or a combination of various intraspecific self-limiting mechanisms, the inevitable population decline follows, with the result that the surviving population is very sparse. The winged predators migrate to areas of higher prey density or prey on other available species. Mammalian predators move, prey more heavily on other species, or, in some cases, die off, as there are fewer prey species in the predators diets in the Arctic than in the Austral region. At such a time there is a low prey stock, apparently of low vitality. The vegetative environment often is not conducive, for a year or more, to an increase of this population. Reproduction for the surviving stock is not at an optimum immediately following the decline, and predation, though greatly reduced, is sufficient to exercise appreciable control for a few years. Eventually, changing environmental conditions allow the reproductive possibilities of the prey species more nearly to be realized and the population once more starts a rise. The timing of these regular periodicities in population changes has been represented by Dymond (1947), "as due to the essential stability of the environment of animals of high biotic potential whose habits tend to insulate them from extreme climatic conditions." A high biotic potential, however, is not prerequisite to attaining high population densities, but only enables species that possess it to reach high densities in less time than is required by those of lower potential. Short-lived environmental changes characteristic of early successional stages on agricultural land can, when favorable, be exploited by highly productive species. When the environment is also somewhat uniform over large areas, as in the Arctic, conditions are ideal for rapid population increases. It has not been the purpose of this discussion to intimate in any way that predation directly occasions the periodicity of population fluctuations or the cyclic decline. The point to be made is that insufficient predation over extensive Arctic areas could well be an environmental feature that enables Arctic prey populations to reach a point where their numbers can "skyrocket."

In more southern latitudes where the fauna and flora are ecologically more complex, periodic fluctuations are not as evident or they are nonexistent. We have shown that in these latitudes raptor predation functions as an important population depressant or regulating mechanism.

Future research should attempt to determine whether predation is as effective as a regulating force in northern latitudes as it is farther south. What major differences exist in the way it operates? If it has a dampening effect on prey populations, is this local or is it widespread? Is a continuous pressure exerted year after year or is it periodic? Is there an increase in predation pressure coinciding with an increase in available prey? If so, how is this response brought about and at what season or seasons of the year? Does reduction of predators by trapping of furbearers and the annual southward migration of a large part of the raptor population materially reduce predation pressure? The predator-prey relationships that involve cyclic populations must be studied more precisely and must be better understood; then such knowledge should be integrated with information concerning other forces that have to do with the build-up, periodicity and die-off. Predation is an integral part of the cyclic phenomenon, and a proper evaluation of its role should be included in any explanation of periodic fluctuations.

RAPTORS AS INDICATORS OF PREY

Raptors are fewer, and usually more readily counted, than their prey. The composition and numbers of a raptor population—both nesting and wintering—are to a trained observer rapid and accurate indicators of the relative abundance of prey populations. They are also, as Errington (1943) has shown, indicators of prey vulnerability. Nest counts and winter censuses of raptors, when properly evaluated, can furnish a wildlife technician with information needed in management. In survey studies, raptor counts can aid materially in acquainting an observer with existing environmental conditions.

THE RELATION OF PREDATION TO THE MANAGEMENT OF DESIRABLE WILDLIFE SPECIES

Predation is one of nature's normal, effective, and continually operating mechanisms for regulating wildlife populations. Man, in managing wildlife, both desirable and undesirable, not only must take account of this function, but must understand the mechanics of it and as far as possible aid nature's efficient regulation to operate smoothly. It behooves us to repeat with somewhat different phrasing the basic concept we have arrived at concerning predation: *That failure of the food supply, adverse weather conditions, or disease sporadically may kill more Bobwhites, rabbits or meadow mice than predation; yet these do not operate continuously or proportionately on the numbers of a collective prey population. The important point is not the number killed, but how they were killed. Predation tends to kill them in almost direct relation to the relative numbers of each species in the prey population; and this steady, proportionate mortality exerts a controlling effect—a control that operates within the limits imposed upon prey populations by all other regulatory factors of the environment.*

It would be well if the sportsman no longer thought of the number of Ring-necked Pheasants taken by a Horned Owl or the Bob-white consumed by a red fox as an indication of whether predation should be tolerated or predators controlled. Similarly, he should realize that merely evaluating the desirable prey species taken, in contrast with the undesirable species, will show only an incomplete picture. We recognize that man wants to increase so-called "desirable" species and decrease the numbers of "undesirable" ones. To what extent can such a policy suc-

The Investigation

PLATE LII. Approximately 1200 ascents of trees were made during the investigation. Horned Owl nest, Section 17, Superior Township, 1942. One owl had left nest, the other flew when the tree was climbed. This date would be considered the brood departure date. A pair of Red-Tailed Hawks built this nest in 1941 and found it usurped by the owls in 1942.

PLATE LIII. Climbing to the nest of a Western Raven. Ecologically a Raven can be classed as a Raptor.

PLATE LV. Nesting hollow of a Long-eared Owl in a narrow-leaf cottonwood. A pair of Screech Owls, Swainson's Hawks, and Sparrow Hawks all nested within a few hundred yards of this site.

PLATE LIV. Climbing to a Screech Owl nest on Western Area. To obtain nesting data, all nest trees were climbed. Eggs and young were counted and information on food habits gathered.

PLATE LVII. View of cultivated land from Blacktail Butte. Individual hawk ranges were easily plotted over this hunting area.

PLATE LVI. Goshawk attacking. While studying and visiting this nest, the hawks struck the authors over a dozen times. A leather jacket was torn, a shirt ripped, and cuts were inflicted on head and hands.

341

PLATE LVIII. Tethering a young Red-tailed Hawk at the base of the nest tree. It is important to remove all debris so the hawk does not entangle itself. Leather jesses, a swivel and a short leash are used.

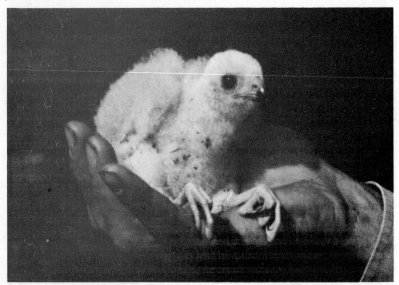

PLATE LIX. Young female Sharp-shinned Hawk at 8 days of age, weight 116 grams. The growth of this and other birds was measured daily and records kept of food consumption.

PLATE LX. Young Sparrow Hawks lowered from nesting hollow in a hat are being measured and weighed to determine growth curves and to gather data helpful in determining sex at an early age.

PLATE LXI. A young male Sparrow Hawk is weighed. Its growth was correlated with food consumption.

PLATE LXII. This Great Gray Owl was raised in captivity and then released within the nesting territory of a Horned Owl. It established a temporary range and was not molested by the Horned Owls.

PLATE LXIII. Bald Eagle returning to its nest with a fish. A tree blind permitted detailed observations.

PLATE LXIV. Both young and adult Horned Owls were fed to determine growth curves and food consumption. This one was completely free for 127 days, returning daily to be fed and weighed.

PLATE LXV. A male Duck Hawk, docile and well adjusted to captivity. Placed on a perch and secured with jesses, swivel and leash its food intake could be measured.

PLATE LXVI. An immature Red-tailed Hawk. Twenty-nine raptors representing eleven species were fed to obtain quantitative data on food requirements. All were handled and cared for using the falconer's technics.

PLATE LXVII. Hundreds of hawks are killed annually at Cape May, New Jersey. Rather high mortality occurs throughout the country during fall migration.

ceed? For example, is it feasible to increase Ring-necked Pheasant populations by eliminating predators and thus predation losses? It is obvious that if man in one way or another temporarily increases a pheasant population, the degree of predation pressure on this population will correspondingly increase. If by any human control predation is decreased, it may be possible to maintain the pheasant population at a higher level and to harvest more birds. Such a gain, however, is fleeting. Great mobility is important in enabling raptor populations to concentrate their pressure. Thus, unless raptors are exterminated over vast expanses, they will concentrate in fall and winter in areas of high prey density. Where a raptor population is greatly reduced, meadow mice and other mice, with their high reproductive powers, tend to increase at a much greater rate than the pheasant. These populations not only adversely alter the vegetative environment, but in winter compete directly with the pheasant for food. They may also attract and hold both bird and mammal predators in greater numbers. While these higher rodent populations decrease relative predation pressure on the pheasant, the environmental disturbances that they cause are adverse to maintenance of a high pheasant population. Eventually, rodent control may have to be undertaken. Thus we have the familiar picture of man attempting to raise a population density by eliminating predators and finding himself bound to a policy of ever greater control—costly control, with the result in the long run, an unbalanced environment where populations of different prey species have greater and greater tendency to fluctuate widely.

How then may we best increase the population density of a desirable species? The answer is quite apparent in the analysis of the dynamics of predation (Figs. 6 and 12). The vulnerability of any prey species varies with a number of conditions that increase risk. Changes in these conditions alter the relationship between prey density and predation pressure. We have even seen that the density relationship between predators and prey can be modified greatly for short periods by changes in one or more types of risk. Changes in the environment—increased cover, better food arrangement, secure travel lanes, better nest sites, or den sites—can reduce greatly certain types of risk and thus decrease the vulnerability of a desirable species to a collective population of either raptors or mammalian predators. In the case of big game mammals, such as deer, elk and antelope, we often have the problem of keeping their numbers within the carrying capacity of a limited winter range. The solution is frequently a decrease in their numbers. When this is brought about, the game enjoys greater security, predation is less, and there is likewise less need of it.

Environmental improvements may be aimed at providing security for particular animals, but such improvements are likewise beneficial to other species. Environmental improvement may raise the threshold of a desirable species against predation and tend simultaneously to decrease intraspecific strife, with the result that fewer losses occur. It is, then, a way to reduce predation's annual harvest of desirable wildlife forms. The results of such environmental improvement are an increase of populations and some increase in predation pressure, but without corresponding losses. In other words, both the carrying capacity and the threshold of security have been raised (Fig. 12). In some cases it may also prove possible to reduce vulnerability through planned breeding that produces faster, stronger, or more hardy species or, in the case of big game, by the reduction of populations in areas where their numbers exceed the carrying capacity of the range.

MANAGEMENT OR CONTROL OF PREDATOR POPULATIONS

From the foregoing data and conclusions it is obvious that general widespread control of raptor populations is undesirable. It is at times best for man's interest to eliminate individual raptors or perhaps occasionally even to reduce the numbers of a particular species. The destruction of a population, however, is entirely unjustified. It would be unwise to provide complete legal protection for the Horned Owl, Cooper's Hawk, and Goshawk, because it then would be difficult to control them in special cases. Man is the most destructive of all predators and his predatory acts will, without special encouragement, be normally sufficient to control these species. Where man has artificially increased local densities of such prey species as pheasant or Bob-white at game farms, and chickens or ducks at poultry establishments, relatively high losses must be expected unless these concentrations are given proper protection. If a farmer keeps large flocks of poultry unfenced, he will suffer rather heavy predation losses. But general destruction of raptors, with the attendant environmental disturbances, in order to permit unwise poultry or game farm-practices, cannot be justified. The kills of larger, more valuable prey are few when compared with the number of prey individuals taken by collective raptor populations. This is because the staple prey of most predators consists of small, numerous, and "undesirable" rodents. Cases of local damage should not be allowed to prejudice our thinking and blind us to the generally beneficial function of raptor predation.

Control of raptors is justified where populations of game birds are very low or are being established in an area. In such instances heavy predation can be anticipated. It is not a case of members of such populations being preyed upon in relation to the population density of other prey forms available. It is a situation where risks are high. A very low prey population or one being artificially established is not secure, since it is not adjusted to the environment. Predation on such populations is likely to be out of proportion to their numbers, with the result that predation has a critical effect, tending to exterminate rather than to regulate.

Actual extinction of animal life has occurred on some oceanic islands, where hogs and rats have been introduced among island forms poorly adapted to cope with predators and with little ability to make necessary adjustments. Furthermore, the introduction of non-predatory animals, such as goats and rabbits, on isolated islands where there have been no predators to check their population growth has resulted in almost complete destruction of the native vegetation and eventual decline of the introduced species.

In such cases the remarkable adaptations of predators to prey, with all of the nicely balanced complexities involved, did not exist—heavy predation pressure is there alien to the balance that has evolved and the sudden introduction of either predation forces or uncontrolled reproductive forces creates an unbalanced situation.

A relatively large proportion of a staple prey population can be killed locally by a population of a single predator species. When this prey, economically or otherwise, happens to be valuable to man and, if because of this value it has been too heavily cropped as a population, or if it has had its environment so altered that it is subject to heavy environmental resistance forces, then predation is very likely to become the condition limiting its population increase. A staple prey is usually a prolific species; thus it still can be sufficiently abundant to bear the brunt of predation and yet exhibit a depressed population level. In such instances, temporary predator control is advisable to make it possible for man to reap a maximum harvest, but it must again be remembered that the fundamental disturbance lies not with predation but with other forces and conditions, frequently land abuse. Predator control then should be a finely calculated emergency operation, employed only until more basic remedies can be applied.

Similarly, when nature's mechanisms for regulating population levels are disrupted by man, or when excess populations of exotic or domestic species are built up, then temporary measures may be necessary to curb an expanding prey population, or to control an artificially created predator-prey relation; but again the major effort should be directed toward moderate control and when possible this should be attained by aiding nature to do the job. Man, by concerted action and persistent effort, can reduce both wild rodent and predator populations to critically low levels over wide areas. The checks and balances inherent in the natural relations between predators and prey and in other regulating forces of the environment are the result of an infinite number of causes and effects that have evolved as animal life evolved. The system may not be perfect as it relates to man, but it operates impersonally and in the state of man's present knowledge is far superior to any widespread control by man himself, which at best considers only a few causes and effects and is based on short-sighted interests.

We must keep in mind that man has altered his environment greatly and will continue to alter it and in doing so will temporarily disturb a natural balance that has evolved. Regardless, however, of how profound his changes, man does not radically deflect or eliminate fundamental natural laws or processes. The force of predation has adjusted itself to man's drastic environmental changes. It may be expected to continue to do so and to function as a density-controlled population regulator, a precise and powerful force tending to restore or establish a balance between various species. This being true, man's objective should be to recognize and utilize the force of predation over vast areas where he himself exerts little or incomplete predatory pressure and to curb it only when there are strong economic reasons for doing so. To argue that man so drastically upsets the equilibrium of nature that he must assume the full burden of controling the levels of all animal species and populations is to overlook the important fact that the complex of population increase and regulating mechanisms has evolved in response to changes —changes perhaps far more drastic and vital than any man has yet caused.

Let us terminate this discussion with a concept of predation as a powerful and complex natural force that should be visualized in its ecological entirety—a force that man should recognize not as an enemy tending to "steal" the rabbits or pheasants he could otherwise harvest, but a force that he has only begun to understand, one that he cannot eliminate but can perhaps best manage if he allows sub-human predation to function harmoniously with human predatory pressure. The former is a natural regulating mechanism, the latter a conscious control or harvest, for man's interests, but both are vital components of a complex whole.

Summary of Chapters

Raptor predation phenomena were studied in Superior Township, Michigan, where predator-prey relationships were interpreted and expressed quantitatively in terms of collective raptor and prey populations and their interrelations. Data essential to analyzing predation were gathered and compared for two full years and some studies were carried through four seasons. Pertinent information on the geology, meteorology, fauna, flora, land use, and history of the region was gathered. This research was then compared with a similar investigation conducted at Moose, Wyoming. The information required to understand the function and dynamics of raptor predation is presented chapter by chapter and then the data and findings are combined to reach the major conclusions.

CHAPTER 2
Hawk Movements and Winter Ranges

1. Hawk populations became stabilized following fall migration and remained more or less numerically constant until spring.

2. Definite and limited fall and winter ranges were established by all the hawks, and thus the population was spatially fixed and did not wander indiscriminately. Three types of movement were recognized and described—migration, drift, and intra-range movement; the latter characterizing the winter population.

3. Daily, weekly, and seasonal ranges of hawks were plotted. Marsh Hawks had the largest seasonal ranges, and American Rough-legged Hawks were second. The largest Red-tailed Hawk ranges covered about four square miles; those of Red-shouldered Hawks and Sparrow Hawks, about two square miles. The average diameter of Cooper's Hawk ranges was between 1.5 and 2 miles. In general, the seasonal range of any species did not exceed a maximum diameter of three miles. The smallest recorded range diameter was one mile.

4. Most hawk flights were under one-quarter mile; the longest flight seldom being more than one mile.

5. The daily ranges of numerous hawks were plotted and the daily activity patterns determined by following both single and paired birds from sunup to sunset. A hunting range generally consisted of two major areas alternately used, morning and afternoon. Perches were used continuously for as long as three hours, but were changed as often as 52 times during a day.

6. Many conditions influenced the size and characteristics of ranges. Large ranges were associated with low density and scattered meadow mouse populations; small ranges with high or vulnerable meadow mouse populations. Within certain bounds, the size of a hawk's winter range was inversely proportional to the food supply.

7. Hawk ranges generally were smaller when the hawk populations were highest. Overlap occurred between ranges of the same and different species when the hawk population and prey densities were high. Ranges tended to be isolated and defended when prey densities were low.

8. Adjustment of each hawk in relation to others was found to be natural, general, and commonly delicate in operation. This movement can be thought of as a form of inter and intra-specific competition, as well as a tolerance of one bird for another. Physical conflict, though it occurred, was the exception rather than the rule.

9. Some hawks wintered as well as nested in the township for a known period of three consecutive years.

10. The great mobility of various species of raptors enables them to concentrate during fall migration in areas of high prey density. The subsequent establishment of definite, relatively small winter ranges makes possible the exertion of a continuous hunting pressure on prey species by the collective members of a raptor population.

CHAPTER 3

HAWK FALL AND WINTER POPULATIONS

1. A car strip census interpreted in terms of known and measured influencing conditions yielded counts that gave a hawk population for both the census strip and the township study area. The census enabled hawk hunting pressure to be measured with almost complete accuracy.

2. The technic of taking a census of a mixed population (6 species of hawks) over an extensive period, of simultaneously determining the association of hawks with habitat, of measuring the activity and distribution of the various members and their numerical and compositional stability, made possible the study of the predatory interrelations of a population unit composed of numerous species.

3. In 1941-42 the fall hawk population of Superior Township was 117 birds of 6 species, an average density of 3.25 hawks per square mile. The winter population was 96 birds of 6 species; an average density of 2.67 hawks per square mile. The spring transition population was 108 hawks, a density of 3 hawks per square mile.

4. In 1947-48 the fall hawk population of the same area was 29 birds of 6 species, an average density of .78 hawks per square mile. The winter population was 27 hawks of 5 species, an average density of .75 hawks per square mile. The spring transition population was 94 hawks, or a density of 2.6 hawks per square mile.

5. An average winter hawk population, determined over a four-year period, was 48 hawks per township or 1.33 hawks per square mile. Under present land use conditions we can normally expect a winter hawk population for southern Michigan of about 24,000 hawks, 50% of which could be Buteos.

6. The movement of even a single individual in the composite hawk population started movement by another member or members of that population. This compensatory movement was largely due to spatial adjustments naturally and readily made by hawks in minimizing competitive hunting pressure. These responses, measured quantitatively, graphically illustrate the delicate and intricate nature of population pressures.

7. In general, the winter population of hawks on any area of land the size of a township will alter little, if any. The number of hawks tends to reflect the carrying capacity of the area for these birds as governed by the density of major prey species.

8. For practical management purposes, a reasonably accurate census of the hawk population per unit of area can be obtained in the farming regions of the north-central states by making three or four consecutive or closely spaced car censuses, supplementing these with random observations on Cooper's Hawks and Sparrow Hawks, interpreting the data and making computations based on them, as indicated in this text.

CHAPTER 4
FALL AND WINTER OWL POPULATIONS

1. Four species of owls wintered—the Horned Owl, the Long-eared Owl, the Screech Owl, and the Short-eared Owl. The populations present during two falls and the winters immediately following were: 1941-Fall, 54; 1942-Winter, 63; 1947-Fall, 35; 1948-Winter, 32. The numbers of Horned Owl and Screech Owl (year-round residents) did not fluctuate noticeably, but those of the Long-eared and Short-eared Owls (migrants) did change. The difference in the populations of the migrants from one year to another reflected the difference in meadow mouse densities.

2. Horned Owls and Screech Owls exhibited little or no movement other than that within relatively small ranges. Both species maintained home ranges where they nested as well as wintered. Winter ranges of Horned Owls seldom exceeded a half-mile in radius. The number, size, and distribution of woodlots appeared to be the major conditions limiting the density and determining the distribution of both wintering and nesting Horned Owls.

3. Long-eared Owls roosted communally, usually in coniferous cover, during the winter and hunted the open meadow mouse habitats. An average of seven hunted the township in 1942, and none was present in 1948.

4. Fourteen Screech Owls were present in the fall and winter of 1941-42; 22 and 19, respectively, in the fall and winter of 1947-48. This was the most difficult owl to count accurately, and the actual numbers present were probably greater than those recorded.

5. The Short-eared Owls, like the Long-eared Owls, migrated into the area in fall and left by early spring. In 1941 the average fall population of Short-eared Owls was 22, the winter, 31. These birds occupied communal roosts on the ground, often in the same field with the Marsh Hawks. The highest count at any one roost was 28. Like the Marsh Hawks they left these roosts to hunt specific open-field areas throughout the township.

6. The total raptor (hawk and owl) populations for the fall and winter periods were: 1941-42—fall, 171; winter, 159; 1947-48—fall, 77; winter, 78. This represents a respective average density of 4.75, 4.14, 2.14, and 2.17 raptors per square mile during these periods.

7. All hawks, with the exception of Marsh Hawks, remained in and hunted definite, localized ranges distributed throughout the area. Owls, with the exception of the Long-eared and Short-eared Owls, did likewise. Each morning a wave of Marsh Hawks spread over the township from a communal roost, and these hawks tended to hunt the areas not in immediate use by hawks of other species. At night the Long-eared and Short-eared Owls repeated the pattern. The result of this distribution and hunting pattern was to limit competition, to exert pressure on all habitats, and to make for hunting efficiency by a mixed population of raptors whose pressure was exerted largely on meadow mice.

8. The hunting pattern of these Michigan winter raptors is believed to be characteristic of other raptor populations in various regions of the country.

CHAPTER 5

PREY POPULATIONS

1. Meadow mouse habitats were rated, then trapped to obtain trap-night indices. From trapping data obtained, the winter meadow mouse population was computed at 303,000 in 1941-42, and 75,000 in 1947-48. Trapping indicated meadow mouse densities as high as 140 per acre in 1941-42, which could well be termed a "mouse year," and scattered concentrations as high as 93 per acre in the winter of 1948.

2. Approximate populations of white-footed mice for the township of 33,000 in 1942 and 27,000 in 1948 were computed on a basis of Burt's (1940) population figures and our trapping results.

3. Detailed Bob-white and Ring-necked Pheasant population data were gathered on an eight-section sample area for two winters, and then projected for the township. The approximate populations following the hunting season were computed to be 1,017 pheasants and 477 Bob-whites in 1942 and 1,000 pheasants and 144 Bob-whites in 1948. The

average number of pheasants per section on the sample area was 28.25 in 1942 and 27.75 in 1948; Bob-whites were 13.25 in 1942 and 4.0 in 1948. During both years there was approximately one pheasant for every 23 acres. In the winter of 1942 there was a ratio of one raptor to 9.4 game birds; in 1948, one to 19.2. The ratio of the raptors (Horned Owl and Cooper's Hawk) which preyed on these species was, however, comparable in the two years (1:71 and 1:60).

4. Two approaches were used in making censuses of rabbits and the populations were rated as being somewhere between the results obtained by the two methods. The population was very low in 1942 (being only 300, compared to 1200 in 1948).

5. The fox squirrel population was much higher in 1948 than in 1942. Following the 1941 hunting season it was computed at about 300, and approximately 1000 in 1948.

6. Estimates of the small-bird populations, based on habitat counts and banding, placed the mixed population at about 23,500 during each winter.

7. This chapter gives an account of the kind and approximate numbers of prey individuals theoretically available to known winter raptor populations. In both study years, meadow mice were the most abundant prey, white-footed mice second, and small birds third. Game birds were fourth in abundance in 1942 and fifth in 1948. Rabbits were fifth in 1942 and fourth in 1948. In both years the fox squirrel populations, following the hunting season, were lower than those of any of the other prey species of which censuses were made.

CHAPTER 6

Winter Food of Raptors

1. Four thousand, seven hundred seventy-six fall and winter pellets dropped by hawks and owls were gathered and analyzed to determine the food of the raptor populations during two winters. The diets are expressed in terms of the number of individual prey animals represented in a pellet sample as well as in terms of the frequency of representation.

2. The Red-tailed Hawks consumed a high proportion of meadow mice (89 and 84 per cent) in both winters. The Red-shouldered Hawk pellets indicated a diet of 94 and 100 per cent meadow mice during the same periods. The American Rough-legged Hawk diet in 1942 was 98 per cent meadow mice. As a group, these three Buteo species consumed 87.4 per cent meadow mice and 5.2 per cent white-footed mice in 1942 and 85.3 and 5.8 per cent, respectively, in 1948. The Buteo population came into an equilibrium with its major food supply so that the relative effect of the Buteos was similar in both winters, although the densities of prey and predator varied.

3. In 1942 the diet of the Marsh Hawks was 93 per cent meadow mice, as compared with 98 per cent in 1948.

4. The Sparrow Hawk diets were 66.7 and 52.3 per cent meadow mice in the two winters.

5. In 1942 meadow mice comprised 97.8 per cent of the diet of the Long-eared Owl. The Long-eared Owl, the Short-eared Owl and the American Rough-legged Hawk did not winter in 1948, when meadow mouse populations were low.

6. Meadow mice comprised 87.7 and 97.6 per cent of the Short-eared Owl diet in 1942 and 1948, respectively, but 31 owls wintered in the former year and none of the few present spent the entire winter in 1948.

7. During these winters meadow mice and white-footed mice combined were 95.3 and 87.2 per cent, respectively, of the Screech Owl diet. In both years white-footed mice were slightly higher in the diet than meadow mice.

8. White-footed mice composed the greatest proportion of individuals in the Horned Owl diet. Meadow mice and white-footed mice combined were 91 per cent of the diet in 1942 and 86 per cent in 1948.

9. Small birds composed the greater proportion of the Cooper's Hawk's winter diet. A tally of kills on an eight-section area revealed that 3.4 per cent of the Bob-white population was taken by this species in 1942 and 12.5 per cent in 1948. No pheasants were known to have been taken in 1942, but 3.4 per cent of the pheasant population was taken in 1948.

10. The diet of each raptor species shows that all species preyed most intensively on the most abundant prey forms in their hunting habitats. Predation by a population of any one hawk or owl species on available prey was largely governed by a density relationship. Meadow mice rated first in the diet of all open-land hunters, white-footed mice first in the diet of all nocturnal woodland hunters, and small birds first in the diet of the only diurnal woodland hunter, the Cooper's Hawk.

CHAPTER 7

THE DYNAMICS OF PREDATION

1. Vulnerability of a prey species is the result of all the physical and biological conditions that cause one species to be preyed on more heavily than another. Prey vulnerability can be separated into two major divisions; *prey density* and *prey risk*. Prey density is the basic condition, the effect of which is modified continually by prey risk. Prey risk includes all conditions operating with prey density to make a prey vulnerable. The two main types of prey risk are: risk originating with the prey and risk originating from the predators.

2. A vulnerability rating determined for the study period for each prey species or group and based on respective representation in the col-

lective raptor diet was: meadow mouse, 1; white-footed mice, 2; small birds, 3; game birds, 4; rabbits, 5; fox squirrels, 6.

3. The vulnerability of any prey species to a raptor population is much influenced by the relative population densities of all other prey species.

4. Raptor hunting pressure on each prey species is roughly in the order of the vulnerability ratings.

5. Through the agency of migration, the population density of restricted feeders (migrants) dropped when meadow mouse vulnerability was low. The populations of the general feeders (year-round residents) remained constant. The latter did not move when meadow mouse vulnerability was low, but their diets changed, indicative of the changing relative vulnerability of the large range of prey that they fed on.

6. Predation is a relation between predator and prey and between predator populations and prey populations, operating so that the prey tend to be taken by the predators generally in proportion to their relative densities, this density relation being modified by variable risks to determine prey vulnerability.

7. The movement of raptors in response to prey fluctuations is an important factor in their ability to balance prey populations.

8. There is a tendency for each component of an entire prey population to provide the percentage of the raptor population diet that it forms of the entire prey population present. This proportionate reduction of prey populations is the key to evaluating predation as a regulatory force, one capable of balancing prey populations.

CHAPTER 8

THE STUDY AREAS AND GENERAL NEST STUDY TECHNICS

1. The physical and biological characteristics of the Western Study Area are treated so that comparisons can be made between predation in Superior Township, Michigan, and predation in a semi-wilderness area at Moose, Wyoming. The nesting studies in Superior Township are a continuation of the fall and winter predation studies already treated.

2. The investigation divided into the following phases: the determination of raptor populations and their productivity; studies of raptor movements, activities, nesting ranges, and interrelationships; studies of prey populations and raptor food habits.

3. The gathering of nesting data involved more than 2000 miles of walking and approximately 1200 ascents of trees. One hundred forty raptor nests were intensively studied on the two areas and 291 additional nests were located and observed in the surrounding vicinities during four breeding seasons.

4. To obtain a count of nesting pairs, to locate renesters, and to distinguish between non-nesting pairs and single birds, it was essential to

record all nests and all raptor observations on an accurate cover map. What contributed most to the accuracy and completeness of the nesting study was the care taken to locate all the raptor nests.

5. It is believed that the nest counting methods were close to 100 percent effective for all species except the Screech Owl and possibly the Barn Owl.

6. A rapid survey of nesting pairs and potential nesters during the early phase of territorial selection and defense can yield a count representing about 85 per cent of the breeding season population.

CHAPTER 9

ACTIVITY AND MOVEMENT

1. The change in numbers and composition of raptor populations, as well as the changes in density relationships between small bird populations and other prey groups, brought about by spring migration, affect predation pressure on all prey populations.

2. Mean monthly temperatures in the Western Study Area in late winter and early spring were comparable to those occurring a month earlier in Superior Township. This temperature difference affected migration dates, nesting dates and the duration of the breeding season.

3. The vital breeding season activities of a raptor population were telescoped in the region where these seasons were shorter.

4. The Horned Owls in the Western Area initiated nesting a month later than those of Superior Township. In both areas the Horned Owl started laying a month before the next raptor species. This has a marked influence on the dominance of the Horned Owl over other raptors.

5. The Horned Owl dominates a nesting population of raptors and decimates the Crow population and the other raptors.

6. The cumulative effect of the interactions of nesting raptor species is measurable and significant; is influenced by the timing of nesting activities and alters predation pressure because of its tendency to reduce raptor productivity.

7. The evolutionary adjustment of a collective raptor population's nesting activities, resulting in a reduction of inter- and intra-specific strife, though perhaps less evident to the observer than conflict, is nevertheless significant in permitting numerous species of predatory birds to live and raise their young in relative harmony and security.

CHAPTER 10

RAPTOR NESTING POPULATIONS

1. In 1942, the Michigan study area (37 square miles) supported 63 pairs of nesting raptors; in 1948 there were 66 pairs and in 1949, 65 pairs, or a density of 1.7, 1.8, and 1.8 pairs, respectively, of nesting raptors per square mile.

2. The Red-shouldered Hawk was the predominant species, comprising 35, 27, and 26 per cent of the population during the three years. The Screech Owl was second in population density, comprising 21, 23, and 22 per cent. The Horned Owl comprised 10 per cent each year.

3. A 12-square-mile semi-wilderness area at Moose, Wyoming, supported 45 nesting pairs of raptors in 1947 and a comparable population density and composition in 1948. There were 3.8 pairs of nesting raptors per square mile, a density more than twice as great as that in southern Michigan.

4. All raptors studied exhibited strong tendencies to reoccupy nesting territories or ranges in consecutive years. On a large land area the raptor population shows the same continuity of occupancy that is exhibited by the individual nesting pairs. In Michigan the nesting pattern of the raptor population of 1947 remained 74 per cent intact in 1948. In the Wyoming study area the nesting pattern remained at least 75 per cent intact from one breeding season through the next.

5. The increase or decrease in any single staple prey species had no immediate effect in altering the raptor nesting density of an area.

6. Over large land areas raptors can maintain an average population density as high as one nesting pair per 0.27 square miles in ideal habitat, and even in areas of intensive land use they can attain an average population density of one nesting pair per 0.56 square mile. They attained a local density as high as one nesting pair per 0.10 square mile.

7. Nesting raptor populations in the study areas tended to regulate the population levels of the prey species.

8. In 1942 the raptor population of Superior Township produced 211 eggs, of which 158 hatched, and of these 126 survived to the fledgling stage; a nesting success of 60 per cent.

In 1948, the population produced 215 eggs, of which 120 hatched, and of these 106 survived to the fledgling stage; a nesting success of 49 per cent.

The nesting success on the Western Study Area was 83 per cent. Of 569 eggs produced in the two study areas during three breeding seasons, 71.9 per cent hatched and 61.5 per cent survived to the fledgling stage; an average productivity of two young per pair of adults.

9. Man was directly and indirectly responsible for nearly half the raptor nest destruction.

10. Raptor population density at the close of the breeding season rose to 7.0 and 6.9 per square mile in Superior Township in 1942 and 1948, respectively. On the Western Area it was 17.4 birds per square mile. The average raptor density for the entire year in Superior Township was 5.0 birds per square mile in 1941-42 and 4.2 birds per square mile in 1947-48. On the Western Area it was 8.7 birds per square mile.

11. The average annual raptor nesting population throughout the hardwood forest region in southern Michigan closely approaches 31,600 nesting pairs; while the average annual post-nesting population closely approximates 113,800 birds of prey.

CHAPTER 11

Nesting Season Ranges of Hawks and Owls

1. In southern Michigan the average size of hawk ranges was:

Red-shouldered Hawk, an area of one-quarter square mile, with a maximum diameter of slightly less than one mile.

Red-tailed Hawk, an area of one and one-half square miles, with a maximum diameter of two miles.

Cooper's Hawk, an area of three-quarters of a square mile, with a maximum diameter of nearly one and one-half miles.

Marsh Hawk, an area of one square mile and a maximum diameter of one and one-half miles.

Sparrow Hawk, an area of one-half square mile and a maximum diameter of one mile.

Immature Red-shouldered Hawk, an area of one-quarter square mile and a maximum diameter of nearly a mile.

Immature Red-tailed Hawk, an area of nearly a square mile and a maximum diameter of one and one-half miles.

2. In Wyoming the average size of hawk ranges was:

Western Red-tailed Hawk, an area of three-quarters of a square mile, with a maximum diameter of one and one-half miles.

Swainson's Hawk, an area of one square mile and a maximum diameter of one and one-half miles.

Western Sparrow Hawk, an area of three-quarters of a square mile, with a maximum diameter of one and one-half miles.

Western Horned Owl, an area of one square mile, with a maximum diameter of one and one-half miles.

Long-eared Owl, an area of one-quarter square mile and a maximum diameter of three-quarters of a mile.

Western Raven, an area of three and one-half square miles and a maximum diameter of three miles.

3. Ranges of most raptors tend to be distinct from those of neighboring pairs of the same species. This isolation is partially due to spatial requirements that vary with circumstances and with species.

4. Overlapping at peripheries of ranges of birds of the same species will normally occur and the degree to which it is exhibited is influenced by the distribution of nest sites and hunting areas and by the abundance of food.

5. A population is always adjusting itself to its members' activities. The tendency toward spatial isolation of nesting ranges of hawks of

the same species gives a certain amount of freedom from these adjustments at a time of the year when breeding activities demand full attention.

6. Localized predation pressure on any prey population is influenced by the range pattern of a nesting population of raptors.

CHAPTER 12

Spring and Summer Food Habits of Raptor Populations

1. During both study years small birds ranked first, meadow mice second and cold-blooded prey groups third in the spring and summer diet of the nesting raptor population in Superior Township.

2. The major prey species taken by the nesting Red-shouldered Hawks during the two study years were: meadow mice, 31.9 and 25.1 per cent; small and medium-sized birds, 20.2 and 29.1 per cent; snakes, 13.8 and 11.3 per cent; frogs, 8.6 and 16.2 per cent; and crawfish, 14.1 and 9.3 per cent. Food habit studies indicated that there was little variation in the diet of the Red-shouldered Hawk population from year to year.

3. Both game birds and game mammals were preyed on more heavily by Red-tailed Hawks than by Red-shouldered Hawks. In southern Michigan these two hawks do not compete seriously with each other for food.

4. Small and medium-sized birds composed 87.7 and 72.0 per cent of the Cooper's Hawk diet in 1942 and 1948, respectively. A large proportion of the small birds taken were fledglings. Mammals composed only 6.8 and 11.8 per cent of the Cooper's Hawk diet. Pheasants were preyed on significantly during both years.

5. Both the Sparrow Hawks and the Marsh Hawks had diets high in meadow mice and small birds.

6. The 1942 Horned Owl diet was 40.5 per cent mammals and 59.6 per cent birds. In 1948 it was 62.1 per cent mammals and 29.7 per cent birds. Pheasants were 20.2 and 13.0 per cent of the Horned Owl's diet in the nesting season in 1942 and 1948, respectively.

7. Meadow mice occurred most frequently in the nesting season diet of the Barn Owl, Long-eared Owl, and Screech Owl.

8. The average minimum adult pheasant mortality in Superior Township from raptors in the spring of the year was estimated to be in the neighborhood of 4 to 5 per cent of the overwintering population.

9. On the Western Study Area in 1947 the Red-tailed Hawks exerted greatest pressure on the Uinta ground squirrel and the meadow mouse. These prey animals represented, respectively, 41.8 and 33.3 per cent of their diet.

10. Meadow mice represented 57.3 per cent of the western Sparrow Hawk diet. The western bird hawks exerted greatest pressure on the

small-bird group, but took more mammals than did the bird hawks in Michigan.

11. Meadow mice and pocket gophers represented 64.5 and 23.5 per cent, respectively, of the Horned Owl diet in Wyoming. Both the Long-eared Owl and the Great Gray Owl exerted greatest pressure on the meadow mice and pocket gophers. Owls exerted much greater pressure on the chiefly nocturnal pocket gopher than did the hawks. This dove-tailing of pressure, resulting from the adaptations of raptor species, enables a collective raptor population to exert a proportional pressure on a wide range of prey forms.

12. In Superior Township, meadow mice were the staple food during the cold months, with small and medium-sized birds assuming this role during the warm months. Fox squirrels, rabbits, Pleasants and Bob-whites were all more heavily preyed on in spring and summer than in fall and winter.

13. The diet of the nesting raptor population in Wyoming more closely parallels the diet of the Michigan winter raptors than it does that of the spring and summer population.

14. In the Western Area a tendency toward balancing prey popula-tions resulted from concerted pressure on the densest prey group, the rodents, during a period of six or seven months, while in the Michigan area a similar tendency resulted from continuous year-round pressure on two major prey groups, rodents and small birds.

CHAPTER 13

THE FUNCTION OF RAPTOR PREDATION

1. An analysis of predation during the breeding season of hawks and owls showed that, as in winter, predation by a collective raptor population was roughly proportional to relative prey densities.

2. Prey risk factors such as protective cover; prey movement, dis-persion, activity and habits; age of prey species; and predator activity all operated to modify the raptor—prey density relation.

3. Taking a broad view of predation, regardless of season, the many prey risk factors are secondary in importance to prey density in deter-mining what raptors feed upon.

4. Pressure by a breeding raptor population is applied in three ways to prey populations: first, by application of raptor pressure at critical periods; second, by application of maximum pressure at periods of max-imum prey density; and third, by continuous proportionate pressure on a collective prey population.

5. During the critical period of early spring, raptors in Superior Township limited prey populations by reducing the breeding stock. Predation was found to be the factor limiting the meadow mouse population.

6. On the Western Area, all grouse represented 1.2 per cent of the total raptor diet; this amounted to a minimum mortality of 7 per cent of the adult grouse population (three species) prior to and during the nesting season.

7. Predation is only one of many regulating forces acting to control the populations of prey species. It is continually operating to lower the rate of prey increase in proportion to prey density and to do this before more drastic but less steadily functioning regulatory forces, such as intraspecific strife, disease, emigration or starvation, become dominant. Prey increases therefore are likely to be checked before they become excessive.

8. Collective raptorial pressure on a collective prey population tends to depress the various species more or less simultaneously toward the threshold of security. Therefore under natural conditions no one prey species can draw enough predation pressure to keep its population at a dangerously low level.

CHAPTER 14

1. Twenty-nine raptors representing eleven species were fed to obtain quantitative data on food requirements.

2. Large raptors (800 to 1000 gms.) ate an average of 10.7% of their average body weights in fall and winter and 7.7% in spring and summer.

3. Medium raptors (200 to 800 gms.) ate an average of 15.8% of their average body weights in fall and winter and 12.5% in spring and summer.

4. Small raptors (100 to 200 gms.) ate an average of 25.3% of their average body weights in fall and winter and 17.0% in spring and summer.

5. In cold weather the large raptors ate 3% more of their average body weights than the same weight class consumed during warm weather. The raptors in the small and medium-sized weight classes ate about 3 and 8 per cent more, respectively, in cold than in warm weather.

6. The percentage of average body weight eaten per day by raptors is closely correlated with the weight of the raptors and with air temperature.

7. It was computed that the raptor population consumed about 2,964,392 grams of food in the fall and winter of 1941-42 and 1,182,633 grams in the fall and winter of 1947-48.

8. During the spring and summer periods of 1941-42 and 1947-48 the populations are calculated to have consumed 2,037,687 and 2,058,-909 grams of food, respectively.

9. On a basis of the average food requirements of raptors and food samples for the populations, it was calculated that the raptor populations consumed more than 90,000 prey individuals during the fall and

winter of 1941-42 and more than 28,000 during the same period of 1947-48.

10. The computed prey numbers calculated to have been consumed during the spring and summer months were approximately 31,000 in 1941-42 and 27,000 in 1947-48.

11. Raptors accounted for approximately 24% of all the major prey species during the fall and winter of 1941-42 and for 20% during 1947-48.

12. A significant number of game individuals were calculated to have been killed by raptor populations during the warm months.

13. Raptor predation is largely a density-regulated mortality factor.

14. The number of major prey animals calculated to have been killed in fall and winter, when compared and related to estimates of the number of individuals of these prey present and available to raptors at this time, shows: that in the case of each major prey species or group, the ratio of the total population of that species or group to the total of the collective prey population approximates the ratio of the kill from that species or group by a collective population of raptors to the total kill from the collective prey population by the same raptor population. This supports an earlier conclusion that raptor predation is determined largely by the relative densities of all the major prey populations.

CHAPTER 15
THE RELATION OF PREDATION PHENOMENA TO WILDLIFE ECOLOGY

1. The basic operation of predation by a collective mammalian population appears to be identical with raptor predation, except that mammals lack the ability to concentrate quickly from distant regions into areas of high prey density.

2. Insufficient predation pressure is one environmental factor that permits Arctic prey populations to reach a point where their numbers can "skyrocket."

3. Predation is a biological process that tends to prevent excessive increases of prey species. In the case of noxious prey species, this constitutes an inexpensive form of control.

4. Widespread and unrestricted persecution of predatory birds is unjustified biologically, although local predator control in special situations may be economically necessary. Predator control should be a finely calculated emergency operation, employed only until more basic remedies, frequently the correction of land abuse, can be applied.

Literature Cited

Allen, D. L. 1943. Michigan Fox Squirrel Management. Game Division, Department of Conservation, Lansing, Michigan: 404 pp.

Bailey, Vernon. 1924. Breeding, Feeding, and other Life Habits of Meadow Mice (Microtus) Jour. Agr. Res., 27:523-536.

Baldwin, P. S., and Charles S. Kendeigh 1938. Variations in the Weight of Birds. Auk, 55:416-467.

Baumgartner, F. M. 1939. Territory and Population in the Great Horned Owl. Auk, 56:274-282.

Bent, A. C. 1937 and 1938. Life Histories of North American Birds of Prey. Part 1. U. S. Nat. Mus. Bull. 167:409 pp. Part 2. U. S. Nat. Mus. Bull. 170:482 pp.

Blair, W. F. 1940. Home Ranges and Populations of the Meadow Vole in Southern Michigan. Jour. Wildl. Mgt., 4:149-161.

Bole, B. P., Jr. 1939. The Quadrat Method of Studying Small Mammal Populations. Cleveland Mus. Nat. Hist. Sci. Publ., 5 (4):15-77.

Burt, W. H. 1940. Territorial Behavior and Populations of Some Small Mammals in Southern Michigan. Misc. Pub. Mus. Zool., Univ. of Mich., No. 45:58 pp.

Burt, W. H. 1946. The Mammals of Michigan. The University of Michigan Press, 288 pp.

Butler, L. 1953. The Nature of Cycles in Populations of Canadian Mammals. Canadian Jour. Zool., 31:242-262.

Cahn, Alvin R., and Jack T. Kemp 1930. On the Food of Certain Owls in East-Central Illinois. Auk, 47:323-328.

Chitty, Dennis 1938. A Laboratory Study of Pellet Formation in the Short-eared Owl (Asio flammeus). Proc. Zool. Soc. London, 108A:271-287.

Craighead, F. C., Jr., and J. J. Craighead 1939. Hawks in the Hand. Houghton Mifflin Company, Boston: pp. 1-290.

Craighead, J. J., and F. C. Craighead, Jr. 1940. Nesting Pigeon Hawks. Wilson Bull., 52:241-248.

Craighead, F. C., Jr., and J. J. Craighead 1949. Nesting Canada Geese on the Upper Snake River. Jour. Wild. Mgt., 13:52-64.

Craighead, F. C., Jr. 1951. A Biological and Economic Evaluation of Coyote Predation. New York Zoological Society & The Conservation Foundation: pp. 1-23.

Daubenmire, R. F. 1948. Plants and Environment. John Wiley and Sons, N. Y.: pp. 1-424.

Dice, L. R. 1941. Methods for Estimating Populations of Mammals. Jour. Wldl. Mgt., 4:398-407.

Dymond, J. R. 1947. Fluctuations in Animal Populations with Special Reference to Those of Canada. Trans. Roy. Soc. Can., Vol. XLI, Section V:1-34.

Elton, Charles. 1942. Voles, Mice, and Lemmings. Oxford Univ. Press: 496 pp.

English, P. F. 1945. Pheasant Mortality. The Ring-necked Pheasant. W. L. McAtee, Editor. Wildl. Mgt. Inst.: pp. 153-154.

Errington, P. L. 1932. Technique of Raptor Food Habits Study. Condor, 34:75-86.

Errington, P. L., and F. N. Hamerstrom, Jr. 1936. The Northern Bob-white's Winter Territory. Agr. Exp. Sta., Iowa State College of Agr. and Mechanic. Arts. Research Bull. 201:443 pp.

Errington, P. L., and F. N. Hamerstrom. 1940. The Great Horned Owl and Its Prey in North-Central United States. Agr. Exp. Sta., Iowa State College of Agr. and Mechanic. Arts. Research Bull. 277:759-850.

Errington, P. L. 1943. An analysis of Mink Predation Upon Muskrats in North-Central United States. Agr. Exp. Sta., Iowa State College of Agr. and Mechanic. Arts. Research Bull. 320:798-924.

Errington, P. L. 1946. Predation and Vertebrate Populations. Quart. Rev. Biol., 21, No. 2:144-177.

Esten, Sidney R. 1931. Bird Weights of 52 Species of Birds. Auk, 48:572-574.

Fitch, H. S. 1947. Predation by Owls in the Sierran Foothills of California. Condor, 49:137-151.

Fryxell, F. M. 1930. Glacial Features of Jackson Hole, Wyoming. Augustana Library Publications, No. 13, Rock Island, Illinois.

Gabrielson, Ira N. 1941. Wildlife Conservation. The Macmillan Co., New York.

George, J. L. 1952. The Birds on a Southern Michigan Farm. 395 pp. Univ. of Michigan Graduate School.

Goodrum, P. D. 1937. Notes on the Gray and Fox Squirrels of eastern Texas. Trans. Second North Amer. Wildl. Conf.: 499-504.

Graham, S. A. 1945. Ecological Classification of Cover Types. Jour. Wildl. Mgt., 7:182-190.

Grinnell, J. 1914. An Account of the Mammals and Birds of the Colorado Valley with Especial Reference to the Distributional Problems Presented. Univ. Calif. Publ. Zool., 12:15-294.

Guérin, G. 1928. La Vie des Chouettes. Régime et Croissance de l'effraye commune, *Tyto alba alba* L., en Vendée. Encyclopedie Ornithologique, Vol. IV, Paris.

Hamilton, W. J., Jr. 1937. The Biology of Microtine Cycles. Jour. Agr. Res., 54, No. 10.

Hamilton, W. J., Jr. 1940. Life and Habits of Field Mice. Sci. Monthly, Vol. 50: 425-434.

Hamilton, W. J., Jr. 1941. Reproduction of the Field Mouse, Microtus pennsylvanicus (Ord). Cornell Univ. Agr. Exp. Sta., Ithaca, New York.

Hammond, M. C., and C. J. Henry, 1949. Success of Marsh Hawk Nests in North Dakota. Auk, 66:271-274.

Haugen, A. O. 1943. Management Studies of the Cottontail Rabbit in Southwestern Michigan. Jour. Wildl. Mgt., 7:102-119.

Hickey, J. J. 1949. Survival Studies of Banded Birds. (Doctoral Thesis) Univ. of Michigan.

Immler, H. R. 1937. Weights of Some Birds of Prey of Western Kansas. Bird Banding, 8:166-169.

Leopold, Aldo. 1937. Game Management. Charles Scribner's Sons. 481 pp.

Lincoln, F. C. 1930. Calculating Waterfowl Abundance on the Basis of Banding Returns. U. S. Dept. Agr. Circ. No. 118.

McAtee, W. L. 1932. Effectiveness in Nature of the So-Called Protective Adaptations in the Animal Kingdom, Chiefly as Illustrated by the Food Habits of Nearctic Birds. Smithsonian Misc. Collections, 85, No. 7:1-201.

McAtee, W. L. 1935. Food Habits of Common Hawks. United States Dept. Agr., Circular No. 370: 1-36.

McDowell, R. D. 1940. The Great Horned Owl. Penna. Game News, Nov.

McDowell, R. D. 1941. The Eastern Goshawk in Pennsylvania. Penna. Game News, Feb.

McDowell, R. D. 1941 (b). The Cooper's Hawk in Pennsylvania. Penna Game News, Apr.

Murie, A. 1934. The Moose of Isle Royale: Univ. Mich. Mus. Zool., Misc. Pub. 25.

Murie, A. 1940. Ecology of the Coyote in the Yellowstone. Fauna of the Natl. Parks of the U. S., Fauna Series No. 4: 1-206.

Murie, A. 1944. The Wolves of Mount McKinley. Fauna of the Natl. Parks of the U. S., Fauna Series No. 5: 1-238.

Murie, O. J. 1935. Food Habits of the Coyote in Jackson Hole, Wyoming. U. S. Dept. of Agr. Cir. No. 362: 1-24.

Murie, O. J. 1945. Birds of Grand Teton National Park. Mimeographed List, Teton Natl. Park, Wyoming.

Nice, M. M. 1937. Studies in the Life History of the Song Sparrow I. Trans. Linnaean Society of New York. Vol. 4: 1-246.

Nice, M. M. 1938. The Biological Significance of Bird Weights. Bird Banding, 9: 1-11.

Nice, M. M. 1943. Studies in the Life History of the Song Sparrow II. Trans. Linnaean Society of New York. Vol. 6: 1-329.

Nicholson, A. J. 1933. The Balance of Animal Populations. Journ. Animal Ecology, 2 (1): 132-178.

Noble, G. K. 1939. The Role of Dominance in the Social Life of Birds. Auk, 56: 263-273.

Peterson, R. T. 1941. A Field Guide to Western Birds. Houghton Mifflin Co., Boston: 1-240.

Piper, S. E. 1909. The Nevada Mouse Plague of 1907-08. U. S. Dept. of Agr. Farmers' Bull., 352: 23.

Poole, Earl L. 1938. Weights and Wing Areas in North American Birds. Auk, 55: 511-517.

Rose Lake Experiment Station Reports; Michigan Department of Conservation
3rd Annual Report, 1941-42.
4th Annual Report, 1942-43.

Rowan, W. 1948. The Ten-Year Cycle. Dept. of Extension, Univ. of Alberta: 1-13 pp.

Russell, J. C., and Frank Leverett. 1915. Geologic Atlas of the United States, Ann Arbor Folio, Michigan. U. S. Geological Survey: 18 pp.

Stewart, R. E. 1949. Ecology of a Nesting Red-shouldered Hawk Population. Wilson Bull., 61:26-35.

Stickel, L. F. 1946. Experimental Analysis of Methods of Measuring Small Mammal Populations. Jour. Wildl. Mgt., 10:150-159.

Stoddard, H. L. 1946. The Bobwhite Quail. Charles Scribner's Sons: 559 pp.

Tinbergen, L. 1946. The Sparrow-hawk (Accipiter nisus L.) as a Predator of Passerine Birds. Ardea, 34:1-213.

Trippensee, R. E. 1934. The Biology and Management of the Cottontail Rabbit—Unpublished thesis—Univ. of Michigan Graduate School.

Uttendorfer, O. 1939. Die Ernahrung der deutschen Raubvogel und Eulen und ihre Bedeutung in der heimischen Natur. Neudamm, Berlin.

Wallace, G. J. 1948. The Barn Owl in Michigan. Michigan State College, Agr. Exp. Sta., East Lansing, Mich., Bull. 208:61 pp.

Watson, Walcott 1935. History of Jackson's Hole, Wyoming, Before the Year 1907. (M. S. Thesis) Harvard University.

Weaver, J. E., and F. C. Clements. 1938. Plant Ecology. McGraw-Hill, N. Y. 601 pp.

Wight, H. M. 1934. The Cover Map Game Census in Pheasant Management. Amer. Game Conf. Trans., 20:329-333.

Yeagley, H. L. 1947. A Preliminary Study of a Physical Basis of Bird Navigation. Jour. of Applied Physics, 18:1035-1063.

Young, Howard. 1949. A Comparative Study of Nesting Birds in a Five-Acre Park. Wilson Bull., 61:36-47.

U. S. Dept. of Agr. 1945. U. S. Census of Agr. 1945.

Flora and Fauna of the Regions Studied

SUPERIOR TOWNSHIP, MICHIGAN

VEGETATION

The country originally was heavily forested with red, white, and black oak; shellbark and bitternut hickory; beech; sugar and red maple; American elm; white and black ash; and basswood. Beech, maple, and basswood are climax species for the area. These forests were cut rapidly until, at present, only 11 per cent of the area is wooded; of this only a small portion carries mature climax stands. The wooded area is divided into widely scattered woodlots, generally located near the centers of the sections.

Upland types predominate, but lowland types present are those of marsh, kettle hole, seepage, flood plain, and lake shore. The upland types of terrestrial origin and the lowland types of aquatic origin are presented in Table 61 after Graham (1945). Six upland types are now present, the area having evolved beyond the early lichen-moss stages of plant succession.

These upland types, numbered in order of their succession, are:

1. Grasses and other perennials.
2. Mixed herbaceous.
3. Shrubs.
4. Intolerant trees.
5. Mid-tolerant trees.
6. Tolerant trees.

All characteristic aquatic successional stages, from open water to the invasion of tolerant trees into mid-tolerant types, are present. The bog series so common in southern Michigan is not represented on the area. The marsh seepage flood plain series are present. The lowland types numbered in order of their succession are:

1–4—Submerged and emergent vegetation.
5—Emergents and sedge-grass.
6—Mixed herbaceous.

7—Swamp shrubs.

8—Intolerant trees.

9—Mid-tolerant trees.

10—Tolerant trees.

For characteristic species in each type, see Table 61. These types were used as a basis for the field maps, but for convenience and simplicity of presentation the types have been grouped into an artificial classification of woodlots, coniferous stands, marsh, grassland, cultivated land, and open water. The basic facts of the predation situation were related to these broad vegetation types and could be correlated with them without recourse to the finer ecological divisions. The percentage of each of these types per section was determined by use of a planimeter on the detailed cover map.

Successional changes rapidly alter vegetation on abandoned farm land (Plate 3) and more slowly change other habitats. Such changes affect animal populations and thus influence predation. Eleven per cent of the township is woodlots; four per cent is marshland; 50 per cent, grassland; 34 per cent, cultivated land; and one per cent, water and buildings. Map I shows the exact relationships of these types of land.

Grasslands

This type is composed of permanent pasture land, timothy and clover fields, wheat and oat stubble containing dense grass and herbaceous growth, unpastured sod grass, and grass-herbaceous mixtures.

Cultivated land

All land supporting grain crops, corn, and soy beans, as well as plowed fields, is considered to be cultivated land. Winter stands of timothy, alfalfa, and clover, though agrciultural crops, functioned ecologically as grasslands and were considered as such.

Marshland

This broad land type included marshes, kettles, seepage areas, and stream floodplains.

Woodlots

Woodlots are of three principal types: (1) the oak-hickory, characteristic of moraines, till plains, and outwash soils; (2) the beech-sugar maple, found on good agricultural land with heavy soil (Plate 4); (3) swamp forest of the wet soils of flood-plains, lake beds, and seepage areas. Typical trees of this type are red maple, American elm, black and white ash.

Fauna

The fauna of the area is typical of southern Michigan. For information concerning mammal species inhabiting the area and for detailed description of them see Burt (1946).

Records of all mammal species seen were kept and information on relative populations of predatory mammals thought to have a direct influence on the ecology of raptor predation was gathered. Actual counts conducted throughout the study periods gave crude population estimates for such species as foxes, raccoons, and skunks (Table 62). These population counts were a result of sight observations, tracking, and field signs. The counts express only minimum populations, but in some cases may be quite close to the actual populations. They serve to give a general quantitative picture of the predaceous fauna of the area, and as such are helpful in evaluating the combined raptor-carnivore predation pressure. Censuses of prey species were made by intensive investigations. The results appear in Chapter 5.

WESTERN STUDY AREA—WYOMING

VEGETATION ZONES OF THE REGION

The Jackson Hole basin and surrounding mountains extend through three life zones—Arctic, Hudsonian, and Canadian.

The vegetation can best be visualized and understood, however, in terms of the formation concept of Weaver and Clements (1948) or of the Life Zones of Daubenmire (1942).

The vegetation of the Jackson Hole basin and surrounding mountain ranges, if classified according to Daubenmire's system, embraces two zones—the sedge-grass zone and the spruce-fir zone. The arborvitae-hemlock zone is not present and the Douglas fir zone is not represented except as local scattered patches of Douglas fir on southern exposures and in the lower end of the valley. Douglas fir occurs as a successional stage following burning of the spruce-fir forests. No ponderosa pine, a tree considered characteristic of successional stages in the Douglas fir zone, is present. Areas of sagebrush and grasses represent the climax vegetation of the sagebrush-grass zone and exist as microclimatic areas within the extensive spruce-fir zone.

Thus extensive associations considered characteristic of three life zones are present within a climatic area embracing only two zones. The sagebrush-grass associations can be considered as microclimatic areas existing in the spruce-fir zone because of a multiplicity of environmental conditions, such as insolation, temperature, soil condition, slope exposure, evaporation, and precipitation, any one of which may act as a controlling condition in local areas.

Sedge-Grass Zone

The sedge-grass zone includes all areas above the limits of forest growth. The vegetation is characterized by dense meadow-like cover dominated by sedges and grasses, open gravelly surfaces supporting chiefly scattered cushion-form plants and boulder fields frequently devoid of any vegetation except lichens and hardy crevice plants.

Spruce-Fir Zone

This zone is the highest forest zone, extending down from upper timberline. It usually occupies about 2000 feet of elevation in the northern Rocky Mountains.

Climax species are Engelmann spruce and alpine fir. Where the climax forests have been destroyed, temporary forests of lodgepole pine, aspen, and Douglas fir exist.

Upper timberline climate usually is too severe for lodgepole pine, aspen, or Douglas fir to colonize burned areas and at such altitudes spruce and fir regenerate directly. The successional stages of which lodgepole pine is characteristic are thus found at lower elevations in the zone.

The predation study area lies within the spruce-fir zone. The vegetation types have evolved largely through the upland succession. Aquatic succession is represented by the Snake River flood plain and seepage area vegetation. Alpine fir and Engelmann spruce characterize the climax type of each of the developmental series.

No attempt will be made to discuss plant succession in the area, but the various major plant types will be described.

VEGETATION TYPES

Although the various successional stages intergrade, definite types, which are generally dominated by one or two species or genera of plants, are often quite marked and quickly recognizable.

Some dominant as well as subdominant plants have been listed under each type. The submerged and floating plant stages of the aquatic succession and the gravel bar of the flood plain were not treated as types. For practical purposes, they were considered as belonging to the larger dry land types within which they were situated.

Spruce-Fir Type

The extent of the spruce-fir type on the study area is limited. The dominant tree species are *Abies lasiocarpa* and *Picea Engelmanni*, with some *Pinus flexilis* and *Pinus albicaulis*. The undergrowth is characterized by such plants as *Vaccinium scoparium, Vaccinium membranaceum, Gaultheria humifusa, Chimaphila umbellata, Ribes petiolare, Rubus parviflorus, Actaea arguta, Mertensia ciliata, Castilleja rhexifolia, Erigeron coulteri, Aquilegia coerulea, Pyrola chlorantha,* and *Saxifraga punctata.*

Note: Authority for scientific names: Davis, R. J. 1952. *Flora of Idaho.* Wm. C. Brown Co., Dubuque, Iowa: 1-828.

Lodgepole Type

The lodgepole pine is a sub-climax species in this region and may dominate areas of porous, well-drained soil for long periods. It is a characteristic tree of the moraines and glacial outwash plains and invades burned-over areas. The lodgepole type generally precedes the spuce-fir climax type. It may consist of nearly pure stands of *Pinus contorta;* a mixture of *Pinus contorta, Populus tremuloides,* and *Populus angustifolia;* or at times *Pseudotsuga taxifolia.* The compostion depends on effects of fire, cutting, grazing, windfall, drifting snow, and exposure. Subdominants of the lodgepole type are *Carex* sp., *Agropyron subsecundum, Phleum pratense, Vaccinium scoparium, Arctostaphylos uva-ursi, Lonicera utahensis, Sorbus scopulina, Ribes cereum, Ribes setosum, Pterospora andromedea, Pachistima myrsinites, Arnica cordifolia, Lupinus parviflorus, Chimaphila umbellata, Corallorhiza striata,* and *Fragaria platypetala.*

Aspen Type

Aspen stands generally are associated with such biological disturbances as fire, cutting, or windfall. Aspen is commonly found with lodgepole pine and cottonwood. Most young aspen stands contain sagebrush as one of the principal subdominants. Subdominants of the aspen type are *Artemisia tridentata, Agropyron subsecundum, Festuca elatior, Lupinus parviflorus, Thalictrum occidentale, Fragaria platypetala, Smilacina stellata, Campanula rotundifolia, Galium boreale, Cerastium arvense, Prunus virginiana, Symphoricarpos albus, Ceanothus velutinus, Valeriana sitchensis, Melica bulbosa, Phleum pratense, Viguiera multiflora, Helianthella quinquenervis,* and species of *Erigeron* and *Solidago.*

Cottonwood Type

The cottonwood type *(Populus angustifolia* and *Populus balsamifera)* is often found in pure stands, but may also contain lodgepole and aspen. It is primarily a transition type evolving through the flood plain succession. In dry regions it may be a subclimax type, but in most of the Jackson Hole region it is succeeded by spruce and fir, either directly or partially through the lodgepole stage before being dominated by the climax species. Subdominants of this type are *Eleagnus canadensis, Rosa Woodsii, Agropyron subsecundum, Rudbeckia occidentalis, Symphoricarpos albus, Glycyrrhiza lepidota, Prunus virginiana, Cornus sericea, Crataegus* sp., *Perideridia gairdneri, Senecio serra,* and *Urtica gracilis.*

Willow-Sedge Type

The willow-sedge type is an easily recognized stage of the aquatic succession. It may vary from a combination of *Salix* sp., *Carex* sp., and *Juncus multicornis,* to pure growths of one or the other accompanied by numerous subdominants. Subdominants of this type are *Potentilla fruticosa, Castilleja* sp., *Pedicularis groenlandica, Mimulus moschatus, Mimulus guttatus, Polygonum bistortoides, Mertensia ciliata, Lonicera involucrata, Alnus tenuifolia, Heracleum lanatum, Sparganium angustifolium, Sagittaria cuneata, Sium suave, Ranunculus aquatilis, Ceratophyllum demersum, Smilacina stellata,* and *Agrostis alba,* and various species of moss.

Sagebrush-Grass Type

The sagebrush-grass type is well represented on the study area and covers extensive areas of the Jackson Hole region, where it is a microclimatic climax type arising through the upland succession. It dominates the dry areas of good soil. The dominant plants of this type are *Artemisia tridentata, Purshia tridentata, Chrysothamnus nauseosus,* and grasses such as *Koeleria cristata, Stipa comata, Stipa lettermani, Sitanion hystrix, Bromus carinatus, Poa pratense, Oryzopsis hymenoides, Elymus cinereus, Hordeum jubatum.* Other common plants are *Balsamorrhiza sagittata, Wyethia amplexicaulis, Gaillardia aristata, Senecio mutabilis, Antennaria* sp., *Eriogonum subalpinum, Rumex paucifolia, Berberis repens, Gilia aggregata, Sedum stenopetalum, Achillea millefolium, Agoseris glauca, Lomatium ambiguum.*

Grass Type

The grass type is one in which grasses and other herbaceous plants predominate. In this region it arises principally as a result of such disturbances as cultivation and irrigation. The grass type, once established, is to some extent a subclimax type. It may hold an area of land for extensive periods, but, if moist enough, it eventually will be invaded by shrubs and trees. On once-irrigated land from which water has been cut off, the pure grass type will be re-invaded by sagebrush. On poor soil, arid sites, and land abused by overgrazing, the grass type contains grass and other herbaceous plants with the latter predominant. Some of the grass species found in this type on the study area are *Phleum pratense, Agropyron subsecundum,* various species of *Poa, Bromus, Stipa,* and *Elymus, Koeleria cristata, Festuca elatior* and *Agrostis alba.* Associated plants are *Taraxacum officinale, Plantago major, Achillea millefolium,* several species of *Carex* and *Sisyrinchium.*

Cultivated Land

The cultivated land in the study area is chiefly hay fields, originally sagebrush, but made productive for crops through irrigation.

FAUNA OF THE AREA

The Wyoming study area differs greatly from Superior Township, not only in its physiography and vegetation, but also in its fauna. A general description of the bird and mammal life is necessary for evaluating the different aspects of predation and will prove helpful in comparing the two regions. Prey species will be treated in greater detail later. No attempt will be made to enumerate all the bird and mammal species present, but the abundance of the more common species will be indicated.

Mammals

Counts were made of a large number of the more conspicuous mammals living on the area during the 1947 period of study and these, together with map plottings, helped to determine approximate populations.

One black bear with two cubs ranged over Blacktail Butte and the river bottom.

Marten, though scarce in summer, ranged occasionally into the area.

Longtail weasels were seldom seen, but winter tracking indicated that they were relatively abundant.

Mink were quite common along the Snake River and the beaver ponds formed in old river channels. Twelve apparently different individuals were observed.

Otter were rare. Only one was seen in the river.

One striped skunk was seen. There were no indications that this mammal was at all abundant.

Numerous badgers were observed. Ten denning pairs (20 individuals) were located on the area. This is considered a rather accurate census of the population.

Three pairs of coyotes denned on the area. The denning areas were widely separated and near the boundary lines.

The yellowbelly marmot, the Uinta ground squirrel, the least chipmunk, and the red squirrel were very common. There was a dense population of marmots on the rocky areas of the Butte and ground squirrels were abundant in the sage flats and hayfields. Chipmunks were well distributed, being commonest in the combination sagebrush, aspen, and cottonwood types. Red squirrels were abundant in all the coniferous stands.

Northern flying squirrels were seen, but did not appear to be abundant.

Northern pocket gophers were very abundant throughout the area, especially in the lodgepole stands of the river bench and on Blacktail Butte.

Beaver utilized bank houses along the Snake River and stick houses in the old river sloughs. Eleven adults and 12 kits were found on the area. Trapping and some losses caused by tularemia had reduced a higher population.

White-footed mice were abundant, especially in the sagebrush habitat.

Bushytail woodrats were found on the cliffs of Blacktail Butte, but did not appear to be particularly abundant.

Redback mice were trapped, but no indication of their abundance was obtained.

Various species of meadow mice were present and a very high population of *Microtus pennsylvanicus* existed in the fall of 1946. The numbers of this mouse had been much reduced by the summer of 1947, but it was still the most abundant mammal on the area. It attained highest densities in the hayfields and in the grasslands on the Butte.

Western jumping mice were trapped, but apparently were not abundant.

Porcupines were occasionally seen.

The whitetail jackrabbit was scarce. Only two were jumped on the area in the entire course of the spring and summer investigation.

Snowshoe hares were also scarce. Only a half-dozen were jumped in spite of the fact that much of their habitat was searched.

In early spring many elk migrated through the area on their way to summer ranges. Small bands spent varying periods on Blacktail Butte and three elk continued to range on and off the area for some time. In late fall, a few elk again banded together on Blacktail Butte.

Mule deer were abundant. Twenty-six adults, at least four of which were bucks, were observed. Three fawns were seen, but more must have been produced. The mule deer congregated on Blacktail Butte in the spring and most of them moved to the river bottom as the season advanced.

Moose were commonly seen. They ranged over Blacktail Butte in spring and returned to the river bottom during summer (Plate 42). Eleven moose were known to have been on the area at one time and though some ranged off and on, 11 is close to the total number. Four of the 11 were bulls.

Birds of the Study Area

A complete list of the summer bird residents will not be given. Olaus J. Murie has prepared a list of the birds of Grand Teton National Park (obtainable at Teton National Park Museum) and additional information is readily obtainable from Peterson, 1941. Almost all the birds that Murie listed for the park and a few additional species have been seen. A selected listing of some of the more common and typical birds will help visualize the avian life. Hawk and owl species are treated later in the text.

The most common ducks were the Mallard, Barrow's Golden-eye, and the American Merganser. All three species nested and raised young on the study area. Several Canada Geese nested along the Snake River and the Harlequin Duck, though rare, was seen occasionally.

Three species of grouse were present and nested. In order of their abundance they were the Gray Ruffed Grouse, the Sage Hen, and Richardson's Grouse.

Sandhill Cranes were observed and a pair nested a mile west of the study area. Wilson's Snipe were heard and less often seen near the beaver ponds and swamps of the river bottom.

The Red-shafted Flicker occurred in fair numbers and the Red-naped Sapsucker was a frequent nester in the aspen groves. Former nesting hollows of this species were utilized by Tree Swallows, Western House Wrens, and Mountain Bluebirds.

Steller's Jay, Clark's Nutcracker, and the Rocky Mountain Jay were conspicuous and noisy residents. All three probably nested, but only the two last named are known definitely to have done so.

The American Magpie was fairly abundant. Sixteen pairs nested and are known to have produced at least twenty-one young birds.

The Western Warbling Vireo was frequently heard in the aspen groves.

The Yellow Warbler, Audubon's Warbler, Macgillivray's Warbler and Yellow-throat were common nesters.

Brewer's Blackbird occurred in greater abundance than did either the Yellow-headed or Red-winged Blackbird.

The Western Tanager was a colorful resident of the lodgepole forests. Much more common in the same habitat was the Pine Siskin. Nests of the Pink-sided Junco were so plentiful that one or more nests were found on almost every trip into lodgepole timber.

The Green-tailed Towhee, the Western Savannah Sparrow, the Western Vesper Sparrow, and Brewer's Sparrow were typical birds of the sagebrush, and common nesters among it.

The foregoing information on the physical and biological characteristics of the Western Area makes clear the major differences existing between it and the Michigan study area. The influences of the two widely different environments on avian predation will be frequently compared to determine whether man-made changes in the environment alter the mechanics of predation or modify its function.

It should be pointed out that the successional stages of the vegetation on an area of land determine the animal species present and the carrying capacity of that area. Each study area as a whole was in relatively stable condition; Superior Township, temporarily arrested in its succession by the activities of man and the Western Area relatively unaltered in its slow progress toward climax conditions. Drastic seasonal land use changes occurred in local habitats, but as land entities the areas changed very little during the years of study.

APPENDIX B

SCIENTIFIC NAMES OF ANIMALS

American Magpie ....Pica pica hudsonia
American Merganser .... Mergus merganser americanus
American Raven ....Corvus corax sinuatus
Arctic fox ....Alopex lagopus
Audubon's Warbler ....Dendroica auduboni
badger ....Taxidea taxus
Bald Eagle ....Haliaeetus leucocephalus
Barn Owl ....Tyto alba pratincola
Barrow's Golden-eye ....Bucephala islandica
bear, black ....Ursus americanus
beaver ....Castor canadensis
bird hawks ....Accipiter
black bear ....Ursus americanus
Blackbird, Brewer's ....Euphagus cyanocephalus
   Red-winged ....Agelaius phoeniceus
   Yellow-headed ....Xanthocephalus xanthocephalus
Bluebird, Eastern ....Sialia sialis
   Mountain ....Sialia currucoides
Blue Jay ....Cyanocitta cristata
Blue Tit ....Parus caeruleus
Bobolink ....Dolichonyx oryzivorus
bog lemming, southern ....Synaptomys cooperi
Brewer's Blackbird ....Euphagus cyanocephalus
Brewer's Sparrow ....Spizella breweri
Broad-winged Hawk ....Buteo platypterus
bull frog ....Rana catesbeiana
Bunting, Indigo ....Passerina cyanea
bushytail woodrat ....Neotoma cinerea
Canada Goose ....Branta canadensis
Canada Lynx ....Lynx canadensis
Cardinal ....Richmondena cardinalis
Chaffinch ....Fringilla coelebs
chipmunk ....Tamias
chipmunk, least ....Eutamias minimus
Clark's Nutcracker ....Nucifraga columbiana
Coal-tit ....Parus ater
Common Goldfinch ....Spinus tristis

Cooper's Hawk ....Accipiter cooperii
Cowbird ....Molothrus ater
coyote ....Canis latrans
Crane, Sandhill ....Grus canadensis
crawfish ....Cambarus
cricket frog ....Acris gryllus
cricket frog, swamp ....Pseudacris nigrita
Crow ....Corvus brachyrhynchos
Cuckoo, Yellow-billed ....Coccyzus americanus
deer ....Cervidae
deer, mule ....Odocoileus hemionus
deer mouse, prairie ....Peromyscus maniculatus bairdii
Domestic Pigeon ....Columba livia
Dove, Mourning ....Zenaidura macroura
Downy Woodpecker ....Dendrocopos pubescens
Duck Hawk ....Falco peregrinus anatum
Duck, Mallard ....Anas platyrhynchos
Duck, Harlequin ....Histrionicus histrionicus
Eagle, Bald ....Haliaeetus leucocephalus
    Golden ....Aquila chrysaëtos canadensis
Eastern Bluebird ....Sialia sialis
eastern wood mouse ....Peromyscus leucopus noveboracensis
elk ....Cervus canadensis
English Sparrow ....Passer domesticus
European Redstart ....Phoenicurus phoenicurus
European Tree Sparrow ....Passer montanus
European Wren ....Troglodytes troglodytes
Falcon, Prairie ....Falco mexicanus
Ferruginous Rough-leg ....Buteo regalis
Flicker ....Colaptes auratus
Flicker, Red-shafted ....Colaptes cafer
flying squirrel, northern ....Glaucomys sabrinus
fox ....Vulpes
fox, Arctic ....Alopex lagopus
fox squirrel ....Sciurus niger
frog, bull ....Rana catesbeiana
    cricket ....Acris gryllus
    green ....Rana clamitans
    leopard ....Rana pipiens
    swamp cricket ....Pseudacris nigrita
    wood ....Rana sylvatica
Golden Eagle ....Aquila chrysaëtos canadensis
Golden-eye, Barrow's ....Bucephala islandica
Goldfinch, Common ....Spinus tristis
Goose, Canada ....Branta canadensis

gopher, northern pocket ....Thomomys talpoides
Goshawk ....Accipiter gentilis
Gray Owl, Great ....Strix nebulosa
Gray Ruffed Grouse ....Bonasa umbellus
Great Gray Owl ....Strix nebulosa
Great Horned Owl ....Bubo virginianus
Great Tit ....Parus major
Green-tailed Towhee ....Chlorura chlorura
Grosbeak, Rose-breasted ....Pheucticus ludovicianus
ground squirrel ....Citellus
ground squirrel, Uinta ....Citellus armatus
Grouse, Gray Ruffed ....Bonasa umbellus
    Richardson's ....Dendragapus obscurus
    Ruffed ....Bonasa umbellus
    Sage ....Centrocercus urophasianus
Hairy Woodpecker ....Dendrocopos villosus
hare, snowshoe ....Lepus americanus
hares and rabbits ....Leporidae
Harlequin Duck ....Histrionicus histrionicus
Hawk, Broad-winged ....Buteo platypterus
    Cooper's ....Accipiter cooperii
    Duck ....Falco peregrinus anatum
    Marsh .... Circus cyaneus hudsonius
    Pigeon ....Falco columbarius
    Red-shouldered ....Buteo lineatus
    Red-tailed ....Buteo jamaicensis
    Rough-legged ....Buteo lagopus s.-johannis
    Sharp-shinned ....Accipiter striatus velox
Hawk, Sparrow ....Falco sparverius
    Sparrow (old world) ....Accipiter nisus
    Swainson's ....Buteo swainsoni
    Western Red-tailed ....Buteo jamaicensis
hawks, bird ....Accipiter
Hen, Sage ....Centrocercus urophasianus
Hermit Thrush ....Hylocichla guttata
Horned Owl, Great ....Bubo virginianus
house mouse ....Mus musculus
House Wren, Western ....Troglodytes aëdon
Ibis ....Threskiornithinae
Indigo Bunting ....Passerina cyanea
jackrabbit ....Lepus sp.
jackrabbit, whitetail ....Lepus townsendi
Jay, Blue ....Cyanocitta cristata
    Rocky Mountain ....Perisoreus canadensis

jumping mice ....Zapodidae
jumping mouse, western ....Zapus princeps
Junco, Pink-sided ....Junco oreganus mearnsi
kinglet ....Regulus
least chipmunk ....Eutamias minimus
least shrew ....Cryptotis parva
lemming ....Lemmus & Dicrostonyx
lemming, southern bog ....Synaptomys cooperi
leopard frog ....Rana pipiens
Lincoln's Sparrow ....Melospiza lincolnii
Long-eared Owl ....Asio otus wilsonianus
longtail weasel ....Mustela frenata
lynx, Canada ....Lynx canadensis
Macgillivray's Warbler ....Oporornis tolmiei
Magpie, American ....Pica pica hudsonia
Mallard Duck ....Anas platyrhynchos
man ....Homo sapiens
marmot ....Marmota
marmot, yellowbelly ....Marmota flaviventris
Marsh Hawk ....Circus cyaneus hudsonius
marten ....Martes americana
Meadowlark ....Sturnella magna
Meadowlark, Western ....Sturnella neglecta
meadow mouse ....Microtus pennsylvanicus
meadow mouse, mountain ....Microtus montanus
Merganser, American ....Mergus merganser americanus
mice, jumping ....Zapodidae
mink ....Mustela vison
moose ....Alces americana
Mountain Bluebird ....Sialia currucoides
mountain meadow mouse ....Microtus montanus
Mourning Dove ....Zenaidura macroura
mouse, eastern wood ....Peromyscus leucopus noveboracensis
    house ....Mus musculus
    meadow ....Microtus pennsylvanicus
    mountain meadow ....Microtus montanus
    prairie deer ....Peromyscus maniculatus bairdii
    redback ....Clethrionomys gapperi
    western jumping ....Zapus princeps
    white-footed ....{ Peromyscus leucopus
                                 { Peromyscus maniculatus
mule deer ....Odocoileus hemionus
northern flying squirrel ....Glaucomys sabrinus
Nutcracker, Clark's ....Nucifraga columbiana
Nuthatch, Red-breasted ....Sitta canadensis

Olive-backed Thrush .... Hylocichla ustulata
Osprey .... Pandion haliaetus carolinensis
otter .... Lutra canadensis
Oven-bird .... Seiurus aurocapillus
Owl, Barn .... Tyto alba pratincola
    Great Gray .... Strix nebulosa
    Great Horned .... Bubo virginianus
    Horned .... Bubo virginianus
    Long-eared .... Asio otus wilsonianus
    Saw-whet .... Aegolius acadicus
    Screech .... Otus asio
    Short-eared .... Asio flammeus
    Snowy .... Nyctea scandiaca
Peeper, spring .... Hyla crucifer
Pine Siskin .... Spinus pinus
Pigeon, Domestic .... Columba livia
Pigeon Hawk .... Falco columbarius
Pigeons .... Columbidae
Pink-sided Junco .... Junco oreganus mearnsi
pocket, gopher, northern .... Thomomys talpoides
porcupine .... Erethizon dorsatum
prairie deer mouse .... Peromyscus maniculatus bairdii
Prairie Falcon .... Falco mexicanus
rabbits and hares .... Leporidae
rabbit, jack .... Lepus townsendi
    snowshoe .... Lepus americanus
    whitetail jack .... Lepus townsendi
Rail, Sora .... Porzana carolina
rat, bushytail wood .... Neotoma cinerea
Raven, American .... Corvus corax sinuatus
redback mouse .... Clethrionomys gapperi
Red-breasted Nuthatch .... Sitta canadensis
Red-headed Woodpecker .... Melanerpes erythrocephalus
Red-naped Sapsucker .... Sphyrapicus varius
Red-shafted Flicker .... Colaptes cafer
Red-shouldered Hawk .... Buteo lineatus
red squirrel .... Tamiasciurus hudsonicus
Redstart, European .... Phoenicurus phoenicurus
Red-tailed Hawk .... Buteo jamaicensis
Red-wing .... Agelaius phoeniceus
Red-winged Blackbird .... Agelaius phoeniceus
Richardson's Grouse .... Dendragapus obscurus
Robin .... Turdus migratorius
Rocky Mountain Jay .... Perisoreus canadensis
Rook .... Corvus frugilegus

Rose-breasted Grosbeak ....Pheucticus ludovicianus
Rough-leg, Ferruginous ....Buteo regalis
Rough-legged Hawk ....Buteo lagopus s.-johannis
Ruffed Grouse ....Bonasa umbellus
Ruffed Grouse, Gray ....Bonasa umbellus
Sage Grouse ....Centrocercus urophasianus
Sage Hen ....Centrocercus urophasianus
Sage Thrasher ....Oreoscoptes montanus
Sandhill Crane ....Grus canadensis
Sapsucker, Red-naped ....Sphyrapicus varius
Saw-whet Owl ....Aegolius acadicus
Screech Owl ....Otus asio
Sharp-shinned Hawk ....Accipiter striatus velox
Short-eared Owl ....Asio flammeus
shorttail shrew ....Blarina brevicauda
shrew, least ....Cryptotis parva
    shorttail ....Blarina brevicauda
Siskin, Pine ....Spinus pinus
skunk, striped ....Mephitis mephitis
snakes ....Thamnophis
Snipe, Wilson's ....Capella gallinago delicata
snowshoe hare ....Lepus americanus
snowshoe rabbit ....Lepus americanus
Snowy Owl ....Nyctea scandiaca
Sora Rail ....Porzana carolina
southern bog lemming ....Synaptomys cooperi
Sparrow, Brewer's ....Spizella breweri
    English ....Passer domesticus
    European Tree ....Passer montanus
Sparrow Hawk ....Falco sparverius
Sparrow Hawk (old world) ....Accipiter nisus
Sparrow, Lincoln's ....Melospiza lincolnii
Sparrows ....Fringillidae
Sparrow, Song ....Melospiza melodia
    Vesper ....Pooecetes gramineus
    Western Savannah ....Passerculus sandwichensis
    Western Vesper ....Pooecetes gramineus
spring peeper ....Hyla crucifer
squirrel, fox ....Sciurus niger
    northern flying ....Glaucomys sabrinus
    red ....Tamiasciurus hudsonicus
squirrel, Uinta ground ....Citellus armatus
squirrels, ground ....Citellus
Starling ....Sturnus vulgaris
Steller's Jay ....Cyanocitta stelleri

striped skunk ....Mephitis mephitis
sucker ....Catostomus
Swainson's Hawk ....Buteo swainsoni
Swallow, Tree ....Iridoprocne bicolor
swamp cricket frog ....Pseudacris nigrita
Tanager, Western ....Piranga ludoviciana
Thrasher, Sage ....Oreoscoptes montanus
Thrush, Hermit ....Hylocichla guttata
 Olive-backed ....Hylocichla ustulata
 Wood ....Hylocichla mustelina
Tit, Blue ....Parus caeruleus
 Great ....Parus major
Titmouse, Tufted ....Parus bicolor
toad, tree ....Hyla versicolor
Towhee ....Pipilo erythrophthalmus
Towhee, Green-tailed ....Chlorura chlorura
Tree Sparrow, European ....Passer montanus
Tree Swallow ....Iridoprocne bicolor
tree toad ....Hyla versicolor
Tufted Titmouse ....Parus bicolor
Uinta ground squirrel ....Citellus armatus
Vesper Sparrow ....Pooecetes gramineus
Vesper Sparrow, Western ....Pooecetes gramineus
Vireo, Western Warbling ....Vireo gilvus
Warbler, Audubon's ....Dendroica auduboni
 Macgillivray's ....Oporornis tolmiei
Warbler, Yellow ....Dendroica petechia
warblers, wood ....Parulidae
Warbling Vireo, Western ....Vireo gilvus
weasel, longtail ....Mustela frenata
weasels ....Mustelidae (in part)
Western House Wren ....Troglodytes aëdon
western jumping mouse ....Zapus princeps
Western Meadowlark ....Sturnella neglecta
Western Red-tailed Hawk ....Buteo jamaicensis
Western Savannah Sparrow ....Passerculus sandwichensis
Western Tanager ....Piranga ludoviciana
Western Vesper Sparrow ....Pooecetes gramineus
Western Warbling Vireo ....Vireo gilvus
Western Yellow-throat ....Geothlypis trichas
white-footed mouse ....{Peromyscus leucopus
           {Peromyscus maniculatus
Wilson's Snipe ....Capella gallinago delicata
wood frog ....Rana sylvatica
wood mouse, eastern ....Peromyscus leucopus noveboracensis

Woodpecker, Downey ....Dendrocopos pubescens
    Hairy ....Dendrocopos villosus
    Red-headed ....Melanerpes erythrocephalus
woodrat, bushytail ....Neotoma cinerea
Wood Thrush ....Hylocichla mustelina
wood warblers ....Parulidae
wren ....Troglodytes sp.
Wren, European ....Troglodytes troglodytes
    Western House ....Troglodytes aëdon
yellowbelly marmot ....Marmota flaviventris
Yellow-headed Blackbird ....Xanthocephalus xanthocephalus
Yellow-throat, Western ....Geothlypis trichas
Yellow Warbler ....Dendroica petechia

SCIENTIFIC NAMES OF PLANTS

This list does not include scientific names of plants that appear in Appendix A without colloquial equivalents.

alpine fir ....Abies lasiocarpa
American elm ....Ulmus americana
apple ....Pyrus (subgen. Malus)
ash, black ....Fraxinus nigra
    white ....Fraxinus americana
aspen ....Populus tremuloides
basswood ....Tilia americana
beech ....Fagus grandifolia
bitterbrush ....Purshia tridentata
bitternut hickory ....Carya cordiformis
black ash ....Fraxinus nigra
black oak ....Quercus velutina
blue grass ....Poa
brome-grass ....Bromus
buttonbush ....Cephalanthus occidentalis
cottonwood ....Populus angustifolia
couch-grass ....Agropyron repens
Douglas fir ....Pseudotsuga taxifolia
elm, American ....Ulmus americana
Engelmann spruce ....Picea Engelmanni
fir, alpine ....Abies lasiocarpa
    Douglas ....Pseudotsuga taxifolia
fungus, heart-rot ....Polyporus benzoinus
goldenrod ....Solidago
grass, blue ....Poa
    orchard ....Dactylis
hawthorn ....Crataegus

heart-rot fungus ....Polyporus benzoinus
hickory, bitternut ....Carya cordiformis
    shellbark ....Carya ovata
lodgepole pine ....Pinus contorta
maple, red ....Acer rubrum
    sugar ....Acer saccharum
meadow-grass ....Poa
oak, black ....Quercus velutina
    red ....Quercus rubra
    white ....Quercus alba
orchard grass ....Dactylis
pine, lodgepole ....Pinus contorta
    ponderosa ....Pinus ponderosa
ponderosa pine ....Pinus ponderosa
ragweed ....Ambrosia
red maple ....Acer rubrum
red oak ....Quercus rubra
redtop ....Agrostis alba
sagebrush ....Artemisia tridentata
shellbark hickory ....Carya ovata
soapberry ....Shepherdia canadensis
spiraea ....Spiraea
spruce, Engelmann ....Picea Engelmanni
sugar maple ....Acer saccharum
sumac ....Rhus
timothy ....Phleum pratense
white ash ....Fraxinus americana
white oak ....Quercus alba

APPENDIX C
Tables 61-100

TABLE 61

SOME PLANTS CHARACTERISTIC OF THE VARIOUS TYPES IN SOUTHERN MICHIGAN

| | Upland types—Terrestrial origin | | | Lowland types—Aquatic origin | | | | Transition types | |
| | | | | Bogs | | Marshes | | | |
	Porous soils P	Nonporous soils A	Rock outcrops R	Seepage BS	Stagnant B	Seepage MS	Stagnant M	Flood plain F	Transition belts E
1	Bare soil	Bare soil	Bare rock	Saturated soil or water	Water	Saturated soil or water	Water	Bare moist soil	Same as corresponding wet or dry land type
2			Crustose lichens		Pond weeds Water lilies		Pond weeds Water lilies		do.
3			Foliose lichens Mosses						do.
4	Annuals Mosses seldom	Annuals grasses and perennials starting	Mosses Lichens Scattered annuals	Sedges Sphagnum seldom	Sedge mat Sphagnum usually Bog plants	Bullrushes Cattails	Bullrushes Cattails	Annuals Perennials starting	do.
5	Various grasses Scattered perennials	Sod grasses Goldenrod Asters	Scattered perennials Mosses persisting	Sedges Leatherleaf Scattered perennials	Sphagnum Leatherleaf Sedges Cranberry	Sedges Grasses Cattails	Sedges Grasses Cattails	Sod grasses Scattered perennials	do.
6	Bluegrass Mixed perennials	Bluegrass Mixed perennials	Grasses Mixed perennials	Mixed perennials	Leatherleaf over sedges and sphagnum	Various mixed perennials	Various mixed perennials	Various mixed perennials	do.

7 Panicled dogwood Rose Rubus	Dogwoods Rose Rubus Witchhazel	Dogwoods Hazel Rubus	Highbush cranberry Dogwoods Willows	Dogwoods Huckleberry Chokeberry Spirea	Elderberry Dogwoods Willows	Dogwoods Buttonbush Willow near margin	Great variety of shrubs	do.
8 Aspen Sassafras Juniper	Aspen Juniper Crataegus Cherry	Crataegus Cherry Aspen	Tamarack Aspen Willows	Tamarack Black spruce	Aspen Willows Black ash Elm	Aspen Dogwoods Black ash Elm	White ash Aspen or other poplar Elm	do.
9 Elm Black oaks Hickories	Black and White oaks Hickories	Black oaks Hickories	Soft maples Swamp white elms Oak	Soft maples Yellow birch Hickories	Soft-maples Elms Hickories	Soft maples Elms Black ash	Soft maples Hickories Oaks Elms	do.
10 Sugar maple Basswood Beech	Sugar maple Basswood Beech	Sugar maple Basswood	Sugar maple Basswood Beech	Sugar maple Basswood Beech	Sugar maple Basswood Beech	Sugar maple Basswood Beech	Sugar maple Basswood Beech	Sugar maple Basswood Beech

After Graham, S. A. 1945.

TABLE 62
APPROXIMATE POPULATIONS OF PREDATORY MAMMALS
SUPERIOR TOWNSHIP—1942 AND 1948

Species	1942	1948
Fox	4–6	14-15 (2 known to have been killed)
Opossum	80–90	Much scarcer than in 1942
Raccoon	36–40	38–44
Skunk	120–160	Much scarcer than in 1942
Badger	2	2–4
Weasel	27–36	Scarcer than in 1942
Mink	8–12	8–12

TABLE 63

METEOROLOGICAL DATA ON CENSUS DAYS
CAR HAWK-CENSUS 1941-42

Census dates	Temperatures in degrees F.				Av. Rel. Hum.	Av. Bar. Pr.	Av. Wind Vel.	General Conditions	No. of Hawks
	Max.	Min.	Av.	Range					
Nov. 10/41	36.7	29.6	33.2	7.1	88	29.26	8	Cloudy	32
Nov. 11/41	34.8	29.8	32.3	5.0	91	29.26	10	Cloudy	54
Nov. 15/41	54.0	37.5	45.8	16.5	85	29.00	6	Clear	47
Dec. 7/41	33.0	19.7	26.4	13.3	84	29.32	4	Clear	41
Dec. 14/41	30.7	22.0	26.3	8.7	89	29.19	9	Clear	34
Jan. 20/42	33.6	29.0	31.3	4.6	80	29.20	8	Cloudy	22
Jan. 21/42	34.5	28.9	31.7	5.6	78	29.17	4	Sun and Cloud	40
Jan. 29/42	33.6	22.9	28.2	10.7	82	29.35	4	Cloudy	29
Feb. 11/42	23.8	16.0	19.9	7.8	84	29.20	4	Clear	32
Feb. 12/42	30.0	7.0	18.5	23.0	76	29.34	7	Clear	35
Feb. 13/42	32.1	11.8	21.9	20.3	80	29.44	4	Clear	35
Feb. 24/42	27.3	12.5	19.9	14.8	82	29.01	8	Clear	34
Feb. 25/42	29.0	10.5	19.8	18.5	94	29.03	7	Sun and Cloud	31
Mar. 19/42	47.8	28.3	38.0	19.5	74	29.08	7	Clear	33
Mar. 24/42	50.4	27.0	38.7	23.4	64	29.27	3	Clear	30
Mar. 31/42	41.0	31.8	36.4	9.2	87	29.17	7	Cloudy rain	15
Apr. 14/42	67.7	35.6	51.6	32.1	68	29.27	7	Clear	27
Apr. 30/42	85.3	58.6	72.0	26.7	61	29.20	5	Clear	12

TABLE 64

METEOROLOGICAL DATA ON CENSUS DAYS
CAR HAWK-CENSUS 1947-48

Census dates	Temperatures in degrees F.			Rel. hum. at beginning of census	Daily av. bar. pr.	Wind velocity at beginning of census	General Conditions	No. of Hawks
	Maximum 6 hrs. prior to census	Minimum 6 hrs. prior to census	Temp. at time of census					
Nov. 19, 1947 ..	44.0	33.6	45	55	29.333	7E	Partly cloudy no snow on ground	9
Nov. 23	38.5	31.2	39	59	29.070	11W	Clear no snow on ground	10
Nov. 26	28.1	24.8	30	51	28.876	16W, NW	Clear snow on ground	8
Nov. 29	25.3	22.2	25	62	29.275	20W, SW	Partly cloudy snow on ground	10
Dec. 6	37.0	33.6	36	64	29.094	12W	Cloudy	7
Dec. 9	29.0	22.0	30	54	29.337	8N	Clear	11
Dec. 19	27.8	17.8	27	57	29.412	12N, NE	Clear snow on ground	10
Jan. 10, 1948 ..	24.1	13.4	24	54	29.454	7NW	Clear ice and snow on ground	9
Jan. 16	23.4	14.7	23	72	29.304	15W, S	Partly cloudy snow on ground	8
Jan. 17	16.2	6.2	14	55	29.452	11NW	Partly cloudy snow on ground	10
Jan. 18	12.4	-1.2	13	48	29.440	6NW	Clear snow on ground	9
Jan. 19	22.9	12.8	23	76	29.218	22SW	Cloudy, snow flurries snow on ground	9
Jan. 28	17.2	-0.5	19	52	29.292	15W, SW	Clear snow on ground	6
Jan. 29	19.3	6.7	23	64	29.110	12W, SW	Clear	11

Feb. 8	15.2	9.8	15	41	29.376	15NW	snow on ground Clear	9
Feb. 9	6.3	0.9	5	80	29.496	6N	snow on ground Clear	9
Feb. 20	20.0	10.2	21	50	29.445	14W, NW	snow on ground Partly cloudy-clearing	12
Feb. 26	45.2	33.6	46	64	29.020	20W	no snow on ground Clear	11
March 24	62.0	43.4	61	21	29.032	8NE	no snow on ground Clear no snow on ground	20

TABLE 65

RECORD OF FOOD OF RED-SHOULDERED HAWKS DURING NESTING PERIOD
Superior Township, 1942
Period March 27 to July 22

Nest No.	Sec. No.	No. of Pellets	No. of non Pellet Food Items	Mammals									
				Rabbit		Fox Squirrel		Muskrat		Rat		Meadow Mouse	
				F.P.	N.P.	F.P.	N.P.	F.P.	N.P.	F.P.	N.P.	F.P.	N.P.
1	2	21	4	..	..	2	..	..	..	..	..	21	1
2	3	4	17	..	..	..	..	..	..	..	..	1	..
3	4	32	10	..	..	3	3	1	..	..	..	27	..
4	7	18	0	1	..	1	..	..	..	..	..	16	..
6	9	28	14	..	..	1	..	..	..	6	..	24	3
9	15	31	5	..	..	1	..	2	..	..	..	28	2
10	16	16	6	..	..	..	..	..	..	..	..	14	2
11	20	40	0	..	..	..	..	..	..	..	..	38	..
12	21	0	3	..	..	..	..	..	..	..	..	..	..
13	22	7	4	..	..	..	..	..	..	..	..	6	..
14	26	23	9	..	..	..	..	..	..	1	..	18	..
15	27	6	8	..	..	..	..	..	..	..	..	3	..
17	28	5	10	..	..	..	..	..	..	..	..	3	1
18	29	0	4	..	..	..	..	..	..	..	..	..	..
19	34	16	6	..	..	..	..	..	..	..	..	11	6
20	36	50	17	2	1	2	1	..	..	..	..	44	..
21	36A	18	0	1	..	..	..	..	..	1	..	13	..
22	36A	19	7	..	..	..	..	..	..	..	..	12	3
23	36A	14	10	..	1	..	..	..	..	..	1	14	..
24	..	12	2	..	..	1	..	..	..	..	..	9	1
Total	..	360	136	4	2	11	4	3	0	8	1	302	19

Percent occurrence in Pellets .... 1.1 1.5 3.0 3.0 .8 0 2.2 .7 83.9 14.0

Total Individuals in Pellet and non-Pellet Material (individuals)

	Fox Squirrel		Muskrat	Rat	Meadow Mouse	
	5	11	3	9	104	5
	1.5	3.3	.9	2.7	31.9	1.5
	5.7			36.1		

F.P. = Frequency of occurrence in pellets.
N.P. = Non-pellet food items.

TABLE 65—*Continued*

RECORD OF FOOD OF RED-SHOULDERED HAWKS DURING NESTING PERIOD
Superior Township, 1942
Period March 27 to July 22

	Birds						Lower Vertebrates				Invertebrates			
	Shrews and Moles		Pheasant		Small & Med. Sized Birds		Snakes		Frogs		Crawfish		Insects	
	F.P	N.P.	F.P	N.P.	F.P	N.P.	F.P	N.P.	F.P	N.P.	F.P	N.P.	F.P	N.P.
1	..	..	..	..	5	2	15	..	..	..	7	1	7	..
2	..	..	..	..	4	..	2	..	..	2	2	15	2	..
3	..	..	..	..	7	..	17	1	..	..	5	6	3	..
4	..	..	..	..	2	..	15	..	..	..	1	..	..	..
6	..	..	..	..	3	1	18	1	..	8	6	1	5	..
9	..	..	..	..	3	1	22	..	..	2	9	..	4	..
10	..	..	..	..	1	..	7	..	..	2	3	2	6	..
11	2	..	..	..	..	..	29	..	..	..	..	..	1	..
12	..	1	..	..	..	..	..	1	..	..	..	1	..	..
13	..	..	..	..	2	2	3	2	..	..	..	..	..	..
14	..	..	..	..	9	1	9	1	..	2	5	5	6	..
15	..	/	..	..	..	2	2	1	..	..	..	5	1	..
17	1	..	1	2	2	2	3	3	..	..	1	2	..	..
18	..	..	..	..	..	..	..	..	..	4	..	..	..	..
19	..	..	..	..	2	..	4	..	..	..	..	..	1	..
20	..	1	..	1	15	7	27	3	..	..	3	3	17	..
21	..	..	..	..	6	..	7	..	..	..	2	..	5	..
22	..	..	..	..	1	1	6	..	..	3	..	..	..	..
23	..	..	..	1	..	2	1	..	..	5	..	..	..	..
24	1	..	..	..	1	1	3	..	..	..	..	..	2	..
Total ...	4	2	1	4	63	22	190	13	0	28	44	41	60	0
Percent occurrence in Pellets	1.1	1.5	.3	3.0	17.5	16.2	52.8	10.0	0	20.6	12.2	30.1	16.6	..

	4		66			45		28		46	Total Indiv.
	1.2		20.2			13.8		8.6		14.1	= 326
		21.4					36.5				

Meadow mice recorded at every nest except one. Garter snakes recorded at every nest except one.
Birds recorded at every nest except three.
Crawfish recorded at every nest except seven.

TABLE 66

RECORD OF FOOD OF RED-SHOULDERED HAWKS DURING NESTING PERIOD
Superior Township, 1948
Period March 28 to July 7

Nest No.	Sec. No.	No. of Pellets	No. of Non-Pellet Food Items	Mammals									
				Rabbit		Fox Squirrel		Muskrat		Rat		Meadow Mouse	
				F.P.	N.P.	F.P.	N.P.	F.P.	N.P.	F.P.	N.P.	F.P.	N.P.
3	5	29	34	..	..	..	..	..	..	..	..	26	6
6	10	14	22	2	1	..	1	..	..	..	..	12	2
11	26	35	13	..	..	..	..	2	..	..	..	28	3
12	27	20	22	..	..	..	..	..	..	3	3	13	..
13	28	41	27	..	..	3	1	..	..	6	1	31	6
14	29	15	18	..	1	..	..	..	..	..	..	12	2
15	30	9	14	..	1	..	..	..	..	3	..	8	2
16	32	5	9	..	..	2	..	1	..	..	..	..	2
17	33	24	11	3	1	..	..	..	..	3	..	20	2
Total No.		192	170	5	4	5	2	3	0	19	4	150	25

Percent occurrence in
Pellets 2.6 2.3   2.6 1.2   1.6 0   7.8 2.3   78.1 14.7

Total Individuals
in Pellet and
non-Pellet
Material   4   3   2   9   62   4

Percent diet
(Individuals)   1.6   1.2   .8   3.6   25.0   1.6

3.6   30.3

F.P. = Frequency of occurrence in pellets.
N.P. = Non-pellet food items.

TABLE 66—*Continued*
RECORD OF FOOD OF RED-SHOULDERED HAWKS DURING NESTING PERIOD
Superior Township, 1948
Period March 28 to July 7

| | Birds | | | | | | Lower Vertebrates | | | | Invertebrates | | | |
| | Shrews and Moles | | Pheasant | | Small & Med. Sized Birds | | Snakes | | Frogs | | Crawfish | | Insects | |
	F.P	N.P.	F.P	N.P.	F.P	N.P.	F.P	N.P.	F.P	N.P.	F.P	N.P.	F.P	N.P.
3	..	..	..	..	10	8	16	3	..	4	6	13	9	..
6	..	1	..	..	3	6	5	3	..	8	..	..	5	..
11	2	..	..	..	22	7	35	..	..	2	1	1	16	..
12	1	1	..	..	4	5	14	..	..	8	8	5	11	..
13	..	..	..	..	16	11	11	1	..	5	4	2	5	..
14	..	..	..	..	5	14	5	1	..	..	..	..	..	..
15	..	1	..	..	5	5	5	..	..	5	..	..	2	..
16	..	..	..	..	4	2	4	..	..	3	1	2	2	..
17	..	..	..	..	8	3	5	..	..	5	..	..	..	..
Total No. .	3	3	0	0	77	61	100	8	0	40	20	23	50	..
Percent occurrence in Pellets	1.6	1.8	0	0	40.1	36.0	52.1	4.7	0	23.5	10.4	13.5	26.0	..

0	72			28	40	23
0	29.1			11.3	16.2	9.3
	29.1				36.8	

Total Indiv. = 247

Meadow mice recorded at all nests.
Small birds recorded at all nests.
Snakes recorded at all nests.
Frogs recorded at every nest but one.
Crawfish recorded at every nest but three.

TABLE 67

RECORD OF FOOD OF RED-TAILED HAWK DURING NESTING PERIOD
Superior Township 1942
Period May 6 to June 8

Nest No.	Sec. No.	No. of Food Items	Mammals					Birds					Reptiles
			Rabbit	Fox Squirrel	Wood Chuck	Ground Squirrel	Weasel	Rat	Meadow Mouse	Pheasant	Small & Medium-Sized Birds	Crow	Garter Snake
24	11	61	6	1	1	1	1	0	26	6	12	2	5
Per Cent Diet			9.8	1.7	1.7	1.7	1.7	0	42.6	9.8	19.6	3.2	8.2
			13.2			46.0				32.6			8.2

TABLE 68

RECORD OF FOOD OF RED-TAILED HAWKS DURING NESTING PERIOD
Superior Township 1948
Period May 1 to July 1

Nest No.	Sec. No.	No. of Food Items	Mammals							Birds		Reptiles
			Rabbit	Fox Squirrel	Muskrat	Racoon	Ground Squirrel	Rat	Meadow Mouse	Pheasant	Small & Medium-Sized Birds	Garter Snake
22	15	73	4	7	11	1	3	0	32	3	12	0
24	33	77	3	1	1	0	0	1	58	1	10	2
Total		150	7	8	12	1	3	1	90	4	22	2
Per Cent Diet			4.7	5.3	8.0	.7	2.0	.7	60.0	2.7	14.7	1.4
			18.7				62.7			17.4		1.4

TABLE 69

RECORD OF FOOD OF COOPER'S HAWK DURING NESTING PERIOD
Superior Township 1942
Period April 20 to July 30

Nest No.	Sec. No.	No. of Food Items	Mammals					Birds			
			Rabbit	Fox Squirrel	Red Squirrel	Meadow Mouse	Moles	Pheasant	Quail	Screech Owl	Small and Medium-Sized Birds
26	10	49		3							46
27	13	34					2	4	1		27
28	16	22				1		1		1	20
31	31	28				1		1	1		25
33	35	32		2		1		2			26
34	36	55			5			1			49
AA											
Total		220	0	5	5	3	2	9	2	1	193
Per Cent Diet				2.2	2.2	1.4	1.0	4.1	1.0	.4	87.7

6.8   93.2

TABLE 70

RECORD OF FOOD OF COOPER'S HAWK DURING NESTING PERIOD
Superior Township 1948
Period May 24 to July 29

Nest No.	Sec. No.	No. of Food Items	Mammals					Birds			
			Rabbit	Fox Squirrel	Meadow Mouse	Chipmunk	Ground Squirrel	Pheasant	Quail	Chicken	Small and Medium-Sized Birds
26	14	19			1			2	1		15
28	26	28						3			25
29	30	114	5	6	6	2	1	18			76
30	34	61			5			13			43
32	36	74	2		4	3		8	2	1	54
AA											
Total		296	7	6	16	5	1	44	3	1	213
Per Cent Diet			2.4	2.0	5.4	1.7	.3	14.9	1.0	.3	72.0

11.8   88.2

TABLE 71
RECORD OF FOOD OF MARSH HAWK DURING NESTING PERIOD
Superior Township 1942
Period June 29 to August 8

Nest No.	Sec. No.	No. of Food Items	Mammals				Birds	
			Rabbit	Ground Squirrel	Rat	Meadow Mouse	Pheasant	Small and Medium-Sized Birds
36	14	51	3	2	2	28	..	16
Per Cent Diet			5.9	3.9	3.9	54.9	0	31.4

TABLE 72
RECORD OF FOOD OF SPARROW HAWKS DURING NESTING PERIOD
Superior Township 1948
Period May 28 to June 17

Nest No.	Sec. No.	No. Pellets	No. Non-Pellet Items	Meadow Mouse		Small Birds		Insects	
				F.P.	N.P.	F.P.	N.P.	F.P.	N.P.
44	15	34	7	34	7	13	0	5	0
Per Cent				100	100	4.0	0	1.5	0
Species Total Pellet & non-Pellet				41		13		No. Individuals = 54	
Per Cent Diet				76.0		24.0			

F.P. = Frequency of occurrence in pellets
N.P. = Non-pellet food items

TABLE 73

RECORD OF FOOD OF GREAT HORNED OWLS DURING NESTING PERIOD
Superior Township 1942
Period February 20 to May 15

Nest No.	Sec. No.	No. of Food Items	Mammals				Birds					
			Rabbit	Rat	Meadow Mouse	White-footed Mice	Pheasant	Chicken	Crow	Rails	Screech Owl	Small and Medium-Sized Birds
45 and 48	5 and 17	.. 99 ..	.. 8 ..	.. 5 ..	.. 21 ..	.. 6 ..	20	2	5	2	1	29
Per Cent Diet			8.1	5.1	21.2	6.1	20.2	2.0	5.1	2.0	1.0	29.3
			40.5				59.6					

TABLE 74

RECORD OF FOOD OF GREAT HORNED OWLS DURING NESTING PERIOD
Superior Township 1948
Period February 20 to May 15

Nest No.	Sec. No.	No. of Food Items	Mammals						Birds						Invertebrates
			Rabbit	Fox Squirrel	Musk-rat	Rat	Meadow Mouse	White-footed Mice	Pheasant	Duck	Crow	Rails	Am. Bittern	Small & Med.-Sized Birds	Crawfish
49	5	71	3	0	1	7	31	1	7	1	3	1	1	8	7
54	20	46	1	0	2	0	28	..	7	..	..	1	..	4	3
55	27	44	1	0	2	3	20	0	7	..	1	..	..	7	3
Total Individuals		161	5	0	5	10	79	1	21	1	4	2	1	19	13
Per Cent Diet			3.1	0	3.1	6.2	49.1	.6	13.0	.6	2.5	1.2	.6	11.8	8.1
			62.1						29.7						8.1

TABLE 75

RECORD OF FOOD OF BARN OWL DURING NESTING PERIOD
SUPERIOR TOWNSHIP, MICHIGAN
1942

Nest No. 58—22 pellets

Prey Species	Per cent of pellets containing prey remains	Number of individuals	Per cent composition of prey species
Meadow Mouse	93.7	28	70.0
White-footed Mice	41.0	12	30.0
Total	..	40	100.0

Long-eared Owl

Nest No. 52—34 pellets

Prey Species	Per cent of pellets containing prey remains	Number of individuals	Per cent composition of prey species
Meadow Mouse	79.4	31	41.3
White-footed Mice	53.0	23	30.7
Bog Lemming	6.0	2	2.7
Shrews	20.6	9	12.0
Small Birds	30.0	10	13.3
Total	..	75	100.0

Screech Owl

Nest No. 56—37 pellets

Prey Species	Per cent of pellets containing prey remains	Number of individuals	Per cent composition of prey species
Meadow Mouse	46.0	17	29.8
White-footed Mice	30.0	13	22.8
Small Birds	27.0	11	19.3
Crawfish	43.2	16	28.0
Insects	21.6	..	..
Total	..	57	99.9

TABLE 76
RECORD OF FOOD OF BUTEOS DURING NESTING PERIOD
Moose, Wyoming
1947

Eight Nests

				Mammals							Birds			Reptiles
	No. of Food Items	Meadow Mouse	Uinta Ground Squirrel	Marmot	Least Chipmunk	Jack Rabbit	Snowshoe Rabbit	Red Squirrel	Pocket Gopher	Weasels	Sage Grouse	Short-Eared Owl	Small & Med.-Sized Birds	Snake
W. Red-Tailed Hawk														
	46	8	26	1	2	1	..	2	..	1	1	..	3	1
	19	8	2	..	..	1	..	..	3	1	1	1	2	..
	26	..	20	..	1	..	..	2	..	..	1	..	2	..
	27	17	3	2	..	1	..	..	2	..	..	..	2	..
	27	17	5	..	..	2	..	..	3	..	..	..	..	..
	32	7	17	5	..	1	1	..	1	..	..	..	..	..
	12	6	6	..	..	..	..	..	..	..	..	..	..	..
Total	189	63	79	8	3	6	1	4	9	2	3	1	9	1
Per Cent Diet	..	33.3	41.8	4.2	1.6	3.2	.5	2.1	4.8	1.1	1.6	.5	4.8	.5 = 100
Swainson's Hawk			*Peromyscus*											
	37	6	16	6	..	1	..	..	2	..	..	..	6	..
Per Cent Diet	..	16.2	43.2	16.2	..	2.7	..	..	5.4	..	..	..	16.2	16.2 = 99.9

Bird Species Identified: Robin, Red-Shafted Flicker, Sparrows, Brewer's Blackbird.

TABLE 77
RECORD OF FOOD OF BIRD HAWKS DURING NESTING PERIOD
Moose, Wyoming
1947

	No. of Food Items	Meadow Mouse	Jumping Mice	Uinta Ground Squirrel	Pocket Gopher	Least Chipmunk	Red Squirrel	Jack Rabbit	Snowshoe Rabbit	Unidentified Mammals	Ruffed Grouse	Small & Med.-Sized Birds	Sparrow Hawk	
Cooper's Hawk	60	5	..	5	..	2	..	..	..	4	1	43	..	
Per Cent Diet ::		8.3	..	8.3	..	3.3	..	..	..	6.6	1.7	71.7	..	99.9
Sharp-shinned Hawk	46	3	..	..	..	..	..	..	..	..	..	43	..	
Per Cent Diet ::		6.5										93.5		100.00
Goshawk	89	1	1	21	1	2	12	1	2	7	6	34	1	
Per Cent Diet		1.1	1.1	23.6	1.1	2.2	13.5	1.1	2.2	7.9	6.7	38.2	1.1	99.9
Total Indiv. in Diet of all Bird Hawks	195	9	1	26	1	4	12	1	2	11	7	120	1	
Per Cent Diet (Bird Hawks) ::		4.6	.5	13.3	.5	2.1	6.2	.5	1.0	5.6	3.6	61.5	.5	99.9

Bird Species Identified: Ruffed Grouse, Red-Naped Sapsucker, Red-Shafted Flicker, Western Tanager, Downy Woodpecker, Hairy Woodpecker, Russet-Backed Thrush, Kinglet, Warbler, Western Yellow Throat, Green-Tailed Towhee, Pink-Sided Junco, Pine Siskin, Lincoln's Sparrow, MacGillivray's Warbler.

TABLE 78

RECORD OF FOOD OF PRAIRIE FALCON DURING NESTING PERIOD
Moose, Wyoming
1947

	No. of Food Items	Meadow Mouse	Uinta Ground Squirrel	Small & Medium Sized Birds	
	100	62	18	20	100
Per Cent Diet	..	62	18	20	100

Bird Species Identified: Mt. Bluebird, Red-Shafted Flicker, Western Meadowlark, Sparrow, Sage Thrasher.

TABLE 79

RECORD OF FOOD OF GREAT HORNED OWL DURING NESTING PERIOD
Moose, Wyoming
1947

Four Nests

No. of Food Items	Mammals							Birds				Fish	
	Meadow Mouse	White-footed Mice	Shrews	Weasles	Red Squirrel	Pocket Gopher	Snow-shoe Rabbit	Ruffed Grouse	Sage Grouse	Mal-lard Duck	Small & Medium Sized Birds	Sucker	
55	34	2	..	..	1	7	3	4	..	..	1	3	
133	77	6	..	2	..	44	1	1	..	1	2	..	
33	20	2	1	..	..	4	4	..	..	..	2	..	
140	102	3	..	..	..	30	2	1	1	1	..	..	
Total 361	233	13	1	2	1	85	10	5	1	2	5	3	
Per Cent Diet	64.5	3.6	.3	.6	.3	23.5	2.8	1.4	.3	.6	1.4	.8	100.1

TABLE 80

RECORD OF FOOD OF LONG-EARED OWL DURING NESTING PERIOD

Moose, Wyoming

1947

Three Nests	No. of Food Items	Meadow Mouse	White-footed Mice	Red-backed Mouse	Unidenti-fied Mice	Shrews	Pocket Gopher	Uinta Ground Squirrel	Red Squirrel	Small & Medium-Sized Birds
	42	6	1	..	24	..	11	..	..	..
	71	37	3	8	3	1	19	..	..	..
	16	10	1	..	..	..	3	..	..	2
Total	129	53	5	8	27	1	33	..	..	2
Per Cent Diet	..	41.1	3.9	6.2	20.9	.8	25.6	..	..	1.6	100.1

TABLE 81
RECORD OF FOOD OF GREAT GRAY OWL DURING NESTING PERIOD
Moose, Wyoming
1947

Per Cent Diet	83	55	1	..	..	..	23	..	3	1
	..	66.3	1.2	..	..	..	27.7	..	3.6	1.2	100.0

TABLE 82
RECORD OF FOOD OF SPARROW HAWKS DURING NESTING PERIOD
Moose, Wyoming

Eight Nests   1947

No. of Pellets	No. of non-Pellet Food Items	Meadow Mouse			White-footed Mice			Uinta Ground Squirrel			Jumping Mice (Mammals)		
		F.P.	I.P.	N.P	F.P.	I.P.	N.P	F.P.	I.P.	N.P	F.P.	I.P.	N.P
16	27	15	5	6	..	..	4	..	..	6	..	..	1
35	27	27	8	13	3	2	3	..	..	..	..	..	..
63	8	39	20	4	11	11	3	..	..	..	..	..	..
36	15	39	13	8	3	3	1	..	..	1	..	..	..
28	1	26	9	..	..	..	..	..	..	..	..	..	..
38	0	28	12	..	..	..	..	..	..	..	..	..	..
54	10	37	16	4	1	1	..	1	1	2	..	..	..
29	0	25	8	..	..	..	..	..	..	..	..	..	..
Total 299	88	236	91	35	18	17	11	1	1	9	..	..	1
Per cent of frequency occurrence ..	..	78.9	..	..	6.0	..	..	.3	..	..	..	..	..
Total No. of Indiv. Represented in Pellet and non-Pellet material ..	..	..	126	..	..	28	..	..	10	..	..	..	1
Percent of diet ..	..	..	57.3	..	..	12.7	..	..	4.5	..	..	..	.5

F.P.—Frequency of occurrence in pellets.
I.P.—Individuals represented in pellets.
N.P.—Individual food items in non-pellet material.

TABLE 82—*Continued*

RECORD OF FOOD OF SPARROW HAWKS DURING NESTING PERIOD
Moose, Wyoming

Eight Nests   1947

									Birds			Invertebrates		
Pocket Gopher			Least Chipmunk			Shrews			Small and Med.-Sized Birds			Insects		
F.P.	I.P.	N.P	F.P.	I.P.	N.P	F.P.	I.P.	N.P	F.P.	I.P.	N.P	F.P.	I.P.	N.P
..	..	2	..	..	..	..	..	..	2	2	8	9	..	..
..	..	1	..	..	4	..	..	..	1	1	6	19	..	..
..	..	..	..	..	..	..	..	..	4	4	1	46	..	..
..	..	..	..	..	..	..	..	..	1	1	5	16	..	..
..	..	..	..	..	..	1	1	..	..	..	1	5	..	..
..	..	..	..	..	..	..	..	..	4	4	..	22	..	..
3	3	..	..	..	..	..	..	..	4	4	4	49	..	..
..	..	..	..	..	..	3	2	..	1	1	..	13	..	..
3	3	3	..	..	4	4	3	..	17	17	25	179	..	..
1.0	..	..	..	..	..	1.3	..	..	5.7	..	..	59.9	..	..
..	6	..	..	..	4	..	3	..	..	42	..	.. = 220		
												Total Indiv.		
..	2.7	..	..	..	1.8	..	1.4	..	..	19.1	..	..	..	..

The number of Meadow Mouse individuals in pellets represents a minimum number computed from osseous remains.

The number of pellets in which only insects were represented or in which they composed the majority of the pellet = 56, or 19 per cent.

TABLE 83

FALL AND WINTER FOOD REQUIREMENTS OF RAPTORS

Raptor Species	Sex	Max. Daily Ration Gms.	Av. Wt. of Food Eaten Per Day, Gms.	Av. Wt. of Raptor Gms.	% of Av. Body Wt. Eaten Per Day	Av. Temp. F. for Feeding Period	No. of Days Fed
Red-Tailed Hawk	F	200	136	1,218	11.2	36.9	68
Red-Tailed Hawk	M	220	117	1,147	10.2	41.0	106
Great Gray Owl	F	162	77	1,045	7.4	14.0	28
Great Horned Owl	F	280	..	..	..	..	..
Prairie Falcon	F	170	122	811	15.0	37.5	118
Duck Hawk	M	170	105	727	14.4	37.1	114
Duck Hawk	M	130	101	716	14.1	41.7	56
Duck Hawk	M	140	106	607	16.9	27.8	35
Barred Owl	F	130	67	625	11.8	39.4	28
Marsh Hawk	F	142	100	526	19.0	41.7	55
Cooper's Hawk	M	100	63	319	19.7	41.7	57
Screech Owl	F	50	40	160	24.8	29.3	55
Screech Owl	F	55	38	148	25.6	29.3	56
Screech Owl	M	60	39	153	25.4	42.5	39

TABLE 84

SPRING AND SUMMER FOOD REQUIREMENTS OF RAPTORS

Raptor Species	Sex	Max. Daily Ration Gms.	Av. Wt. of Food Eaten Per Day, Gms.	Av. Wt. of Raptor Gms.	% of Av. Body Wt. Eaten Per Day	Av. Temp. F. for Feeding Period	No. of Days Fed
Great Horned Owl	F	210	85	1,210	7.0	64.4	67
Great Horned Owl	M	190	82	1,108	7.4	78.4	26
Red-Tailed Hawk	M	183	73	855	8.6	55.1	29
Prairie Falcon	F	163	103	801	12.9	67.1	16
Duck Hawk	M	120	83	721	11.5	67.1	15
Red-shouldered Hawk ..	F	109	71	636	11.2	60.0	59
Red-shouldered Hawk ..	M	110	58	519	11.2	61.2	86
Cooper's Hawk	F	110	79	453	17.3	70.0	8
Cooper's Hawk	F	100	70	437	16.0	74.0	20
Cooper's Hawk	M	60	47	288	16.2	70.0	8
Marsh Hawk*	M	70	42	343	12.1	69.6	17
Sharp-shinned Hawk ...	F	59	39	167	23.3	55.1	30
Sparrow Hawk	F	34	21	112	18.4	70.8	27
Sparrow Hawk	F	36	18	107	16.9	70.8	30
Sparrow Hawk	M	30	20	91	22.3	61.2	40
Screech Owl	M	50	14	134	10.3	82.9	29

* Gained 100 gms.

TABLE 85

DIFFERENCES IN PER CENT OF BODY WEIGHT CONSUMED BY RAPTORS DURING COLD AND WARM WEATHER

FALL AND WINTER FOOD REQUIREMENTS			SPRING AND SUMMER FOOD REQUIREMENTS		
Large Raptors Wt. Class 800-1200 gms.	% of Av. Body Wt. Eaten Per Day	Av. Temp. F.	*Large Raptors* Wt. Class 800-1200 gms.	% of Av. Body Wt. Eaten Per Day	Av. Temp. F.
Red-tailed Hawk	10.2	41.0	Great Horned Owl	7.0	64.4
Red-tailed Hawk	11.2	36.9	Great Horned Owl	7.4	78.4
Average	10.7	38.9	Red-tailed Hawk	8.6	55.1
			Average	7.7	65.9
Medium Raptors Wt. Class 200-800 gms.			*Medium Raptors* Wt. Class 200-800 gms.		
Duck Hawk*	14.4	37.1	Duck Hawk*	11.5	67.1
Duck Hawk	14.1	41.7	Prairie Falcon*	12.9	67.1
Duck Hawk	16.9	22.8	Marsh Hawk	12.1	69.6
Prairie Falcon*	15.0	37.5	Red-shouldered Hawk	11.2	61.2
Cooper's Hawk	19.7	41.7	Red-shouldered Hawk	11.2	60.0
Marsh Hawk	19.0	41.7	Cooper's Hawk	16.0	74.0
Barred Owl	11.8	39.4	Average	12.5	66.5
Average	15.8	37.4			
Small Raptors Wt. Class 100-200 gms.			*Small Raptors* Wt. Class 100-200 gms.		
Screech Owl	25.6	29.3	Sparrow Hawk	18.4	70.8
Screech Owl	25.4	42.5	Sparrow Hawk	16.9	70.8
Screech Owl	24.8	29.3	Sparrow Hawk	22.3	61.2
Average	25.3	33.7	Screech Owl	10.3	82.9
			Average	17.0	71.4

* Same individual birds fed summer and winter.

TABLE 86
DETERMINATION OF RAPTOR-DAYS—FALL AND WINTER

Species	Raptors 1941-1942 Fall	Winter	Spring Trans.	Raptors 1947-1948 Fall	Winter	Spring Trans.	Raptor-Days 1941-1942 Fall	Winter	Spring Trans.	Total	Raptor-Days 1947-1948 Fall	Winter	Spring Trans.	Total
Unidentified Buteos	26	13	8	3	0	0	2,548	1,105	112	3,765	294	0	0	294
Red-tailed Hawk	7	17	14	6	12	15	686	1,445	196	2,327	588	1,020	210	1,818
Am. Rough-legged Hawk	30	13	7	2	0	2	2,940	1,105	98	4,143	196	0	28	224
Great Horned Owl	11	11	11	13	13	16	1,078	935	154	2,167	1,274	1,105	224	2,603
Total Large Raptors	74	54	40	24	25	33	7,252	4,590	560	12,402	2,352	2,125	462	4,939
Red-shouldered Hawk	2	1	38	1	3	34	196	85	532	813	98	255	476	829
Short-eared Owl	22	31	10	0	0	4	2,156	2,635	140	4,931	0	0	56	56
Long-eared Owl	7	7	2	0	0	1	686	595	28	1,309	:	:	14	14
Marsh Hawk	38	37	20	7	1	18	3,724	3,145	280	7,149	686	85	252	1,023
Cooper's Hawk	9	10	16	6	6	16	882	850	224	1,956	588	510	224	1,322
Total Medium Raptors	78	86	86	14	10	73	7,644	7,310	1,204	16,158	1,372	850	1,022	3,244
Sparrow Hawk	5	5	5	4	5	9	490	425	70	985	392	425	126	943
Screech Owl	14	14	26	22	19	30	1,372	1,190	364	2,926	2,156	1,615	420	4,191
Total Small Raptors	19	19	31	26	24	39	1,862	1,615	434	3,911	2,548	2,040	546	5,134
Grand Total	171	159	157	64	59	145	16,758	13,515	2,198	32,471	6,272	5,015	2,030	13,317

Note: The number of hunting days used to calculate raptor-days was:

Fall—98
Winter—85
Spring Transition—14

TABLE 87

DETERMINATION OF RAPTOR-DAYS—SPRING AND SUMMER

Species	Raptors				No. Hunting Days				Raptor-Days					
	1941-42		1947-48		1941-42		1947-48		1941-42			1947-48		
	Spring	Summer	Spring	Summer	Spring	Summer	Spring	Summer	Spring	Summer	Total	Spring	Summer	Total
Red-tailed Hawk	5	6	15	19	39	128	..	..	195	768	963	585	2,432	3,017
Great Horned Owl	11	17	16	19	14	153	..	..	154	2,601	2,755	224	2,907	3,131
Total Large Raptors	16	23	31	38	..	..	..	..	349	3,369	3,718	809	5,339	6,148
Red-shouldered Hawk	40	77	34	60	55	112	..	..	2,200	8,624	10,824	1,870	6,720	8,590
Long-eared Owl	2	6	1	1	34	133	..	..	68	798	866	34	133	167
Marsh Hawk	14	30	18	18	82	85	..	..	1,148	2,550	3,698	1,476	1,530	3,006
Cooper's Hawk	16	31	16	30	77	90	..	..	1,232	2,790	4,022	1,232	2,700	3,932
Barn Owl	2	2	1	0	44	123	..	..	88	246	334	44	0	44
Total Medium Raptors ...	74	146	70	109	..	..	..	..	4,736	15,008	19,744	4,656	11,083	15,739
Sparrow Hawk	7	15	9	19	61	106	..	..	427	1,590	2,017	549	2,014	2,563
Screech Owl	26	57	30	70	45	122	..	..	1,170	6,954	8,124	1,350	8,540	9,890
Total Small Raptors	33	72	39	89	..	..	..	..	1,597	8,544	10,141	1,899	10,554	12,453
Grand Total	123	241	140	236	..	..	..	..	6,682	26,921	33,603	7,364	26,976	34,340

TABLE 88

RAPTOR PRESSURE EXPRESSED AS RAPTOR WEIGHT IN GRAMS—FALL AND WINTER

	1941-1942			1947-1948		
Raptor Species	Av. No. Raptors	Av. Wt.° from Field Records	Total Raptor Wt.	Av. No. Raptors	Av. Wt.° from Field Records	Total Raptor Wt.
Wt. Class—Large						
Unidentified Buteos	16	1,139	18,224	1	1,139	1,139
Red-tailed Hawk	13	1,126	14,638	11	1,126	12,386
Am. Rough-legged Hawk	17	1,152	19,584	2	1,152	2,304
Great Horned Owl	11	1,505	16,555	14	1,505	21,070
Total	57		69,001	28		36,899
Wt. Class—Medium						
Red-shouldered Hawk ..	14	625	8,750	13	625	8,125
Short-eared Owl	21	340	7,140	1.3	340	442
Long-eared Owl	5	245	1,225	0.3	245	73.5
Marsh Hawk	32	521	16,672	9	521	4,689
Cooper's Hawk	12	470	5,640	9	470	4,230
Total	84		39,427	32.6		17,559.5
Wt. Class—Small						
Sparrow Hawk	5	114	570	6	114	684
Screech Owl	18	172	3,096	24	172	4,128
Total	23		3,666	30		4,812
Grand Total			112,094			59,271

Note: Barn Owl on area only 14 days, so not recorded here.

Av. Wt. of raptors of each weight class on study area is obtained by dividing the total raptor weight of each weight class by the av. number of raptors in that class.
°See table 99.

TABLE 89

RAPTOR PRESSURE EXPRESSED AS RAPTOR WEIGHT IN GRAMS—SPRING AND SUMMER

Raptor Species	1941-1942			1947-1948		
	Av. No. Raptors	Av. Wt. from Field Records	Total Raptor Wt.	Av. No. Raptors	Av. Wt. from Field Records	Total Raptor Wt.
Wt. Class—Large						
Red-tailed Hawk	5.5	1,126	6,193	17	1,126	19,142
Great Horned Owl	14	1,505	21,070	17.5	1,505	26,338
Total	19.5		27,263	34.5		45,480
Wt. Class—Medium						
Red-shouldered Hawk ..	58.5	625	36,563	47	625	29,375
Long-eared Owl	4	245	980	1	245	245
Marsh Hawk	22	521	11,462	18	521	9,378
Cooper's Hawk	23.5	470	11,045	23	470	10,810
Barn Owl	2	437	874	0.5	437	219
Total	110.0		60,924	89.5		50,027
Wt. Class—Small						
Sparrow Hawk	11	114	1,254	14	114	1,596
Screech Owl	41.5	172	7,138	50	172	8,600
Total	52.5		8,392	64		10,196
Grand Total			96,579			105,703

Note: Total raptor weights are an expression of both raptor weight and growth. See page 316.

TABLE 90

Raptor Pressure Expressed as Grams of Food Consumed—Fall and Winter

Raptor Group	Total Raptor-Days		Computed Av. Wt. in grams of Food Eaten per Day by Raptors on Study Area		Computed Food Consumed Gms.	
	1941-42	1947-48	1941-42	1947-48	1941-42	1947-48
Large						
Unidentified Buteos	3,765	294			489,450	41,454
Red-tailed Hawk	2,327	1,818			302,510	256,338
Am. Rough-legged Hawk ..	4,143	224			538,590	31,584
Great Horned Owl	2,167	2,603			281,710	367,023
Total Large Raptors ...	12,402	4,939	130	141	1,612,260	696,399
Medium						
Red-shouldered Hawk	813	829			60,162	70,465
Short-eared Owl	4,931	56			364,894	4,760
Long-eared Owl	1,309	14			96,866	1,190
Marsh Hawk	7,149	1,023			529,026	86,955
Cooper's Hawk	1,956	1,322			144,744	112,370
Total Medium Raptors .	16,158	3,244	74	85	1,195,692	275,740
Small						
Sparrow Hawk	985	943			39,400	38,663
Screech Owl	2,926	4,191			117,040	171,831
Total Small Raptors ...	3,911	5,134	40	41	156,440	210,494
Grand Total	32,471	13,317			°2,964,392	°1,182,633

Note: Computed food consumed = Total No. of Raptor-days x computed Av. Wt. of food eaten per day.

° These figures represent the amount of food that the respective raptor populations are presumed to have consumed during the fall and winter periods.

TABLE 91

RAPTOR PRESSURE EXPRESSED AS GRAMS OF FOOD CONSUMED—SPRING AND SUMMER

Raptor Group	Total Raptor-Days		Computed Av. Wt. in grams of Food Eaten per Day by Raptors on Study Area		Computed Food Consumed Gms.	
	1941-42	1947-48	1941-42	1947-48	1941-42	1947-48
Large						
Red-tailed Hawk	963	3,017			104,004	304,717
Great Horned Owl	2,755	3,131			297,540	316,231
Total Large Raptors ...	3,718	6,148	108	101	401,544	620,948
Medium						
Red-shouldered Hawk ...	10,824	8,590			746,856	601,300
Long-eared Owl	866	167			59,754	11,690
Marsh Hawk	3,698	3,006			255,162	210,420
Cooper's Hawk	4,022	3,932			277,518	275,240
Barn Owl	334	44			23,046	3,080
Total Medium Raptors ..	19,744	15,739	69	70	1,362,336	1,101,730
Small						
Sparrow Hawk	2,017	2,563			54,459	69,201
Screech Owl	8,124	9,890			219,348	267,030
Total Small Raptors ...	10,141	12,453	27	27	273,807	336,231
Grand Total	33,603	34,340			*2,037,687	*2,058,909

* These figures represent the amount of food that the respective raptor populations are presumed to have consumed during the spring and summer periods.

TABLE 92

RELATION OF RAPTOR WEIGHT TO DAILY RATION—FALL AND WINTER

	1941-1942			1947-1948		
Raptor Group	% of Av. Body Wt. Eaten/ Day	Av. Wt. in grams of Rap- tors on Study Area, from Wt. Records (See Table 88)	Computed Av. Wt. in grams of Food Eaten/Day by Raptors on Study Area	% of Av. Body Wt. Eaten/ Day	Av. Wt. in grams of Rap- tors on Study Area, from Wt. Records (See Table 88)	Computed Av. Wt. in grams of Food Eaten/Day by Raptors on Study Area
Large	10.7	1,211	130	10.7	1,318	141
Medium	15.8	469	74	15.8	539	85
Small	25.3	159	40	25.3	160	41

TABLE 93

RELATION OF RAPTOR WEIGHT TO DAILY RATION—SPRING AND SUMMER

	1941-1942			1947-1948		
Raptor Group	% of Av. Body Wt. Eaten/ Day	Av. Wt. in grams of Rap- tors on Study Area, from Wt. Records (See Table 89)	Computed Av. Wt. in grams of Food Eaten/Day by Raptors on Study Area	% of Av. Body Wt. Eaten/ Day	Av. Wt. in grams of Rap- tors on Study Area, from Wt. Records (See Table 89)	Computed Av. Wt. in grams of Food Eaten/Day by Raptors on Study Area
Large	7.7	1,398	108	7.7	1,318	101
Medium	12.5	554	69	12.5	559	70
Small	17.0	160	27	17.0	159	27

TABLE 94

NUMERICAL EXPRESSION OF RAPTOR PREDATION DURING FALL AND WINTER SUPERIOR TOWNSHIP, MICHIGAN

	1	2	3		4	
Prey Species	Av. Wt. of Prey in Grams	Av. Max. Raptor Ration, Grams	No. of Prey Individuals in Food Sample		Total Prey Wt. in Grams Represented by Food Sample	
			1941-42	1947-48[a]	1941-42	1947-48
Meadow Mouse	34	....	10,104	897	343,536	30,498
White-footed Mouse	14	....	976	348	12,764	4,872
Other Mice, Rats, Shrews and Moles ...	17	....[a]	302	16	5,134	272
Weasels	..	230[a]	1	2	230	460
Muskrat	..	230	3	2	690	460
Rabbit	..	230	18[a]	22	4,140	5,060
Fox Squirrel	..	230	2	2	460	460
Pheasant	..	230	4	7	920	1,610
Bob-white	165	....	5	4	825	660
Small and Medium-sized Birds	41	....	130	139	5,330	5,699
Other Items	126[1]	....	24	118	3,024	14,868
TOTALS	.	....	11,569	1,557	377,053	64,919

TABLE 94

NUMERICAL EXPRESSION OF RAPTOR PREDATION DURING FALL AND WINTER
SUPERIOR TOWNSHIP, MICHIGAN

	5		6		7	
	Per Cent of Prey Weight in Diet		Computed No. of Grams of Prey Animals Consumed by the Raptor Populations		Computed No. of Prey Animals Consumed by the Raptor Populations	
Prey Species	1941-42	1947-48	1941-42	1947-48	1941-42	1947-48
Meadow Mouse	91.11	46.98	2,700,858	555,600	79,437	16,341
White-footed Mouse	3.39	7.50	100,492	88,697	7,178	6,336
Other Mice, Rats, Shrews and Moles	1.36	0.42	40,316	4,968	2,372	292
Weasels	0.06	0.71	1,778	8,397	8	37
Muskrat	0.18	0.71	5,336	8,397	23	37
Rabbit	1.10	7.79	32,608	92,127	142	401
Fox Squirrel	0.12	0.71	3,557	8,397	16	37
Pheasant	0.25	2.48	7,410	29,329	32	128
Bob-white	0.22	1.02	6,522	12,063	40	73
Small and Medium-sized Birds	1.41	8.78	41,798	103,835	1,019	2,533
Other Items	0.80	22.90	23,715	270,823	188	2,149
TOTALS	100.00	100.00	2,964,392[5]	1,182,633[6]	90,455	28,364

[1] 126 grams "other items" = an average wt. for miscellaneous items.
[2] 230 grams = the average maximum ration for large raptors.
[3] Nine rabbits recorded in the Marsh Hawk diet were excluded as unreliable data.
[4] The Horned Owl diet for 1947-48 was weighted.
[5] and [6] These figures were calculated from the quantitative feeding experiment. See Table 90.

TABLE 95

NUMERICAL EXPRESSION OF RAPTOR PREDATION DURING SPRING AND SUMMER
SUPERIOR TOWNSHIP, MICHIGAN

	1	2	3		4	
			No. of Prey Individuals in Food Sample		Total Prey Wt. in Grams Represented by Food Sample	
Prey Species	1/2 Av. Wt. of Small Prey, Gms.	1/3 Av. Wt. of Large Prey, Gms.	1941-42	1947-48	1941-42	1947-48
Meadow Mouse	17	....	258	288	4,386	4,896
White-footed Mouse	7	....	54	1	378	7
Other Mice, Rats, Shrews and Moles	9	....	34	24	306	216
Weasels		200	1	0	200	0
Muskrat		700	3	19	2,100	13,300
Rabbit		500	22	23	11,000	11,500
Fox Squirrel		260	17	17	4,420	4,420
Pheasant		360	79[2]	69	28,440	24,840
Bob-white		54	2	3	108	162
Small and Medium-sized Birds	21	....	337	339	7,077	7,119
Snakes	32	....	50	30	1,600	960
Frogs °	27[1]	....	56[3]	80[3]	1,512	2,160
Crawfish	7[1]	....	62	36	434	252
Other Items	126[1]	....	22	19	2,772	2,394
TOTALS	....	....	997	948	64,733	72,226

TABLE 95

NUMERICAL EXPRESSION OF RAPTOR PREDATION DURING SPRING AND SUMMER
SUPERIOR TOWNSHIP, MICHIGAN

	5		6		7	
Prey Species	Per Cent of Prey Weight in Diet		Computed No. of Grams of Prey Animals Consumed by the Raptor Populations		Computed No. of Prey Animals Consumed by the Raptor Populations	
	1941-42	1947-48	1941-42	1947-48	1941-42	1947-48
Meadow Mouse	6.78	6.78	138,155	139,594	8,127	8,211
White-footed Mouse	.58	.01	11,819	206	1,688	29
Other Mice, Rats, Shrews and Moles ...	.47	.30	9,577	6,177	1,064	686
Weasels	.31	0	6,317	0	32	0
Muskrat	3.24	18.41	66,022	379,045	94	541
Rabbit	16.99	15.92	346,203	327,778	692	656
Fox Squirrel	6.83	6.12	139,174	126,005	535	485
Pheasant	43.93	34.39	895,156	708,059	2,487	1,967
Bob-white	.17	.22	3,464	4,530	64	84
Small and Medium-sized Birds	10.93	9.86	222,719	203,008	10,606	9,667
Snakes	2.47	1.33	50,331	27,383	1,573	856
Frogs [*]	2.34	2.99	47,682	61,561	1,766	2,280
Crawfish	.67	.35	13,653	7,206	1,950	1,029
Other Items	4.29	3.32	87,417	68,357	694	543
TOTALS	100.00	100.00	2,037,687[4]	2,058,909[4]	31,372	27,034

[1] Adult weights of these prey were used. See text.
[2] Forty pheasants were added to make this figure comparable to the 1947-48 data. Explained in text.
[3] The number of frogs was doubled. Explained in text.
[4] These figures were calculated from the quantitative feeding experiment. See Table 91.
[*] Frogs = average weight of green and leopard frogs together.

TABLE 96

PERCENTAGES OF PREY POPULATIONS COMPUTED TO HAVE BEEN KILLED BY RAPTORS
IN SUPERIOR TOWNSHIP FALL AND WINTER

Prey Species or Groups	1941-42			1947-48		
	Prey Population	Computed Number Killed	Per Cent of Each Prey Population Killed	Prey Population	Computed Number Killed	Per Cent of Each Prey Population Killed
Meadow Mouse ..	303,000	79,437	26	75,000	16,341	22
White-footed Mice	33,000	7,178	22	27,000	6,336	24
Small Birds	23,000	1,019	4	23,000	2,533	11
Game Birds	1,500	72	5	1,100	201	18
Rabbit	300	142	47	1,200	401	33
Fox Squirrel	300	16	5	1,000	37	4
Totals	361,100	87,864	24	128,300	25,849	20

TABLE 97

RAPTOR PREDATION ON RING-NECKED PHEASANTS IN TERMS OF THE PRODUCTIVE POTENTIAL OF THE PHEASANT POPULATION—SUPERIOR TOWNSHIP, 36 SQ. MI. 1941-42

		Per Cent of Potential Fall Population
1941-42 Winter population	1,008	
Calculated fall and winter kill by raptors	32	
* Other mortality	Nil	
Late winter population	976	
Average early spring pre-nesting mortality (6.3% of late winter population)	61	
Breeding population	915	
Sex ratio observed, 1 M to 4 F	.80	
Maximum clutch successfully incubated	20	
Biotic potential, 20 x .80	16.00	
Productive capacity of birds on area, 16.00 x 915	14,640	
Total potential fall population, 915 + 14,640	15,555	100.00
Potential fall population calculated lost to raptors	2,487	16.00
Expected 1942 fall population (approx. double winter population)	2,000	12.9
Potential fall population lost to other resistance factors	11,068	71.1

* Note: No mortality from causes other than Raptor Predation was noted in the winter of 1941-42.

TABLE 98

RELATION OF THE PERCENTAGE OF SPECIES IN TOTAL PREY POPULATION TO THE PERCENTAGE OF SPECIES IN TOTAL PREY KILL

Prey Species or Groups	1941-42		1947-1948	
	Per Cent Representation of Prey in the Total Prey Population	Per Cent Representation of Prey in the Total Prey Kill	Per Cent Representation of Prey in the Total Prey Population	Per Cent Representation of Prey in the Total Prey Kill
Meadow Mouse	83.9	90.4	58.4	63.2
White-footed Mouse	9.1	8.1	21.0	24.5
Small Birds	6.4	1.2	17.9	9.8
Game Birds	0.4	0.08	0.9	0.8
Rabbit	0.08	0.16	0.9	1.6
Fox Squirrel	0.08	0.02	0.9	0.14

Note: Compare with Table 36.

TABLE 99

Raptor Weights

Species	Sex	No. of Records	Av. Wt. in Grams
Great Horned Owl	M	895	1,304
	F	772	1,706
	Com.	1,667	1,505
Am. Rough-legged Hawk	M	11	1,027
	F	17	1,278
	Com.	28	1,152
Red-tailed Hawk	M	108	1,028
	F	100	1,224
	Com.	208	1,126
Great Gray Owl	M	..	..
	F	7	1,084
	Com.	7	1,084
Swainson's Hawk	M	5	908
	F	7	1,069
	Com.	12	988
Goshawk	M	62	860
	F	114	1,095
	Com.	176	978
Duck Hawk	M	65	719
	F	1	1,223
	Com.	66	971
Prairie Falcon	M	..	..
	F	34	801
	Com.	34	801
Red-shouldered Hawk	M	25	550
	F	24	701
	Com.	49	625
Barred Owl	M	..	..
	F	9	625
	Com.	9	625
Marsh Hawk	M	30	472
	F	13	570
	Com.	43	521
Cooper's Hawk	M	34	380
	F	143	561
	Com.	177	470
Broad-winged Hawk	M	14	420
	F	13	490
	Com.	27	455
Barn Owl	M	1	385
	F	4	489
	Com.	5	437
Short-eared Owl	M	1	262
	F	1	417
	Com.	2	340
Long-eared Owl	M	1	230
	F	2	259
	Com.	3	245

TABLE 99 (Continued)

RAPTOR WEIGHTS

Species	Sex	No. of Records	Av. Wt. in Grams
Screech Owl	M	26	167
	F	21	177
	Com.	47	172
Pigeon Hawk	M	5	162
	F	..	..
	Com.	5	162
Sharp-shinned Hawk	M	98	102
	F	92	179
	Com.	190	140
Sparrow Hawk	M	50	109
	F	67	119
	Com.	117	114

SOURCE OF DATA

The raptor weights presented here have been accumulated from varied sources. Many were contributed by the authors, but the majority were obtained from unpublished records of the Penna. Game Comm., made available by Roger M. Latham. Other sources of data were R. D. McDowell, 1940 and 1941; Ralph H. Immler, 1937; Sidney R. Esten, 1931; Earl L. Poole, 1938; and unpublished records of the Museum of Vertebrate Zoology, Univ. of Michigan.

TABLE 100

AVERAGE WEIGHTS OF MAJOR PREY SPECIES AND PREY GROUPS

Species	Sex	No. of Records	Av. Wt. in Grams
Meadow Mouse		561	34.2
White-footed Mouse		56	14.1
Other Mice and Shrews			
House Mouse		16	
Bog Lemming		15	
Short-tailed Shrew		18	17.4
Masked Shrew		2	
Least Shrew		10	
		—	
Total		61	
Small Birds as Winter Residents	Com.	4,607	41.15
Ring-necked Pheasant	M	63	1,293
	F	72	973
	Com.	135	1,133
Bob-white Quail	M	475	165.1
	F	413	164.8
	Com.	888	164.9
Cottontail Rabbit	M	172	1,433
	F	174	1,458
	Com.	346	1,445
Fox Squirrel	M	..	760
	F	..	771
	Com.	409	764
Crawfish		4	6.5
Garter Snake (Adults and Young) ..		9	64
Wood Frog		65	6.1
Green Frog		18	28.7
Leopard Frog		56	24.9

SOURCE OF DATA

The average weight for small birds was computed from data of P. S. Baldwin and C. S. Kendeigh, 1938, and M. M. Nice, 1938. Data for pheasant weights were obtained from unpublished records of H. M. Wight and from the 3rd Annual Report of the Rose Lake Experiment Station, 1941-42. The Bob-white Quail weights of H. L. Stoddard (1946) were used; Cottontail Rabbit weights were obtained from R. E. Trippensee (1934) and the 4th Annual Report of the Rose Lake Wildlife Experiment Station (1942-43). Source of Fox Squirrel data was D. L. Allen, 1943.

INDEX

431

A CATALOG OF SELECTED
DOVER BOOKS
IN ALL FIELDS OF INTEREST

A CATALOG OF SELECTED DOVER BOOKS IN ALL FIELDS OF INTEREST

DRAWINGS OF REMBRANDT, edited by Seymour Slive. Updated Lippmann, Hofstede de Groot edition, with definitive scholarly apparatus. All portraits, biblical sketches, landscapes, nudes. Oriental figures, classical studies, together with selection of work by followers. 550 illustrations. Total of 630pp. 9⅛ × 12¼.
21485-0, 21486-9 Pa., Two-vol. set $25.00

GHOST AND HORROR STORIES OF AMBROSE BIERCE, Ambrose Bierce. 24 tales vividly imagined, strangely prophetic, and decades ahead of their time in technical skill: "The Damned Thing," "An Inhabitant of Carcosa," "The Eyes of the Panther," "Moxon's Master," and 20 more. 199pp. 5⅜ × 8½. 20767-6 Pa. $3.95

ETHICAL WRITINGS OF MAIMONIDES, Maimonides. Most significant ethical works of great medieval sage, newly translated for utmost precision, readability. Laws Concerning Character Traits, Eight Chapters, more. 192pp. 5⅜ × 8½.
24522-5 Pa. $4.50

THE EXPLORATION OF THE COLORADO RIVER AND ITS CANYONS, J. W. Powell. Full text of Powell's 1,000-mile expedition down the fabled Colorado in 1869. Superb account of terrain, geology, vegetation, Indians, famine, mutiny, treacherous rapids, mighty canyons, during exploration of last unknown part of continental U.S. 400pp. 5⅜ × 8½. 20094-9 Pa. $6.95

HISTORY OF PHILOSOPHY, Julián Marías. Clearest one-volume history on the market. Every major philosopher and dozens of others, to Existentialism and later. 505pp. 5⅜ × 8½. 21739-6 Pa. $8.50

ALL ABOUT LIGHTNING, Martin A. Uman. Highly readable non-technical survey of nature and causes of lightning, thunderstorms, ball lightning, St. Elmo's Fire, much more. Illustrated. 192pp. 5⅜ × 8½. 25237-X Pa. $5.95

SAILING ALONE AROUND THE WORLD, Captain Joshua Slocum. First man to sail around the world, alone, in small boat. One of great feats of seamanship told in delightful manner. 67 illustrations. 294pp. 5⅜ × 8½. 20326-3 Pa. $4.50

LETTERS AND NOTES ON THE MANNERS, CUSTOMS AND CONDITIONS OF THE NORTH AMERICAN INDIANS, George Catlin. Classic account of life among Plains Indians: ceremonies, hunt, warfare, etc. 312 plates. 572pp. of text. 6⅛ × 9¼. 22118-0, 22119-9 Pa. Two-vol. set $15.90

ALASKA: The Harriman Expedition, 1899, John Burroughs, John Muir, et al. Informative, engrossing accounts of two-month, 9,000-mile expedition. Native peoples, wildlife, forests, geography, salmon industry, glaciers, more. Profusely illustrated. 240 black-and-white line drawings. 124 black-and-white photographs. 3 maps. Index. 576pp. 5⅜ × 8½. 25109-8 Pa. $11.95

CATALOG OF DOVER BOOKS

HOW TO WRITE, Gertrude Stein. Gertrude Stein claimed anyone could understand her unconventional writing—here are clues to help. Fascinating improvisations, language experiments, explanations illuminate Stein's craft and the art of writing. Total of 414pp. 4⅜ × 6⅜. 23144-5 Pa. $5.95

ADVENTURES AT SEA IN THE GREAT AGE OF SAIL: Five Firsthand Narratives, edited by Elliot Snow. Rare true accounts of exploration, whaling, shipwreck, fierce natives, trade, shipboard life, more. 33 illustrations. Introduction. 353pp. 5⅜ × 8½. 25177-2 Pa. $7.95

THE HERBAL OR GENERAL HISTORY OF PLANTS, John Gerard. Classic descriptions of about 2,850 plants—with over 2,700 illustrations—includes Latin and English names, physical descriptions, varieties, time and place of growth, more. 2,706 illustrations. xlv + 1,678pp. 8½ × 12¼. 23147-X Cloth. $75.00

DOROTHY AND THE WIZARD IN OZ, L. Frank Baum. Dorothy and the Wizard visit the center of the Earth, where people are vegetables, glass houses grow and Oz characters reappear. Classic sequel to *Wizard of Oz.* 256pp. 5⅜ × 8. 24714-7 Pa. $4.95

SONGS OF EXPERIENCE: Facsimile Reproduction with 26 Plates in Full Color, William Blake. This facsimile of Blake's original "Illuminated Book" reproduces 26 full-color plates from a rare 1826 edition. Includes "The Tyger," "London," "Holy Thursday," and other immortal poems. 26 color plates. Printed text of poems. 48pp. 5¼ × 7. 24636-1 Pa. $3.50

SONGS OF INNOCENCE, William Blake. The first and most popular of Blake's famous "Illuminated Books," in a facsimile edition reproducing all 31 brightly colored plates. Additional printed text of each poem. 64pp. 5¼ × 7. 22764-2 Pa. $3.50

PRECIOUS STONES, Max Bauer. Classic, thorough study of diamonds, rubies, emeralds, garnets, etc.: physical character, occurrence, properties, use, similar topics. 20 plates, 8 in color. 94 figures. 659pp. 6⅛ × 9¼. 21910-0, 21911-9 Pa., Two-vol. set $14.90

ENCYCLOPEDIA OF VICTORIAN NEEDLEWORK, S. F. A. Caulfeild and Blanche Saward. Full, precise descriptions of stitches, techniques for dozens of needlecrafts—most exhaustive reference of its kind. Over 800 figures. Total of 679pp. 8⅛ × 11. Two volumes. Vol. 1 22800-2 Pa. $10.95
Vol. 2 22801-0 Pa. $10.95

THE MARVELOUS LAND OF OZ, L. Frank Baum. Second Oz book, the Scarecrow and Tin Woodman are back with hero named Tip, Oz magic. 136 illustrations. 287pp. 5⅜ × 8½. 20692-0 Pa. $5.95

WILD FOWL DECOYS, Joel Barber. Basic book on the subject, by foremost authority and collector. Reveals history of decoy making and rigging, place in American culture, different kinds of decoys, how to make them, and how to use them. 140 plates. 156pp. 7⅞ × 10¾. 20011-6 Pa. $7.95

HISTORY OF LACE, Mrs. Bury Palliser. Definitive, profusely illustrated chronicle of lace from earliest times to late 19th century. Laces of Italy, Greece, England, France, Belgium, etc. Landmark of needlework scholarship. 266 illustrations. 672pp. 6⅛ × 9¼. 24742-2 Pa. $14.95

THE BOOK OF BEASTS: Being a Translation from a Latin Bestiary of the Twelfth Century, T. H. White. Wonderful catalog real and fanciful beasts: manticore, griffin, phoenix, amphivius, jaculus, many more. White's witty erudite commentary on scientific, historical aspects. Fascinating glimpse of medieval mind. Illustrated. 296pp. 5⅜ × 8¼. (Available in U.S. only) 24609-4 Pa. $5.95

FRANK LLOYD WRIGHT: ARCHITECTURE AND NATURE With 160 Illustrations, Donald Hoffmann. Profusely illustrated study of influence of nature—especially prairie—on Wright's designs for Fallingwater, Robie House, Guggenheim Museum, other masterpieces. 96pp. 9¼ × 10¾. 25098-9 Pa. $7.95

FRANK LLOYD WRIGHT'S FALLINGWATER, Donald Hoffmann. Wright's famous waterfall house: planning and construction of organic idea. History of site, owners, Wright's personal involvement. Photographs of various stages of building. Preface by Edgar Kaufmann, Jr. 100 illustrations. 112pp. 9¼ × 10. 23671-4 Pa. $7.95

YEARS WITH FRANK LLOYD WRIGHT: Apprentice to Genius, Edgar Tafel. Insightful memoir by a former apprentice presents a revealing portrait of Wright the man, the inspired teacher, the greatest American architect. 372 black-and-white illustrations. Preface. Index. vi + 228pp. 8¼ × 11. 24801-1 Pa. $9.95

THE STORY OF KING ARTHUR AND HIS KNIGHTS, Howard Pyle. Enchanting version of King Arthur fable has delighted generations with imaginative narratives of exciting adventures and unforgettable illustrations by the author. 41 illustrations. xviii + 313pp. 6⅛ × 9¼. 21445-1 Pa. $5.95

THE GODS OF THE EGYPTIANS, E. A. Wallis Budge. Thorough coverage of numerous gods of ancient Egypt by foremost Egyptologist. Information on evolution of cults, rites and gods; the cult of Osiris; the Book of the Dead and its rites; the sacred animals and birds; Heaven and Hell; and more. 956pp. 6⅛ × 9¼. 22055-9, 22056-7 Pa., Two-vol. set $20.00

A THEOLOGICO-POLITICAL TREATISE, Benedict Spinoza. Also contains unfinished *Political Treatise*. Great classic on religious liberty, theory of government on common consent. R. Elwes translation. Total of 421pp. 5⅜ × 8½. 20249-6 Pa. $6.95

INCIDENTS OF TRAVEL IN CENTRAL AMERICA, CHIAPAS, AND YUCATAN, John L. Stephens. Almost single-handed discovery of Maya culture; exploration of ruined cities, monuments, temples; customs of Indians. 115 drawings. 892pp. 5⅜ × 8½. 22404-X, 22405-8 Pa., Two-vol. set $15.90

LOS CAPRICHOS, Francisco Goya. 80 plates of wild, grotesque monsters and caricatures. Prado manuscript included. 183pp. 6⅝ × 9⅝. 22384-1 Pa. $4.95

AUTOBIOGRAPHY: The Story of My Experiments with Truth, Mohandas K. Gandhi. Not hagiography, but Gandhi in his own words. Boyhood, legal studies, purification, the growth of the Satyagraha (nonviolent protest) movement. Critical, inspiring work of the man who freed India. 480pp. 5⅜ × 8½. (Available in U.S. only) 24593-4 Pa. $6.95

PLANTS OF THE BIBLE, Harold N. Moldenke and Alma L. Moldenke. Standard reference to all 230 plants mentioned in Scriptures. Latin name, biblical reference, uses, modern identity, much more. Unsurpassed encyclopedic resource for scholars, botanists, nature lovers, students of Bible. Bibliography. Indexes. 123 black-and-white illustrations. 384pp. 6 × 9. 25069-5 Pa. $8.95

FAMOUS AMERICAN WOMEN: A Biographical Dictionary from Colonial Times to the Present, Robert McHenry, ed. From Pocahontas to Rosa Parks, 1,035 distinguished American women documented in separate biographical entries. Accurate, up-to-date data, numerous categories, spans 400 years. Indices. 493pp. 6½ × 9¼. 24523-3 Pa. $9.95

THE FABULOUS INTERIORS OF THE GREAT OCEAN LINERS IN HISTORIC PHOTOGRAPHS, William H. Miller, Jr. Some 200 superb photographs capture exquisite interiors of world's great "floating palaces"—1890's to 1980's: *Titanic, Ile de France, Queen Elizabeth, United States, Europa,* more. Approx. 200 black-and-white photographs. Captions. Text. Introduction. 160pp. 8⅞ × 11¼. 24756-2 Pa. $9.95

THE GREAT LUXURY LINERS, 1927–1954: A Photographic Record, William H. Miller, Jr. Nostalgic tribute to heyday of ocean liners. 186 photos of Ile de France, Normandie, Leviathan, Queen Elizabeth, United States, many others. Interior and exterior views. Introduction. Captions. 160pp. 9 × 12. 24056-8 Pa. $9.95

A NATURAL HISTORY OF THE DUCKS, John Charles Phillips. Great landmark of ornithology offers complete detailed coverage of nearly 200 species and subspecies of ducks: gadwall, sheldrake, merganser, pintail, many more. 74 full-color plates, 102 black-and-white. Bibliography. Total of 1,920pp. 8⅜ × 11¼. 25141-1, 25142-X Cloth. Two-vol. set $100.00

THE SEAWEED HANDBOOK: An Illustrated Guide to Seaweeds from North Carolina to Canada, Thomas F. Lee. Concise reference covers 78 species. Scientific and common names, habitat, distribution, more. Finding keys for easy identification. 224pp. 5⅜ × 8½. 25215-9 Pa. $5.95

THE TEN BOOKS OF ARCHITECTURE: The 1755 Leoni Edition, Leon Battista Alberti. Rare classic helped introduce the glories of ancient architecture to the Renaissance. 68 black-and-white plates. 336pp. 8⅜ × 11¼. 25239-6 Pa. $14.95

MISS MACKENZIE, Anthony Trollope. Minor masterpieces by Victorian master unmasks many truths about life in 19th-century England. First inexpensive edition in years. 392pp. 5⅜ × 8½. 25201-9 Pa. $7.95

THE RIME OF THE ANCIENT MARINER, Gustave Doré, Samuel Taylor Coleridge. Dramatic engravings considered by many to be his greatest work. The terrifying space of the open sea, the storms and whirlpools of an unknown ocean, the ice of Antarctica, more—all rendered in a powerful, chilling manner. Full text. 38 plates. 77pp. 9¼ × 12. 22305-1 Pa. $4.95

THE EXPEDITIONS OF ZEBULON MONTGOMERY PIKE, Zebulon Montgomery Pike. Fascinating first-hand accounts (1805–6) of exploration of Mississippi River, Indian wars, capture by Spanish dragoons, much more. 1,088pp. 5⅜ × 8½. 25254-X, 25255-8 Pa. Two-vol. set $23.90

ILLUSTRATED DICTIONARY OF HISTORIC ARCHITECTURE, edited by Cyril M. Harris. Extraordinary compendium of clear, concise definitions for over 5,000 important architectural terms complemented by over 2,000 line drawings. Covers full spectrum of architecture from ancient ruins to 20th-century Modernism. Preface. 592pp. 7½ × 9⅜. 24444-X Pa. $14.95

THE NIGHT BEFORE CHRISTMAS, Clement Moore. Full text, and woodcuts from original 1848 book. Also critical, historical material. 19 illustrations. 40pp. 4⅝ × 6. 22797-9 Pa. $2.25

THE LESSON OF JAPANESE ARCHITECTURE: 165 Photographs, Jiro Harada. Memorable gallery of 165 photographs taken in the 1930's of exquisite Japanese homes of the well-to-do and historic buildings. 13 line diagrams. 192pp. 8⅜ × 11¼. 24778-3 Pa. $8.95

THE AUTOBIOGRAPHY OF CHARLES DARWIN AND SELECTED LETTERS, edited by Francis Darwin. The fascinating life of eccentric genius composed of an intimate memoir by Darwin (intended for his children); commentary by his son, Francis; hundreds of fragments from notebooks, journals, papers; and letters to and from Lyell, Hooker, Huxley, Wallace and Henslow. xi + 365pp. 5⅜ × 8. 20479-0 Pa. $5.95

WONDERS OF THE SKY: Observing Rainbows, Comets, Eclipses, the Stars and Other Phenomena, Fred Schaaf. Charming, easy-to-read poetic guide to all manner of celestial events visible to the naked eye. Mock suns, glories, Belt of Venus, more. Illustrated. 299pp. 5¼ × 8¼. 24402-4 Pa. $7.95

BURNHAM'S CELESTIAL HANDBOOK, Robert Burnham, Jr. Thorough guide to the stars beyond our solar system. Exhaustive treatment. Alphabetical by constellation: Andromeda to Cetus in Vol. 1; Chamaeleon to Orion in Vol. 2; and Pavo to Vulpecula in Vol. 3. Hundreds of illustrations. Index in Vol. 3. 2,000pp. 6⅛ × 9¼. 23567-X, 23568-8, 23673-0 Pa., Three-vol. set $36.85

STAR NAMES: Their Lore and Meaning, Richard Hinckley Allen. Fascinating history of names various cultures have given to constellations and literary and folkloristic uses that have been made of stars. Indexes to subjects. Arabic and Greek names. Biblical references. Bibliography. 563pp. 5⅜ × 8½. 21079-0 Pa. $7.95

THIRTY YEARS THAT SHOOK PHYSICS: The Story of Quantum Theory, George Gamow. Lucid, accessible introduction to influential theory of energy and matter. Careful explanations of Dirac's anti-particles, Bohr's model of the atom, much more. 12 plates. Numerous drawings. 240pp. 5⅜ × 8½. 24895-X Pa. $4.95

CHINESE DOMESTIC FURNITURE IN PHOTOGRAPHS AND MEASURED DRAWINGS, Gustav Ecke. A rare volume, now affordably priced for antique collectors, furniture buffs and art historians. Detailed review of styles ranging from early Shang to late Ming. Unabridged republication. 161 black-and-white drawings, photos. Total of 224pp. 8⅜ × 11¼. (Available in U.S. only) 25171-3 Pa. $12.95

VINCENT VAN GOGH: A Biography, Julius Meier-Graefe. Dynamic, penetrating study of artist's life, relationship with brother, Theo, painting techniques, travels, more. Readable, engrossing. 160pp. 5⅜ × 8½. (Available in U.S. only) 25253-1 Pa. $3.95

SUNDIALS, Albert Waugh. Far and away the best, most thorough coverage of ideas, mathematics concerned, types, construction, adjusting anywhere. Over 100 illustrations. 230pp. 5⅜ × 8½. 22947-5 Pa. $4.00

PICTURE HISTORY OF THE NORMANDIE: With 190 Illustrations, Frank O. Braynard. Full story of legendary French ocean liner: Art Deco interiors, design innovations, furnishings, celebrities, maiden voyage, tragic fire, much more. Extensive text. 144pp. 8⅜ × 11¾. 25257-4 Pa. $9.95

THE FIRST AMERICAN COOKBOOK: A Facsimile of "American Cookery," 1796, Amelia Simmons. Facsimile of the first American-written cookbook published in the United States contains authentic recipes for colonial favorites—pumpkin pudding, winter squash pudding, spruce beer, Indian slapjacks, and more. Introductory Essay and Glossary of colonial cooking terms. 80pp. 5⅜ × 8½. 24710-4 Pa. $3.50

101 PUZZLES IN THOUGHT AND LOGIC, C. R. Wylie, Jr. Solve murders and robberies, find out which fishermen are liars, how a blind man could possibly identify a color—purely by your own reasoning! 107pp. 5⅜ × 8½. 20367-0 Pa. $2.00

THE BOOK OF WORLD-FAMOUS MUSIC—CLASSICAL, POPULAR AND FOLK, James J. Fuld. Revised and enlarged republication of landmark work in musico-bibliography. Full information about nearly 1,000 songs and compositions including first lines of music and lyrics. New supplement. Index. 800pp. 5⅜ × 8¼. 24857-7 Pa. $14.95

ANTHROPOLOGY AND MODERN LIFE, Franz Boas. Great anthropologist's classic treatise on race and culture. Introduction by Ruth Bunzel. Only inexpensive paperback edition. 255pp. 5⅜ × 8½. 25245-0 Pa. $5.95

THE TALE OF PETER RABBIT, Beatrix Potter. The inimitable Peter's terrifying adventure in Mr. McGregor's garden, with all 27 wonderful, full-color Potter illustrations. 55pp. 4¼ × 5½. (Available in U.S. only) 22827-4 Pa. $1.75

THREE PROPHETIC SCIENCE FICTION NOVELS, H. G. Wells. *When the Sleeper Wakes, A Story of the Days to Come* and *The Time Machine* (full version). 335pp. 5⅜ × 8½. (Available in U.S. only) 20605-X Pa. $5.95

APICIUS COOKERY AND DINING IN IMPERIAL ROME, edited and translated by Joseph Dommers Vehling. Oldest known cookbook in existence offers readers a clear picture of what foods Romans ate, how they prepared them, etc. 49 illustrations. 301pp. 6⅛ × 9¼. 23563-7 Pa. $6.00

SHAKESPEARE LEXICON AND QUOTATION DICTIONARY, Alexander Schmidt. Full definitions, locations, shades of meaning of every word in plays and poems. More than 50,000 exact quotations. 1,485pp. 6½ × 9¼. 22726-X, 22727-8 Pa., Two-vol. set $27.90

THE WORLD'S GREAT SPEECHES, edited by Lewis Copeland and Lawrence W. Lamm. Vast collection of 278 speeches from Greeks to 1970. Powerful and effective models; unique look at history. 842pp. 5⅜ × 8½. 20468-5 Pa. $10.95

A CONCISE HISTORY OF PHOTOGRAPHY: Third Revised Edition, Helmut Gernsheim. Best one-volume history—camera obscura, photochemistry, daguerreotypes, evolution of cameras, film, more. Also artistic aspects—landscape, portraits, fine art, etc. 281 black-and-white photographs. 26 in color. 176pp. 8⅜ × 11¼. 25128-4 Pa. $12.95

THE DORÉ BIBLE ILLUSTRATIONS, Gustave Doré. 241 detailed plates from the Bible: the Creation scenes, Adam and Eve, Flood, Babylon, battle sequences, life of Jesus, etc. Each plate is accompanied by the verses from the King James version of the Bible. 241pp. 9 × 12. 23004-X Pa. $8.95

HUGGER-MUGGER IN THE LOUVRE, Elliot Paul. Second Homer Evans mystery-comedy. Theft at the Louvre involves sleuth in hilarious, madcap caper. "A knockout."—Books. 336pp. 5⅜ × 8½. 25185-3 Pa. $5.95

FLATLAND, E. A. Abbott. Intriguing and enormously popular science-fiction classic explores the complexities of trying to survive as a two-dimensional being in a three-dimensional world. Amusingly illustrated by the author. 16 illustrations. 103pp. 5⅜ × 8½. 20001-9 Pa. $2.00

THE HISTORY OF THE LEWIS AND CLARK EXPEDITION, Meriwether Lewis and William Clark, edited by Elliott Coues. Classic edition of Lewis and Clark's day-by-day journals that later became the basis for U.S. claims to Oregon and the West. Accurate and invaluable geographical, botanical, biological, meteorological and anthropological material. Total of 1,508pp. 5⅜ × 8½.
21268-8, 21269-6, 21270-X Pa. Three-vol. set $25.50

LANGUAGE, TRUTH AND LOGIC, Alfred J. Ayer. Famous, clear introduction to Vienna, Cambridge schools of Logical Positivism. Role of philosophy, elimination of metaphysics, nature of analysis, etc. 160pp. 5⅜ × 8½. (Available in U.S. and Canada only) 20010-8 Pa. $2.95

MATHEMATICS FOR THE NONMATHEMATICIAN, Morris Kline. Detailed, college-level treatment of mathematics in cultural and historical context, with numerous exercises. For liberal arts students. Preface. Recommended Reading Lists. Tables. Index. Numerous black-and-white figures. xvi + 641pp. 5⅜ × 8½.
24823-2 Pa. $11.95

28 SCIENCE FICTION STORIES, H. G. Wells. Novels, *Star Begotten* and *Men Like Gods,* plus 26 short stories: "Empire of the Ants," "A Story of the Stone Age," "The Stolen Bacillus," "In the Abyss," etc. 915pp. 5⅜ × 8½. (Available in U.S. only)
20265-8 Cloth. $10.95

HANDBOOK OF PICTORIAL SYMBOLS, Rudolph Modley. 3,250 signs and symbols, many systems in full; official or heavy commercial use. Arranged by subject. Most in Pictorial Archive series. 143pp. 8⅛ × 11. 23357-X Pa. $5.95

INCIDENTS OF TRAVEL IN YUCATAN, John L. Stephens. Classic (1843) exploration of jungles of Yucatan, looking for evidences of Maya civilization. Travel adventures, Mexican and Indian culture, etc. Total of 669pp. 5⅜ × 8½.
20926-1, 20927-X Pa., Two-vol. set $9.90

AMERICAN CLIPPER SHIPS: 1833–1858, Octavius T. Howe & Frederick C. Matthews. Fully-illustrated, encyclopedic review of 352 clipper ships from the period of America's greatest maritime supremacy. Introduction. 109 halftones. 5 black-and-white line illustrations. Index. Total of 928pp. 5⅜ × 8½.
25115-2, 25116-0 Pa., Two-vol. set $17.90

TOWARDS A NEW ARCHITECTURE, Le Corbusier. Pioneering manifesto by great architect, near legendary founder of "International School." Technical and aesthetic theories, views on industry, economics, relation of form to function, "mass-production spirit," much more. Profusely illustrated. Unabridged translation of 13th French edition. Introduction by Frederick Etchells. 320pp. 6⅛ × 9¼. (Available in U.S. only) 25023-7 Pa. $8.95

THE BOOK OF KELLS, edited by Blanche Cirker. Inexpensive collection of 32 full-color, full-page plates from the greatest illuminated manuscript of the Middle Ages, painstakingly reproduced from rare facsimile edition. Publisher's Note. Captions. 32pp. 9⅜ × 12¼. 24345-1 Pa. $4.50

BEST SCIENCE FICTION STORIES OF H. G. WELLS, H. G. Wells. Full novel *The Invisible Man,* plus 17 short stories: "The Crystal Egg," "Aepyornis Island," "The Strange Orchid," etc. 303pp. 5⅜ × 8½. (Available in U.S. only)
21531-8 Pa. $4.95

AMERICAN SAILING SHIPS: Their Plans and History, Charles G. Davis. Photos, construction details of schooners, frigates, clippers, other sailcraft of 18th to early 20th centuries—plus entertaining discourse on design, rigging, nautical lore, much more. 137 black-and-white illustrations. 240pp. 6⅛ × 9¼.
24658-2 Pa. $5.95

ENTERTAINING MATHEMATICAL PUZZLES, Martin Gardner. Selection of author's favorite conundrums involving arithmetic, money, speed, etc., with lively commentary. Complete solutions. 112pp. 5⅜ × 8½. 25211-6 Pa. $2.95

THE WILL TO BELIEVE, HUMAN IMMORTALITY, William James. Two books bound together. Effect of irrational on logical, and arguments for human immortality. 402pp. 5⅜ × 8½. 20291-7 Pa. $7.50

THE HAUNTED MONASTERY and THE CHINESE MAZE MURDERS, Robert Van Gulik. 2 full novels by Van Gulik continue adventures of Judge Dee and his companions. An evil Taoist monastery, seemingly supernatural events; overgrown topiary maze that hides strange crimes. Set in 7th-century China. 27 illustrations. 328pp. 5⅜ × 8½. 23502-5 Pa. $5.00

CELEBRATED CASES OF JUDGE DEE (DEE GOONG AN), translated by Robert Van Gulik. Authentic 18th-century Chinese detective novel; Dee and associates solve three interlocked cases. Led to Van Gulik's own stories with same characters. Extensive introduction. 9 illustrations. 237pp. 5⅜ × 8½.
23337-5 Pa. $4.95

Prices subject to change without notice.
Available at your book dealer or write for free catalog to Dept. GI, Dover Publications, Inc., 31 East 2nd St., Mineola, N.Y. 11501. Dover publishes more than 175 books each year on science, elementary and advanced mathematics, biology, music, art, literary history, social sciences and other areas.